1983

Axiomatic Theory of Sets and Classes

Axiomatic Theory of Sets and Classes

MURRAY EISENBERG
University of Massachusetts, Amherst

HOLT, RINEHART AND WINSTON, INC.
New York Chicago San Francisco Atlanta Dallas
Montreal Toronto London Sydney

Library of Congress Catalog Card Number: 70–128362
AMS 1970 Subject Classifications: 04—01, 04A05, 04A10, 04A25, 02K15, 02—01
SBN: 03-083632-8
Printed in the United States of America
1 2 3 4 22 1 2 3 4 5 6 7 8 9

To my teachers

Bernard S. Warshaw and Walter H. Gottschalk

Preface

This book presents in an axiomatic setting the fundamental concepts of set theory and logic. It is designed as a text to be used in a one-semester course for advanced undergraduate and beginning graduate students. Although calculus and a little algebra are useful for understanding some of the examples and exercises, no specific subject matter is prerequisite. What is required of the reader is a willingness to grapple with some very general and abstract (and therefore very simple) ideas.

The choice of topics and their manner of exposition are dictated by three primary aims. The first is to present Cantor's classical theory of infinite sets from a wholly modern point of view. The second is to introduce those set-theoretic tools now required for advanced study in all branches of mathematics. The third is to explain systematically what the most basic and general objects of mathematics really are and why they behave as they do.

To achieve the third aim especially, an axiomatic approach has of necessity been adopted. Despite the rigor this entails, excessive formalism is avoided, and proofs are written in the standard informal style of mathematical discourse. Because this is neither a text on "foundations" nor a treatise for specialists in set theory, just one set of axioms is presented that is adequate for describing what every mathematician should know about set theory. An axiom system of the von Neumann-Bernays-Gödel-

Morse-Kelley type has been chosen because it accommodates more naturally than does a Zermelo-Fraenkel-Skolem system the "big" collections mathematicians routinely study. The axiom of choice is accepted without special apology as just one more axiom in this system needed by the working mathematician.

To understand any real mathematics one must be able to read and even write proofs, yet the student taking a course in set theory typically has an inadequate comprehension of logical principles. Hence Chapters 2–3 are devoted to formulating explicitly common techniques of proof. These are derived as consequences of an axiomatized logic which is an integral part of the entire system in that the only statements it treats concern the objects of set theory proper.

In accordance with the needs of modern mathematics, a map always has a codomain as well as a domain. The map-theoretic viewpoint is emphasized throughout the text, and commutative diagrams are used wherever possible. The axiom of choice itself is introduced in order to guarantee the existence of right inverses of surjective maps.

Owing to the cumulative nature of the subject matter, the chapters should be read in order, except that Chapter 15 may be read immediately after Chapter 11. The Appendix may be read immediately after Chapter 15. The following topics may be omitted without loss of continuity: detailed derivation of proof rules in Chapters 2–3, simultaneous and primitive recursion and generalized sums and products in Chapter 10, calculations involving finite sets in 14.5–14.12, ordinal arithmetic in 18.16–18.20, and the proof of equipollence of proper classes in 18.21–18.27.

The exercises range in difficulty from the routine to the challenging. They are designed not only to test comprehension of the ideas presented in the text, but also to present extensions and applications of these ideas.

All definitions, examples, theorems, and so on, within a chapter are numbered consecutively, so that 8.11 refers to item 11 in Chapter 8. Exercises are labeled alphabetically, so that 8 C is the third exercise in Chapter 8.

The Bibliography lists only those books and articles referred to in the text or suggested for further reading. References to the Bibliography are indicated by numbers enclosed in brackets.

This book grew out of lecture notes for a course I taught at the University of Massachusetts during 1967–1968. The efforts of many people made possible the development of those notes into printed form. I especially wish to express my gratitude to Mr. David Sibley, who assisted in preparing the lecture notes; to my students in Mathematics 371/671, who pointed out defects in fact and exposition; to Professor Walter H. Gottschalk, who suggested improvements in a preliminary version; to Professor Trevor J. McMinn and Mr. William Shelden, who critically read the entire manuscript; to Professor Victor Klee, the publisher's editorial consultant, who suggested numerous improvements in the text; to Miss Rosemarie Fischer, who ably and speedily typed the manuscript; to the staff of Holt, Rinehart and Winston, and particularly to Mr. Jack S. Murphy and Mrs. Marilyn Genzer, who courteously and efficiently handled the editing; and above all to my wife, who encouraged and aided me at every stage of writing this book.

Murray Eisenberg

Amherst, Massachusetts
September 1970

Contents

Axiomatic Theory of Sets and Classes

1. The Formalization of Set Theory

INTRODUCTION

Many familiar mathematical entities are collections of objects—"sets" as we informally call such collections here. The Euclidean plane is the set consisting of all its points. Each point of the plane is, according to Descartes' treatment of geometry, an ordered set of two real numbers, its coordinates. Each real number is, according to Dedekind's construction of real numbers as "cuts," a set of rational numbers. Even a real-valued function defined on the set of all real numbers is a set, namely, its graph. Indeed, all mathematical entities can be represented as sets.

In a series of papers initiated in 1874, G. Cantor [6], [7] developed an intuitively grounded general theory of sets treating particularly those sets having infinitely many members. (For an idea of the spirit of Cantor's work, see [17].) Using this theory he proved some theorems which startled the mathematical community. One of these concerns the algebraic numbers.

Recall that an "algebraic number" is a real number which is a solution of some polynomial equation

$$a_0x^n + a_1x^{n-1} + \cdots + a_{n-1}x + a_n = 0$$

having integers $(\ldots, -2, -1, 0, 1, 2, \ldots)$ as its coefficients. Among the algebraic numbers are the natural numbers $0, 1, 2, \ldots$, for each natural number k is a solution of the equation $x - k = 0$. Of course, most algebraic numbers are not natural numbers; each rational number is algebraic, as are such irrational numbers as $\sqrt{2}$, $\sqrt[3]{5}$, and so on. Cantor showed that there are nonetheless just as many natural numbers as there are algebraic numbers, in the following sense. It is possible to construct a one-to-one correspondence between the set of all natural numbers and the set of all algebraic numbers. In other words, one can make correspond to each natural number one and only one algebraic number in such a way that each algebraic number is put into correspondence with exactly one natural number.

Mathematicians did not work long with Cantor's ideas before a crisis developed. It was possible to formulate certain statements about sets which are contradictory, that is, both the statements and their denials are provable.

One of the first, and certainly the simplest, of these contradictions was discovered by Bertrand Russell in 1900. Russell reasoned as follows. Given a set S and object x, the rules of logic dictate that either x is a member of S or else x is not a member of S. In particular, a set S either is a member of itself or else is not a member of itself. Of course, most sets one thinks of are not members of themselves. For example, the entire set of real numbers is not itself a real number and so is not a member of itself.

Russell considered the set $\mathcal{R}$ consisting of all sets which are *not* members of themselves: A set S belongs to $\mathcal{R}$ if and only if S is not a member of S. The question then arises, "Is $\mathcal{R}$ a member of itself?" If $\mathcal{R}$ is a member of $\mathcal{R}$, then by definition of $\mathcal{R}$, $\mathcal{R}$ is not a member of $\mathcal{R}$. However, if $\mathcal{R}$ is not a member of $\mathcal{R}$, then again by the definition of $\mathcal{R}$, $\mathcal{R}$ is a member of $\mathcal{R}$. Hence $\mathcal{R}$ is a member of $\mathcal{R}$ if and only if $\mathcal{R}$ is not a member of $\mathcal{R}$. This is a contradiction!

The appearance of such paradoxes as Russell's precipitated a search for a rigorous foundation of set theory which would avoid contradiction. Cantor's vague pseudo-definition of "set" to mean "a collection into a whole of definite, well-distinguished objects of our intuition or thought" would no longer suffice. What was needed instead was a precisely stated system of axioms saying enough about the behavior of sets to capture the intuitive meaning of "set" and yet, hopefully, to so delimit this concept as to avoid paradox.

The earliest attempt to axiomatize Cantor's "naive" theory was that of G. Frege in 1893–1903. Frege included a so-called axiom of abstraction, which asserts the existence, for any given property, of a set whose members are precisely those objects having the property. Of course, Russell's paradox is an immediate consequence of that axiom.

The various axiom systems currently and widely in use by mathematicians may be divided into two kinds according to the way they resolve the difficulties of the axiom of abstraction. The first is due to E. Zermelo, A. A. Fraenkel, and T. Skolem. There a given property determines only the set of those objects which have the property and which are also members of some set already known to exist. Thus the Zermelo-Fraenkel-Skolem system simply prohibits the formation of such gigantic collections as Russell's $\mathcal{R}$. (For a good account of this system, see Suppes [38].)

The second kind of system is due essentially to J. von Neumann [28], [29], [30], P. Bernays [1], and K. Gödel [11]. Here the primitive notion corresponding to the intuitive idea of a collection of objects is that of a *class*. The axioms specify the behavior of classes. A *set* is by definition a class which is a member of some class. For a given property, one is guaranteed the existence of a class whose members are exactly those classes which are actually sets and which have the property. Russell's collection turns out to be a class which is not a set.

The axiom system we shall present in detail is of the second kind and is closely related to the Morse-Kelley system (described in the Appendix to Kelley [18]), a derivative of the von Neumann-Bernays-Gödel system. For discussions of other axiom systems as well as relationships between various systems, the reader is referred to Cohen [8], Fraenkel and Bar-Hillel [10], Hatcher [15], Morse [26], and Quine [32].

THE FORMAL LANGUAGE OF SET THEORY

To construct the theory of sets we start from scratch, taking nothing for granted. The very language used to convey the theory must be prescribed in detail. This rigid, unambiguous language—the *formal language*—consists of certain successions of symbols together with very strict mechanical rules prescribing certain combinations of symbols to be "true" statements.

To describe this formal language and to make statements about it, we shall use the ordinary English language (supplemented by some special symbols) which in this role may be called the *metalanguage*. (In general, the prefix 'meta-' is applied to a language or theory which describes or studies another language or theory. Thus *metamathematics* is the theory which studies the formal structure of mathematics.)

The formal language is specified by several kinds of rules:

(i) Rules for writing the symbols that constitute the alphabet of the language.

(ii) Rules for forming those combinations of symbols that serve as the nouns and sentences of the language.

(iii) Rules for calling certain of these sentences 'true'. Some rules declare special sentences, the axioms, to be true by fiat. Others allow additional sentences, the theorems, to be labeled as true because they can be deduced from the axioms.

We now proceed to elaborate these rules. The alphabet of the formal language consists of the following *symbols*:

(1) Letters, namely any upper or lower case letter of the English alphabet, together with any number of copies of the symbol '′' (prime). Thus 'X', 'X'', 'X''' are three different letters. (Allowing the use of primes insures that the supply of letters is inexhaustible.)
(2) Logical connectives: $\neg$ (negation sign), $\vee$ (disjunction sign), $\forall$ (universal quantifier), and ι (descriptor).
(3) Equality sign: $=$.
(4) Elementhood sign: $\in$ (not to be confused with the Greek epsilon, ϵ).
(5) Parentheses: (,). (Parentheses serve as punctuation marks.)

The standard intuitive interpretations of these symbols are given in the next section.

Any succession of symbols of the formal language written one after another is said to be a *string*. A string may consist of just a single symbol. Thus 'x'', '$\in$', and '$x \in X$' are strings.

In the preceding sentence mention was made of three specific strings. Now an individual string is an expression in the formal language, *not* an expression in the metalanguage. Hence we had to use *names* of the three strings, not the three strings themselves. This was done in accordance with the following convention:

In the metalanguage, names of strings of the formal language are formed by use of single quotation marks. The one-symbol strings

$$\neg, \vee, \forall, \iota, =, \in, (,)$$

which are not letters may be used as names of themselves.

Although the distinction between things and names of things is often violated in informal mathematical discourse, it is important to make the distinction when laying foundations. Otherwise, confusion may result from the fact that the things our metalanguage talks about are themselves linguistic entities. The distinction is discussed in detail in Quine [31, Sections 4–6] and Suppes [37, Chapter 6] and is dramatized in the following conversation reported in Lewis Carroll's *Through the Looking-Glass*:

> "The name of the song is called '*Haddocks' Eyes*,' " [the White Knight said].
>
> "Oh, that's the name of the song, is it?" Alice said, trying to feel interested.
>
> "No, you don't understand," the Knight said, looking a little vexed. "That's what the name is *called*. The name really *is* '*The Aged Aged Man*.' "

> "Then I ought to have said 'That's what the *song* is called'?" Alice corrected herself.
>
> "No, you oughtn't: that's quite another thing! The *song* is called '*Ways and Means*': but that's only what it's *called*, you know!"
>
> "Well, what *is* the song, then?" said Alice, who was by this time completely bewildered.
>
> "I was coming to that," the Knight said. "The song really *is* . . .". . . then, slowly beating time with one hand . . . he began [to sing].

For the sake of brevity, it is convenient to use in the metalanguage certain characters ("metavariables") which denote ambiguously the names of arbitrary, unspecified strings and which may be replaced by the names of specific strings. We shall so use boldface letters '**P**', '**Q**', '**x**', '**y**', and so on. For example,

'x' does not appear in '$y = z$',

'y' does not appear in '$a \in x$'

are both instances of the metamathematical assertion

The letter **x** does not appear in **P**.

If **P** and **Q** are names of strings, then we write '**PQ**' for a name of the string formed by writing the string called **Q** to the immediate right of the string called **P**. For example, if **P** is '$(x \in$' and **Q** is '$X)$', then **PQ** is a name of '$(x \in X)$'. Similarly, if **P** is '$x \in X$', then $\neg(\mathbf{P})$ denotes '$\neg(x \in X)$'.

A letter **x** is said to be *bound in* a string **S** if either of the strings

$$(\forall \mathbf{x}), \qquad (\iota \mathbf{x})$$

appears in **S**. Otherwise, **x** is said to be *free in* **S**. In particular, **x** is free in **S** if **x** does not appear in **S**. For example, in

$$x = (\iota y)((\forall z)(z \in x))$$

the letters 'x' and 'w' are free but the letters 'y' and 'z' are bound.

TERMS AND FORMULAS

We are about to state the rules for constructing those strings—*terms* and *formulas*—that are to be regarded as meaningful. Intuitively, a term denotes an object the formal language talks about, and a formula denotes an assertion, not necessarily true, about such objects. The precise definitions of 'term' and 'formula' are given by the eight rules below. The rules are syntactical, that is, they allow one to determine whether a given string is a term or a formula by its form alone.

(C1) Each letter is a term.
(C2) If $\mathbf{X}$ and $\mathbf{Y}$ are terms, then $\mathbf{X} = \mathbf{Y}$ is a formula.
(C3) If $\mathbf{x}$ and $\mathbf{X}$ are terms, then $\mathbf{x} \in \mathbf{X}$ is a formula.
(C4) If $\mathbf{P}$ is a formula, then $\neg(\mathbf{P})$ is a formula.
(C5) If $\mathbf{P}$ and $\mathbf{Q}$ are formulas, then $(\mathbf{P}) \vee (\mathbf{Q})$ is a formula.
(C6) If $\mathbf{P}$ is a formula and $\mathbf{x}$ is a letter which is *free* in $\mathbf{P}$, then $(\forall\mathbf{x})(\mathbf{P})$ is a formula.
(C7) If $\mathbf{P}$ is a formula and $\mathbf{x}$ is a letter which is *free* in $\mathbf{P}$, then $(\iota\mathbf{x})(\mathbf{P})$ is a term.
(C8) The only terms and formulas are those given by rules (C1)–(C7).

Some authors call 'well-formed formula' ('wff' for short) what we have called 'formula'.

Particular terms and formulas are constructed by repeated applications of (C2)–(C7) starting with letters.

EXAMPLES

(1) By (C1) the three letters 'x', 'Y', and 'z' are terms. By (C2) and (C3) the two strings

$$x \in Y, \qquad x = z$$

are formulas. By (C5) the string

$$(x \in Y) \vee (x = z)$$

is a formula. By (C7) the string

$$(\iota x)((x \in Y) \vee (x = z))$$

is a term.

(2) The string '$\neg(x)$' is neither a term nor a formula. In fact, each term either is a letter or begins with the string '$(\iota$'; the only formulas beginning with '$\neg($' are given by (C4), but the letter 'x' is not a formula.

(3) The string

$$(\forall x)(x = (\iota x)(x \in y))$$

is obviously not a term. It is not a formula, for the letter 'x' is bound in

$$(\iota x)(x \in y),$$

and rule (C6) does not permit a formula to be obtained by prefixing '$(\forall x)$' to a formula in which 'x' is already bound.

To guide the intuitive understanding of particular terms and formulas, we tabulate below typical renderings into ordinary language of the strings constructed in accordance with rules (C2)–(C7). Later, when formal language yields to a more informal style of discourse, the phrases in the right column of this table will often replace the symbolism in the left column.

$\mathbf{X} = \mathbf{Y}$	**X** equals **Y**
$\mathbf{x} \in \mathbf{X}$	**x** is an element of **X** **x** is a member of **X** **x** belongs to **X** **x** is in **X**
$\neg(\mathbf{P})$	not **P** it is not the case that **P**
$(\mathbf{P}) \vee (\mathbf{Q})$	**P** or **Q** (either) **P** or **Q** or both **P** and **Q** **P** and/or **Q**
$(\forall \mathbf{x})(\mathbf{P})$	for all **x**, **P** for every **x**, **P** for each **x**, **P** for arbitrary **x**, **P** for any **x**, **P**
$(\iota \mathbf{x})(\mathbf{P})$	the **x** such that **P** the unique **x** for which **P**

In this table the ordinary language renderings of the symbolism must not be construed as defining the symbolism. In fact, we have not and shall not define such elementary formulas as '$X = Y$' and '$x \in X$', for to do so would require reference to still prior notions. Rather, equality and elementhood are "primitive," undefined notions of set theory. Whatever meaning the symbolism has is expressed by the axioms yet to be stated. Of course, the axioms will reflect the standard interpretations of the symbolism.

We can now explain in more detail how terms and formulas are to be regarded. Terms are of two kinds, those of the form $(\iota \mathbf{x})(\mathbf{P})$ and those that are single letters (*variables*). A term of the form $(\iota \mathbf{x})(\mathbf{P})$ names or designates some one specific class; such a term functions as does a proper noun ('George Washington') or definite description ('the first president of the U.S.') in ordinary language. A term which is a single letter ambiguously denotes an unspecified class; such a term functions somewhat as does a pronoun ('it', 'he') or common noun ('a thing', 'a man').

Formulas are analogs in the formal language of declarative sentences in ordinary language. Formulas too come in two kinds, those containing no free letters (*closed* formulas) and those containing free letters (*open* formulas). A formula containing no free letters makes a definite statement about whose truth or falsity it makes sense to inquire; such a formula makes an assertion about the specific classes named in it (compare 'the rose I hold is red') or about all classes (compare 'every thing which is a rose is red'). A formula containing free letters makes no statement, because of the ambiguity in its free variables (compare 'it is red'); such a formula is "statement-like," for it is converted into a formula making a statement if its free variables are replaced by terms denoting specific classes.

In ordinary language, the sentence

A thing which is a rose is red

at least sometimes means

Every thing which is a rose is red

(but in certain contexts might very well mean 'At least one rose is red'). Likewise, our axioms will guarantee, a closed formula $(\forall \mathbf{x})(\mathbf{P})$, where $\mathbf{x}$ is free in $\mathbf{P}$, is true exactly when the open formula $\mathbf{P}$ is true. Hence we sometimes speak informally—and inaccurately—about open formulas as if they, too, made statements.

Some terms denote classes, and some formulas make statements about classes, but what are classes? Classes are not expressions in the formal language, and the word 'class' is not an expression in the formal language or in the metalanguage. Rather, classes are the abstract, mental constructs mathematicians deal with. In so far as we can systematically study these intellectual objects by means of a precise language, we do so through terms and formulas.

Just as letters and formulas of the forms $\mathbf{X} = \mathbf{Y}$ and $\mathbf{x} \in \mathbf{X}$ are the elementary units from which all terms and formulas are built, so abbreviations of whole terms and formulas are the units out of which names of more complex terms and formulas are constructed. These abbreviations, which introduce new symbolism into the metalanguage, are called *definitions*. Our first definition introduces the *implication sign* (or *conditional*) '$\Rightarrow$'.

DEFINITION

Suppose $\mathbf{P}$ and $\mathbf{Q}$ are formulas. Then

$$(\mathbf{P}) \Rightarrow (\mathbf{Q})$$

is an abbreviation for the formula

$$(\neg(\mathbf{P})) \vee (\mathbf{Q}).$$

The formula $(\mathbf{P}) \Rightarrow (\mathbf{Q})$ is called an *implication,* **P** is called its *hypothesis* (or *antecedent*), and **Q** its *conclusion* (or *consequent*).

Typical renderings of $(\mathbf{P}) \Rightarrow (\mathbf{Q})$ into ordinary language are:

P implies **Q**	**P** is sufficient for **Q**
if **P**, then **Q**	**Q** is necessary for **P**
P only if **Q**	**Q** provided that **P**.

Additional renderings are obtained by replacing 'if' above by 'when' or by 'in case'.

Through definitions we shall introduce later the additional logical signs '&' and '$\Leftrightarrow$', used in the forms $(\mathbf{P})\ \&\ (\mathbf{Q})$ and $(\mathbf{P}) \Leftrightarrow (\mathbf{Q})$, and the sign '$\exists$', used in the form $(\exists \mathbf{x})(\mathbf{P})$.

In the metalanguage as well as in the informal style in which we shall actually develop set theory, parentheses are omitted whenever feasible. For example, we understand

$$(\forall \mathbf{x})(\mathbf{P}) \vee (\exists \mathbf{x})(\mathbf{Q})$$

to mean

$$((\forall \mathbf{x})(\mathbf{P})) \vee ((\exists \mathbf{x})(\mathbf{Q})),$$

'$\neg\mathbf{P}$' to mean '$\neg(\mathbf{P})$', '$\mathbf{P} \Rightarrow \mathbf{Q}$' to mean '$(\mathbf{P}) \Rightarrow (\mathbf{Q})$', and so on. Of course, parentheses cannot be omitted at will if ambiguity is to be avoided. For example, '$\neg\mathbf{P} \Rightarrow \mathbf{Q}$' could signify either '$(\neg\mathbf{P}) \Rightarrow \mathbf{Q}$' or '$\neg(\mathbf{P} \Rightarrow \mathbf{Q})$'.

When parentheses are missing we usually rely on context plus familiar usages of ordinary language to indicate precisely which fully punctuated strings of the formal language are intended. To avoid genuine ambiguity, we assign to the logical and set-theoretic signs increasing "strengths" in the order listed here:

$$=,\ \in,\ \neg,\ \vee,\ \&,\ \Rightarrow,\ \Leftrightarrow.$$

To interpret expressions from which parentheses are missing, we agree that *the sign with the greater strength reaches further.* Specific illustrations best explain this somewhat vaguely stated rule.

EXAMPLES

(1) '$\neg\mathbf{P} \Rightarrow \mathbf{Q}$' means '$(\neg\mathbf{P}) \Rightarrow \mathbf{Q}$'.
(2) '$x = y \Rightarrow x \in X$' means '$(x = y) \Rightarrow (x \in X)$'.
(3) '$\mathbf{P} \vee \mathbf{Q} \Leftrightarrow \mathbf{R}$' means '$(\mathbf{P} \vee \mathbf{Q}) \Leftrightarrow \mathbf{R}$'.

THEOREMS AND AXIOMS

We now formulate the rules for calling statements "true."

Let **C** be a formula and $\mathfrak{A}$ be a given list of formulas. Suppose **C** is the final entry in a list $\mathfrak{D}$ of formulas so formed that for each entry **S** in $\mathfrak{D}$ at least one of the following holds:

(D1) **S** is in the given list $\mathfrak{A}$.
(D2) A formula **R** and the formula $\mathbf{R} \Rightarrow \mathbf{S}$ both appear above **S** as entries in $\mathfrak{D}$.

Suppose also that no letter which is bound in **C** is free in any entry of $\mathfrak{A}$ (the reason for this restriction will become apparent in Chapter 3). Then one calls $\mathfrak{D}$ a *deduction of* **C** *from* $\mathfrak{A}$, and one calls the entries in $\mathfrak{A}$ the *premises* (or *assumptions* or *hypotheses*) and **C** the *conclusion of* $\mathfrak{D}$.

To illustrate the idea of a deduction, let **P**, **Q**, and **R** be any three specific formulas having no bound letters. For example, one could take **P** to be '$x \in X$', **Q** to be '$x \in Y$', and **R** to be '$x \in Z$'. Here is a deduction of **R** from the three premises **P**, $\mathbf{P} \Rightarrow \mathbf{Q}$, $\mathbf{Q} \Rightarrow \mathbf{R}$:

$$\mathbf{P}$$

$$\mathbf{P} \Rightarrow \mathbf{Q}$$

$$\mathbf{Q}$$

$$\mathbf{Q} \Rightarrow \mathbf{R}$$

$$\mathbf{R}$$

The conclusion of a deduction with premises $\mathfrak{A}$ is a formula which is in $\mathfrak{A}$, or is implied by a formula in $\mathfrak{A}$, or is implied by a formula which in turn is implied by a formula in $\mathfrak{A}$, or .... Such a conclusion may be regarded as "conditionally true," subject only to the truth of the premises from which it has been deduced. In order to regard some things as unconditionally true we must start by accepting certain formulas as absolutely true. These formulas we postulate are the *axioms*. They are conceived as expressing the fundamental truths of set theory. We shall discuss shortly which formulas are actually axioms.

If one can write a deduction of a formula **C** whose premises are the entries in a list $\mathfrak{B}$ together with certain axioms, then one says that **C** is *deducible from* $\mathfrak{B}$ and indicates this metamathematically by writing

$$\mathfrak{B} \vdash \mathbf{C}.$$

We emphasize that '$\mathfrak{B} \vdash \mathbf{C}$' does not necessarily mean that every premise of the deduction in question appears in $\mathfrak{B}$, only that every premise which is not an axiom does so appear. Thus we are regarding axioms as universally available to serve as premises, and so we can suppress explicit mention of axiomatic premises.

A deduction of a formula **C** all of whose premises are axioms is called a (formal) *proof of* **C**. When it is possible to write a proof of **C**, and only then, one says that **C** is *true*, calls **C** a *theorem*, and writes

$$\vdash \mathbf{C}.$$

The notation '$\vdash$**C**' is consistent with the notation introduced above, for '$\vdash$**C**' means there is a deduction of **C** whose premises are axioms plus the entries in a list having no entries.

It is to be emphasized we use 'true' synonymously with 'provable'. ('True' is also used differently in metamathematics to mean 'valid in any interpretation'.)

Following the usual practice, we call subsidiary theorems *lemmas*, *propositions*, and *corollaries*.

If the negation $\neg$**C** of a formula **C** is true, then **C** is said to be *false*.

According to the definition of 'proof', in order to write a proof of a given formula, one must proceed from scratch, oblivious to all theorems already proved, even though earlier proofs may appear intact as part of the new proof. In practice, of course, such duplication of effort is avoided, and short-cuts are taken. These are based on *proof rules*, metamathematical assertions that certain formulas are theorems once it is already known that others are. Proof rules embody the common modes of formal mathematical reasoning. Here is the first.

1.1. PROOF RULE (MODUS PONENS)

Suppose $\vdash$**P** and $\vdash \mathbf{P} \Rightarrow \mathbf{Q}$. Then $\vdash$**Q**.

Justification

Form a list $\mathfrak{D}$ of formulas as follows. First write the entries in a proof of **P**, then the entries in a proof of $\mathbf{P} \Rightarrow \mathbf{Q}$, and finally the formula **Q**. Then $\mathfrak{D}$ is a deduction of **Q** from the axioms which are the premises of the given proof of **P** and of the given proof of $\mathbf{P} \Rightarrow \mathbf{Q}$. □

The symbol '□' appearing immediately above is a substitute for 'Q.E.D.' and always indicates the end of a line of reasoning.

There are thirteen rules for describing which formulas are axioms. Twelve of these, called *axiom schemes*, say that any formula having thus-and-such a form is an axiom. The thirteenth simply declares eight specific formulas to be axioms. These eight are the *explicit axioms*. In contrast, any formula determined to be an axiom by one of the axiom schemes is an *implicit axiom*.

Because the formulations of the explicit axioms and the axiom schemes would be unduly complicated without many abbreviations and would be difficult to grasp without much commentary, we shall introduce them only gradually. The axiom schemes are stated in Chapters 2–4; the first explicit axiom is given in Chapter 4, the last not until Chapter 9. Readers wishing a preview may consult the summaries at the end of the book.

We are hardly about to attempt the tedious task of writing all of set theory in the formal language and constructing formal proofs of all the theorems. We will rest content with indicating how, in principle, this could be done. We shall be adopting an informal style approximating more and more closely normal mathematical discourse, with proofs informally employing standard modes of reasoning as embodied in our proof rules. Always before us, however, is the possibility of casting proofs in a strictly formal form, their validity susceptible to being checked mechanically.

EXERCISES

A. Write as an unabbreviated formula in the formal language: $(\forall x)(x \in X \Rightarrow x \in Y)$.

B. Determine which letters are bound and which are free in the following strings:
   (a) $(\forall x)((x \in X) \vee (x \in Y)) \vee (\forall x)(x \in Z)$.
   (b) z.
   (c) $X = (\iota X)(X = X)$.
   (d) $(x = y) \Rightarrow (\forall y)(y = x)$.
   (e) $(\iota x)(y)$.
   (f) $(x \in y) \Rightarrow (x \in z) \Rightarrow (y \in z)$.
   (g) $(\forall y)(\neg(y = y) \Rightarrow (\forall y)(y = y))$.
   (h) $(\iota x)((\forall y)(\neg(y \in x)))$.
   (i) $\neg(\forall x)(\neg(\forall y)(\neg(y \in x)))$.
   (j) $\neg(\neg(\forall x)(x \in X \Rightarrow x \in Y) \vee \neg(\forall x)(x \in Y \Rightarrow x \in X)) \Rightarrow X = Y$.

C. (a) Determine which of the strings in Exercise B are terms and which are formulas.
   (b) Where feasible render these terms and formulas into ordinary language.

D. Symbolize the following informal expressions:
   (a) If for every y, $y = y$, then $x = x$.
   (b) The x for which x is not equal to x.
   (c) The y such that it is not the case that y equals y.
   (d) A necessary condition for x to equal y is that y be equal to x.
   (e) In order that $Y = X$, it suffices that $X = Y$.

E. Restore missing parentheses to the following formulas:
   (a) $x \in y \vee x \in z \Rightarrow x \in X$.
   (b) $(\forall \mathbf{x})(\mathbf{P} \,\&\, \mathbf{Q}) \Rightarrow (\forall \mathbf{x})(\mathbf{P}) \,\&\, (\forall \mathbf{x})(\mathbf{Q})$.
   (c) $\neg\neg x \in y \Leftrightarrow x \in y$.

F. (a) Justify the following proof rule: If **P**, **Q**, **R**, **S** are formulas and if

$$\vdash \mathbf{P},\ \vdash \mathbf{P} \Rightarrow \mathbf{Q},\ \vdash \mathbf{Q} \Rightarrow \mathbf{R},\ \vdash \mathbf{R} \Rightarrow \mathbf{S},$$

then $\vdash \mathbf{S}$.

(b) Generalize both (a) and 1.1.

G. To check mechanically whether a given term or formula has the correct number of parentheses, one counts the left parentheses '(' and then counts the right parentheses ')' and checks that the two counts give the same number. Explain why this procedure works.

2. Propositional Calculus

Four axiom schemes will be presented here governing the meaning of the logical connectives '$\neg$' ('not') and '$\vee$' ('or'); they are formulated with the aid of the connective '$\Rightarrow$' ('implies') defined earlier. In addition, the new connectives '&' ('and') and '$\Leftrightarrow$' ('is equivalent to') will be defined. Most of this chapter is devoted to elaborating the *propositional calculus,* that system of proof rules resulting solely from the first four axiom schemes.

In his previous mathematical studies the reader has doubtless used, even if implicitly, the rules of the propositional calculus. Our aim here is to state these rules quite explicitly and to convince the reader of the possibility of deriving them from just four simple axiom schemes. Hence we suggest that the stated rules not be committed to memory and that the details of their derivation not be mastered.

IMPLICATION

The four axiom schemes below were first formulated in 1910 by Russell and Whitehead [39]. Actually, Russell and Whitehead included a fifth scheme, but this follows from the others—see 2.14(2).

AXIOM SCHEME 1

Let **P** be a formula. Then

$$\mathbf{P} \vee \mathbf{P} \Rightarrow \mathbf{P}$$

is an axiom.

AXIOM SCHEME 2

Let **P** and **Q** be formulas. Then

$$\mathbf{P} \Rightarrow \mathbf{P} \vee \mathbf{Q}$$

is an axiom.

AXIOM SCHEME 3

Let **P** and **Q** be formulas. Then

$$\mathbf{P} \vee \mathbf{Q} \Rightarrow \mathbf{Q} \vee \mathbf{P}$$

is an axiom.

AXIOM SCHEME 4

Let **P**, **Q**, and **R** be formulas. Then

$$(\mathbf{P} \Rightarrow \mathbf{Q}) \Rightarrow (\mathbf{R} \vee \mathbf{P} \Rightarrow \mathbf{R} \vee \mathbf{Q})$$

is an axiom.

To shorten the exposition, we make the following conventions for the remainder of this chapter: *The letters* '**P**', '**Q**', '**R**', *and* '**S**' *stand for formulas, and* '$\mathcal{A}$' *for a list of formulas.*

Applications of 1.1 to the four axiom schemes immediately yield new proof rules. These are rather uninteresting and are stated only for later use.

2.1. PROOF RULE

Suppose $\mathcal{A} \vdash \mathbf{P} \vee \mathbf{P}$. Then $\mathcal{A} \vdash \mathbf{P}$.

2.2. PROOF RULE

Suppose $\mathcal{A} \vdash \mathbf{P}$. Then $\mathcal{A} \vdash \mathbf{P} \vee \mathbf{Q}$.

2.3. PROOF RULE

Suppose $\mathcal{A} \vdash \mathbf{P} \vee \mathbf{Q}$. Then $\mathcal{A} \vdash \mathbf{Q} \vee \mathbf{P}$.

2.4. PROOF RULE

Suppose $\mathcal{A} \vdash \mathbf{P} \Rightarrow \mathbf{Q}$. Then

$$\mathcal{A} \vdash \mathbf{R} \vee \mathbf{P} \Rightarrow \mathbf{R} \vee \mathbf{Q}.$$

Using Axiom Scheme 4, we can write a one-step proof of the theorem

$$(*) \quad (x = y \Rightarrow x = z) \Rightarrow ((x = w \Rightarrow x = y) \Rightarrow (x = w \Rightarrow x = z))$$

as follows:

$$(x = y \Rightarrow x = z) \Rightarrow ((\neg x = w \vee x = y) \Rightarrow (\neg x = w \vee x = z)).$$

This statement is the implicit axiom obtained by taking **P** in Axiom Scheme 4 to be '$x = y$', **Q** to be '$x = z$', and **R** to be '$x = w$'. If we replace '$x = y$', '$x = z$', '$x = w$' in this proof by '$x \in X$', '$x \in Y$', '$x \in Z$', respectively, we obtain a proof of the new theorem

$$(**) \quad (x \in X \Rightarrow x \in Y) \Rightarrow ((x \in Z \Rightarrow x \in X) \Rightarrow (x \in Z \Rightarrow x \in Y)).$$

Theorems (*) and (**) have the same form, as do the indicated proofs of them. In order to avoid such wasteful duplication of patterns of proof, we want to state a rule guaranteeing we can write a proof of any formula having the same form as (*) and (**). Such a rule is called a *metatheorem.*

A metatheorem bears the same relation to a theorem as does an axiom scheme to an implicit axiom. We call the argument justifying a metatheorem a *metaproof*; it shows how to construct a proof of any particular formula having the form indicated by the metatheorem.

2.5. METATHEOREM

$$\vdash (\mathbf{P} \Rightarrow \mathbf{Q}) \Rightarrow ((\mathbf{R} \Rightarrow \mathbf{P}) \Rightarrow (\mathbf{R} \Rightarrow \mathbf{Q})).$$

Metaproof

Substitution of $\neg\mathbf{R}$ for **R** in Axiom Scheme 4 gives the axiom

$$(\mathbf{P} \Rightarrow \mathbf{Q}) \Rightarrow (\neg\mathbf{R} \vee \mathbf{P} \Rightarrow \neg\mathbf{R} \vee \mathbf{Q}).$$

The stated theorem is just an abbreviation for this axiom. □

Metatheorems furnish new proof rules.

2.6. PROOF RULE (SYLLOGISM)

Suppose

$$\mathbf{P} \Rightarrow \mathbf{Q},\ \mathbf{Q} \Rightarrow \mathbf{R}$$

are both deducible from $\mathcal{A}$. Then so is

$$\mathbf{P} \Rightarrow \mathbf{R}.$$

Justification

By 2.5,

$$\vdash (\mathbf{Q} \Rightarrow \mathbf{R}) \Rightarrow ((\mathbf{P} \Rightarrow \mathbf{Q}) \Rightarrow (\mathbf{P} \Rightarrow \mathbf{R})).$$

Since $\mathcal{A} \vdash \mathbf{Q} \Rightarrow \mathbf{R}$, 1.1 yields

$$\mathcal{A} \vdash (\mathbf{P} \Rightarrow \mathbf{Q}) \Rightarrow (\mathbf{P} \Rightarrow \mathbf{R}).$$

Since also $\mathcal{A} \vdash \mathbf{P} \Rightarrow \mathbf{Q}$, a second use of 1.1 yields the desired result. □

As an example of 2.6, suppose one has already proved: If a function f is differentiable, then f is continuous; if f is continuous, then f is integrable. Then one may conclude: If f is differentiable, then f is integrable. Of course, 2.6 may be applied several times in the course of proving an implication.

2.7. METATHEOREM

$\vdash \mathbf{P} \Rightarrow \mathbf{P}$.

Metaproof

By Axiom Schemes 1 and 2,

$$\vdash \mathbf{P} \vee \mathbf{P} \Rightarrow \mathbf{P}, \vdash \mathbf{P} \Rightarrow \mathbf{P} \vee \mathbf{P}.$$

Now use 2.6. ∎

To illustrate the economy of using metatheorems and proof rules to demonstrate a formula is true without actually writing a formal proof of it, let us give a formal proof of the instance

$$x \in X \Rightarrow x \in X$$

of 2.7. We abbreviate '$\neg x \in X$' by '$x \notin X$'.

$$x \in X \vee x \in X \Rightarrow x \in X$$
$$(x \in X \vee x \in X \Rightarrow x \in X) \Rightarrow$$
$$(x \notin X \vee (x \in X \vee x \in X) \Rightarrow x \notin X \vee x \in X)$$
$$x \notin X \vee (x \in X \vee x \in X) \Rightarrow x \notin X \vee x \in X$$
$$x \notin X \vee (x \in X \vee x \in X)$$
$$x \notin X \vee x \in X.$$

Here the first formula is an implicit axiom obtained from Axiom Scheme 1, and the second an implicit axiom obtained from Axiom Scheme 4. The third formula appears after the first two by virtue of condition (D2) for a deduction. The fourth formula is obtained from Axiom Scheme 2, and the fifth appears after the third and fourth by virtue of condition (D2).

2.8. METATHEOREM (LAW OF THE EXCLUDED MIDDLE)

$$\vdash \mathbf{P} \vee \neg \mathbf{P}.$$

Metaproof

By 2.7, $\vdash \neg \mathbf{P} \vee \mathbf{P}$. Now use Axiom Scheme 3 together with 1.1. ∎

For a given $\mathbf{P}$, 2.8 asserts simply the truth of the compound formula $\mathbf{P} \vee \neg \mathbf{P}$, and not the truth of either of the components $\mathbf{P}$, $\neg \mathbf{P}$ of this formula. That is, one can *not* draw from 2.8 the metamathematical inference that either $\mathbf{P}$ is true or else $\mathbf{P}$ is false. In fact, some celebrated conjectures have for centuries resisted all efforts to determine their truth or falsity.

For example, in 1637, P. Fermat asserted that for each integer $n > 2$ the equation $x^n + y^n = z^n$ has no solution in nonzero integers x, y, z, but to date this "Fermat's last theorem" has been neither proved nor disproved.

The possibility of some formula being neither true nor false is intimately related to the possibility of some formula being both true and false. A celebrated "incompleteness theorem" of K. Gödel says that a formal system such as ours, whose axioms are rich enough for the development of arithmetic, must have a formula which is simultaneously true and false unless it has one which is neither. For an exposition of these metamathematical considerations the reader is referred to Hatcher [15], Kleene [19], or Lyndon [23].

2.9. METATHEOREM (LAWS OF DOUBLE NEGATION)

(1) $\vdash \mathbf{P} \Rightarrow \neg\neg\mathbf{P}$.
(2) $\vdash \neg\neg\mathbf{P} \Rightarrow \mathbf{P}$.

Metaproof

(1) The formula $\mathbf{P} \Rightarrow \neg\neg\mathbf{P}$ is $(\neg\mathbf{P}) \vee (\neg\neg\mathbf{P})$, and this is true by 2.8.

(2) Replacement of $\mathbf{P}$ by $\neg\mathbf{P}$ in (1) yields $\vdash \neg\mathbf{P} \Rightarrow \neg\neg\neg\mathbf{P}$.

By 2.4,

$$\vdash \mathbf{P} \vee \neg\mathbf{P} \Rightarrow \mathbf{P} \vee \neg\neg\neg\mathbf{P}.$$

By 2.8 and 1.1,

$$\vdash \mathbf{P} \vee \neg\neg\neg\mathbf{P}.$$

Hence by 2.3,

$$\vdash \neg\neg\neg\mathbf{P} \vee \mathbf{P},$$

and this last statement is (2). □

According to 2.9(1), $\neg(\neg\mathbf{P})$ is true, that is, $\neg\mathbf{P}$ is false, in case $\mathbf{P}$ is true. By the very definition of 'false', $\neg\mathbf{P}$ is true in case $\mathbf{P}$ is false. These facts may be summarized in the following *truth table* for negation.

$\mathbf{P}$	$\neg\mathbf{P}$
true	false
false	true

This table does not tell the whole story about the truth or falsity of $\neg\mathbf{P}$, only its truth or falsity *given* the truth or falsity of $\mathbf{P}$. It conceals the possibility that $\mathbf{P}$ is neither true nor false (see the discussion of Fermat's last theorem above) as well as the possibility that **P** be both true and false (see the discussion of contradictions below)! Because of these considerations, we have assigned to truth tables a decidedly minor role in our treatment of the propositional calculus.

2.10. METATHEOREM

(1) $\vdash (\mathbf{P} \Rightarrow \mathbf{Q}) \Rightarrow (\neg\mathbf{Q} \Rightarrow \neg\mathbf{P})$.
(2) $\vdash (\neg\mathbf{Q} \Rightarrow \neg\mathbf{P}) \Rightarrow (\mathbf{P} \Rightarrow \mathbf{Q})$.

Metaproof

(1) By Axiom Scheme 3,

$$\vdash (\mathbf{P} \Rightarrow \mathbf{Q}) \Rightarrow (\mathbf{Q} \vee \neg\mathbf{P}).$$

By 2.9,

$$\vdash \mathbf{Q} \Rightarrow \neg\neg\mathbf{Q}.$$

By 2.4,

$$\vdash (\mathbf{Q} \vee \neg\mathbf{P}) \Rightarrow (\neg\neg\mathbf{Q} \vee \neg\mathbf{P}),$$

that is,

$$\vdash (\mathbf{Q} \vee \neg\mathbf{P}) \Rightarrow (\neg\mathbf{Q} \Rightarrow \neg\mathbf{P}).$$

Now use 2.6. □

The implication $\neg\mathbf{Q} \Rightarrow \neg\mathbf{P}$ is called the *contrapositive* of the implication $\mathbf{P} \Rightarrow \mathbf{Q}$. From 2.10 one concludes:

2.11. PROOF RULE

If the contrapositive of $\mathbf{P} \Rightarrow \mathbf{Q}$ is deducible from $\mathcal{A}$, so is $\mathbf{P} \Rightarrow \mathbf{Q}$. If $\mathbf{P} \Rightarrow \mathbf{Q}$ is deducible from $\mathcal{A}$, so is its contrapositive.

In particular, an implication is true precisely when its contrapositive is true. This frequently allows one to prove an implication by proving instead its contrapositive, an implication which may be easier to handle. For example, let n be an integer, and recall that 'n is even' means 'n is not odd'. To prove the implication:

if n^2 is odd, then n is odd

one proves its contrapositive:

if n is even, then n^2 is even.

The reader is cautioned against confusing the contrapositive $\neg\mathbf{Q} \Rightarrow \neg\mathbf{P}$ of the implication $\mathbf{P} \Rightarrow \mathbf{Q}$ with the *converse*

$$\mathbf{Q} \Rightarrow \mathbf{P}$$

of $\mathbf{P} \Rightarrow \mathbf{Q}$. The converse of a true implication may be false. For example, the implication

if n is a natural number divisible by 4,
then n is an even natural number

is true, but its converse

if n is an even natural number, then n is
a natural number divisible by 4

is false.

The next proof rule justifies informal proofs which take the form: "We have one of the two cases **P** or **Q**. On the one hand, **P** implies **R**; on the other hand, **Q** implies **R**. Hence **R** is true in any case."

2.12. PROOF RULE (METHOD OF CASES)

Suppose

$$\mathbf{P} \vee \mathbf{Q}, \qquad \mathbf{P} \Rightarrow \mathbf{R}, \qquad \mathbf{Q} \Rightarrow \mathbf{R}$$

are deducible from $\mathcal{A}$. Then **R** is deducible from $\mathcal{A}$.

Justification

By 2.4,

$$\mathcal{A} \vdash \mathbf{Q} \vee \mathbf{P} \Rightarrow \mathbf{Q} \vee \mathbf{R},$$

$$\mathcal{A} \vdash \mathbf{R} \vee \mathbf{Q} \Rightarrow \mathbf{R} \vee \mathbf{R}.$$

By Axiom Scheme 3,

$$\mathcal{A} \vdash \mathbf{P} \vee \mathbf{Q} \Rightarrow \mathbf{Q} \vee \mathbf{P},$$

$$\mathcal{A} \vdash \mathbf{Q} \vee \mathbf{R} \Rightarrow \mathbf{R} \vee \mathbf{Q}.$$

By Axiom Scheme 1,

$$\mathcal{A} \vdash \mathbf{R} \vee \mathbf{R} \Rightarrow \mathbf{R}.$$

Use of 2.6 four times shows that

$$\mathcal{A} \vdash \mathbf{P} \vee \mathbf{Q} \Rightarrow \mathbf{R},$$

and one concludes by using 1.1. □

Rule 2.12 is frequently applied in the following special form based on 2.8.

2.13. PROOF RULE

Suppose both the formulas

$$\mathbf{P} \Rightarrow \mathbf{Q}, \qquad \neg\mathbf{P} \Rightarrow \mathbf{Q}$$

are deducible from $\mathcal{A}$. Then so is **Q**.

To illustrate 2.13, we give the standard proof that the square of a real number $x \neq 0$ is positive: If $x > 0$, then $x^2 = x \cdot x > 0$ (because the product of any two positive numbers is positive). If, however, $x \ngtr 0$, then $x < 0$, $-x > 0$, and so $x^2 = (-x)^2 > 0$.

Rule 2.12 has an obvious extension to situations where there are three (or more) cases to consider. Here it is convenient to be able to write $\mathbf{P} \vee \mathbf{Q} \vee \mathbf{R}$ as an abbreviation for both $\mathbf{P} \vee (\mathbf{Q} \vee \mathbf{R})$ and $(\mathbf{P} \vee \mathbf{Q}) \vee \mathbf{R}$. The next result guarantees that $\mathbf{P} \vee (\mathbf{Q} \vee \mathbf{R})$ is true precisely when $(\mathbf{P} \vee \mathbf{Q}) \vee \mathbf{R}$ is true, so that the abbreviation is unambiguous.

2.14. METATHEOREM

(1) $\vdash \mathbf{P} \vee (\mathbf{Q} \vee \mathbf{R}) \Rightarrow \mathbf{Q} \vee (\mathbf{P} \vee \mathbf{R})$.
(2) $\vdash \mathbf{P} \vee (\mathbf{Q} \vee \mathbf{R}) \Rightarrow (\mathbf{P} \vee \mathbf{Q}) \vee \mathbf{R}$.
(3) $\vdash (\mathbf{P} \vee \mathbf{Q}) \vee \mathbf{R} \Rightarrow \mathbf{P} \vee (\mathbf{Q} \vee \mathbf{R})$.

Metaproof

(1) By Axiom Schemes 2 and 3,

$$\vdash \mathbf{R} \Rightarrow \mathbf{P} \vee \mathbf{R},$$

so by 2.4,

$$\vdash \mathbf{Q} \vee \mathbf{R} \Rightarrow \mathbf{Q} \vee (\mathbf{P} \vee \mathbf{R}).$$

By 2.4 again,

$$\vdash \mathbf{P} \vee (\mathbf{Q} \vee \mathbf{R}) \Rightarrow \mathbf{P} \vee (\mathbf{Q} \vee (\mathbf{P} \vee \mathbf{R})).$$

Using Axiom Scheme 2, we see then that

$$(*) \qquad \vdash \mathbf{P} \vee (\mathbf{Q} \vee \mathbf{R}) \Rightarrow (\mathbf{Q} \vee (\mathbf{P} \vee \mathbf{R})) \vee \mathbf{P}.$$

Next,

$$\vdash \mathbf{P} \Rightarrow \mathbf{R} \vee \mathbf{P},$$

$$\vdash \mathbf{P} \vee \mathbf{R} \Rightarrow \mathbf{Q} \vee (\mathbf{P} \vee \mathbf{R}),$$

$$\vdash \mathbf{P} \Rightarrow \mathbf{Q} \vee (\mathbf{P} \vee \mathbf{R})$$

(why?), and application of 2.4 to the last assertion yields

$$(**) \qquad (\mathbf{Q} \vee (\mathbf{P} \vee \mathbf{R})) \vee \mathbf{P} \Rightarrow (\mathbf{Q} \vee (\mathbf{P} \vee \mathbf{R})) \vee (\mathbf{Q} \vee (\mathbf{P} \vee \mathbf{R})).$$

Assertion (1) now follows from (*) and (**).

(2) One first shows

$$\vdash \mathbf{P} \vee (\mathbf{Q} \vee \mathbf{R}) \Rightarrow \mathbf{P} \vee (\mathbf{R} \vee \mathbf{Q})$$

and then uses (1) with **P**, **Q**, **R** in (1) replaced by **P**, **R**, **Q**, respectively.

(3) is left as an exercise. □

CONJUNCTION AND CONTRADICTION

2.15. DEFINITION

The formula

$$\neg(\neg \mathbf{P} \vee \neg \mathbf{Q})$$

is abbreviated by

$$\mathbf{P} \,\&\, \mathbf{Q}$$

and is read as

P and **Q**

both **P** and **Q**

and so on. The ampersand '&' (for which the symbol '$\wedge$' is sometimes used) is called the *conjunction sign.*

That the intuitive meaning of '**P** and **Q**' is conveyed by this definition will be evident from 2.17 and 2.19.

2.16. METATHEOREM

$$\vdash \mathbf{P} \,\&\, \mathbf{Q} \Rightarrow \mathbf{Q} \,\&\, \mathbf{P}.$$

Metaproof

Use Axiom Scheme 3 and 2.11. □

2.17. METATHEOREM

$$\vdash \mathbf{P} \,\&\, \mathbf{Q} \Rightarrow \mathbf{P}, \qquad \vdash \mathbf{P} \,\&\, \mathbf{Q} \Rightarrow \mathbf{Q}.$$

Metaproof

The second assertion follows from the first by 2.16. To prove $\mathbf{P} \,\&\, \mathbf{Q} \Rightarrow \mathbf{P}$, we prove its contrapositive. By Axiom Scheme 2,

$$\vdash \neg \mathbf{P} \Rightarrow \neg \mathbf{P} \vee \neg \mathbf{Q}.$$

By 2.9,

$$\vdash \neg \mathbf{P} \vee \neg \mathbf{Q} \Rightarrow \neg\neg(\neg \mathbf{P} \vee \neg \mathbf{Q}),$$

that is,

$$\vdash \neg \mathbf{P} \vee \neg \mathbf{Q} \Rightarrow \neg(\mathbf{P} \,\&\, \mathbf{Q}).$$

Hence

$$\vdash \neg \mathbf{P} \Rightarrow \neg(\mathbf{P} \,\&\, \mathbf{Q}). \quad \square$$

According to 2.17, **P** and **Q** are both true in case **P** & **Q** is true.

The next rule expresses the ideas that anything implies a true statement, and that a false statement implies anything. (These properties of implication are quite reasonable: see the discussion of the truth table for $\Rightarrow$ following 2.23.)

2.18. PROOF RULE

(1) If $\mathbf{Q}$ is deducible from $\mathcal{A}$, so is $\mathbf{P} \Rightarrow \mathbf{Q}$.
(2) If $\neg\mathbf{P}$ is deducible from $\mathcal{A}$, so is $\mathbf{P} \Rightarrow \mathbf{Q}$.

Justification

Use 2.2 and 2.3. □

2.19. PROOF RULE

Suppose $\mathbf{P}$ and $\mathbf{Q}$ are both deducible from $\mathcal{A}$. Then $\mathbf{P} \,\&\, \mathbf{Q}$ is deducible from $\mathcal{A}$.

Justification

By 2.9,

$$\vdash \neg(\mathbf{P} \,\&\, \mathbf{Q}) \Rightarrow \neg\mathbf{P} \vee \neg\mathbf{Q}.$$

Since $\mathcal{A} \vdash \mathbf{P}$, by 2.18(1),

$$\mathcal{A} \vdash \neg\mathbf{P} \vee \neg\mathbf{Q} \Rightarrow \mathbf{P},$$

and hence

$$\mathcal{A} \vdash \neg\mathbf{P} \Rightarrow \mathbf{P} \,\&\, \mathbf{Q}.$$

Similarly,

$$\mathcal{A} \vdash \neg\mathbf{Q} \Rightarrow \mathbf{P} \,\&\, \mathbf{Q}.$$

By 2.12,

$$\mathcal{A} \vdash \neg(\mathbf{P} \,\&\, \mathbf{Q}) \Rightarrow \mathbf{P} \,\&\, \mathbf{Q},$$

and the desired result follows. □

In particular, $\mathbf{P} \,\&\, \mathbf{Q}$ is true in case $\mathbf{P}$ and $\mathbf{Q}$ are both true simultaneously. Hence to prove the conjunction $\mathbf{P} \,\&\, \mathbf{Q}$ one need only prove separately first $\mathbf{P}$ and then $\mathbf{Q}$.

The conjunction $\mathbf{P} \,\&\, \mathbf{Q}$ is false in case $\mathbf{Q}$ is false, for in this case $\neg\mathbf{Q}$ is true, $\neg\mathbf{P} \vee \neg\mathbf{Q}$ is true, and $\mathbf{P} \,\&\, \mathbf{Q}$, which is $\neg(\neg\mathbf{P} \vee \neg\mathbf{Q})$, is false. Similarly $\mathbf{P} \,\&\, \mathbf{Q}$ is false in case $\mathbf{P}$ is false.

These comments on the truth and falsity of $\mathbf{P} \,\&\, \mathbf{Q}$ may be summarized in the following truth table for conjunction.

P	Q	P & Q
true	true	true
true	false	false
false	true	false
false	false	false

2.20. PROOF RULE

If

$$\mathbf{P} \Rightarrow \mathbf{Q}, \qquad \mathbf{P} \Rightarrow \mathbf{R}$$

are both deducible from $\mathcal{A}$, so is

$$\mathbf{P} \Rightarrow \mathbf{Q} \,\&\, \mathbf{R}.$$

Justification

It suffices to show

$$\mathcal{A} \vdash (\neg\mathbf{Q} \vee \neg\mathbf{R}) \Rightarrow \neg\mathbf{P}. \quad \square$$

A formula of the form $\mathbf{P} \,\&\, \neg\mathbf{P}$ for some formula $\mathbf{P}$ is called a *contradiction*. The next proof rule is often used informally in the form: "We wish to prove $\mathbf{P}$. Just suppose $\mathbf{P}$ is false. Then the statements $\mathbf{Q}$ and $\neg\mathbf{Q}$ are both true. Thus we have reached a contradiction, which is absurd. Hence $\mathbf{P}$ is in fact true."

2.21. PROOF RULE (METHOD OF CONTRADICTION; METHOD OF INDIRECT PROOF; REDUCTIO AD ABSURDUM)

Let $\mathbf{C}$ be a contradiction $\mathbf{Q} \,\&\, \neg\mathbf{Q}$, and suppose

$$\neg\mathbf{P} \Rightarrow \mathbf{C}$$

is deducible from $\mathcal{A}$. Then $\mathbf{P}$ is deducible from $\mathcal{A}$.

Justification

We have

$$\mathcal{A} \vdash \neg\mathbf{P} \Rightarrow \mathbf{Q}, \qquad \mathcal{A} \vdash \neg\mathbf{P} \Rightarrow \neg\mathbf{Q},$$

hence

$$\mathcal{A} \vdash \neg\mathbf{Q} \Rightarrow \mathbf{P}, \qquad \mathcal{A} \vdash \mathbf{Q} \Rightarrow \mathbf{P}.$$

Reversing the roles of $\mathbf{P}$ and $\mathbf{Q}$ in 2.13, we see that $\mathcal{A} \vdash \mathbf{P}$. $\square$

When using 2.21, to prove $\neg\mathbf{P} \Rightarrow \mathbf{Q} \,\&\, \neg\mathbf{Q}$ one often proceeds by employing 2.20, proving separately the implications $\neg\mathbf{P} \Rightarrow \mathbf{Q}$ and $\neg\mathbf{P} \Rightarrow \neg\mathbf{Q}$.

To illustrate the method of contradiction, we show that the absolute value function f, defined by $f(x) = |x|$, does not have a derivative at 0. Recall first that $|0| = 0$, $|x| = x$ in case $x > 0$, and $|x| = -x$ in case $x < 0$, so that $|x|/x = 1$ if $x > 0$ and $|x|/x = -1$ if $x < 0$. On the one hand, if it is not the case that f does not have a derivative at 0, then f has a derivative at 0,

$$\begin{aligned} f'(0) &= \lim_{x\to 0} \frac{f(x) - f(0)}{x - 0} = \lim_{x\to 0} \frac{|x|}{x} = \lim_{x\to 0^+} \frac{|x|}{x} \\ &= \lim_{x\to 0^+} 1 = 1, \end{aligned}$$

so $f'(0) = 1$. On the other hand, if it is not the case that f does not have a derivative at 0, then

$$f'(0) = \lim_{x \to 0} \frac{f(x) - f(0)}{x - 0} = \lim_{x \to 0} \frac{|x|}{x} = \lim_{x \to 0^-} \frac{|x|}{x}$$
$$= \lim_{x \to 0^-} (-1) = -1 \neq 1,$$

so $f'(0) \neq 1$.

From 2.18(2) one obtains:

2.22. PROOF RULE

If some contradiction is deducible from $\mathcal{A}$, then every formula is deducible from $\mathcal{A}$.

In particular, if some contradiction is true, then every formula is true, and in this case our theory is said to be *inconsistent*. Of course an inconsistent theory has no mathematical interest.

To prove a contradiction is to show that some formula is simultaneously true and false. There is no guarantee whatever that our axioms do not lead to a contradiction (see the reference following 2.8 to Gödel's incompleteness theorem). Hence one has no right to claim that a true formula is not false and a false formula is not true!

2.23. PROOF RULE

Suppose **P** and $\neg$**Q** are both deducible from $\mathcal{A}$. Then $\neg$(**P** $\Rightarrow$ **Q**) is deducible from $\mathcal{A}$.

Justification

According to 2.19, **P** & $\neg$**Q**, that is,

$$\neg(\neg \mathbf{P} \vee \neg\neg \mathbf{Q})$$

is deducible from $\mathcal{A}$. Hence $\neg(\neg$**P** $\vee$ **Q**) is deducible from $\mathcal{A}$. ▯

It follows from 2.23 that **P** $\Rightarrow$ **Q** is false in case **P** is true and **Q** is false. By 2.18, **P** $\Rightarrow$ **Q** is true in case both **P** and **Q** are true, in case **P** is false and **Q** is true, and in case both **P** and **Q** are false. These observations yield the truth table for implication.

P	Q	P $\Rightarrow$ Q
true	true	true
true	false	false
false	true	true
false	false	true

Our definition of '$\mathbf{P} \Rightarrow \mathbf{Q}$' as '$\neg\mathbf{P} \vee \mathbf{Q}$' was designed to produce this truth table. The last two lines there say that $\mathbf{P} \Rightarrow \mathbf{Q}$ is true whenever **P** is false. Beginners often find this mathematical usage of 'if . . . then' perplexing. To call 'if **P**, then **Q**' true when **P** and **Q** are both false should cause no confusion, for such usage occurs even in ordinary discourse. Consider, for example: "If Hubert H. Humphrey had been elected president in 1968, then Edmund Muskie would have been elected vice-president in 1968." (The reader may be interested in the discussion of such "contrary-to-fact conditionals" found in Goodman [12].) The case of **P** false and **Q** true requires further comment.

Rather than argue by analogy with ordinary discourse, we offer a mathematical justification for the consequences of our definition of implication. The implication

> if n is an integer divisible by 4, then n is
> an even integer

is certainly to be considered true, and it seems entirely reasonable to call true any implication which results from replacing the letter 'n' by a particular integer. Thus, both

> if 6 is an integer divisible by 4, then 6 is
> an even integer

and

> if 5 is an integer divisible by 4, then 5 is
> an even integer

are to be regarded as true.

EQUIVALENCE

2.24. DEFINITION

The formula

$$(\mathbf{P} \Rightarrow \mathbf{Q}) \ \& \ (\mathbf{Q} \Rightarrow \mathbf{P})$$

is abbreviated by

$$\mathbf{P} \Leftrightarrow \mathbf{Q}.$$

Some informal renderings of $\mathbf{P} \Leftrightarrow \mathbf{Q}$ are

P is equivalent to **Q**
P if and only if **Q**
P iff **Q**
P is necessary and sufficient for **Q**
P precisely when **Q**.

The symbol '$\Leftrightarrow$' is the *equivalence sign* (or *biconditional*).

According to 2.19, one may prove an equivalence $\mathbf{P} \Leftrightarrow \mathbf{Q}$ by proving separately the implication $\mathbf{P} \Rightarrow \mathbf{Q}$ and its converse $\mathbf{Q} \Rightarrow \mathbf{P}$. Thus to prove

> a necessary and sufficient condition for a triangle to be isosceles is that two of its angles be equal

one first proves *necessity*—'if a triangle is isosceles, then two of its angles are equal'—and then proves *sufficiency*—'if a triangle has two equal angles, then it is isosceles'.

In terms of equivalence, 2.9 and 2.14 say

$$\vdash \mathbf{P} \Leftrightarrow \neg\neg\mathbf{P}$$

$$\vdash \mathbf{P} \vee (\mathbf{Q} \vee \mathbf{R}) \Leftrightarrow (\mathbf{P} \vee \mathbf{Q}) \vee \mathbf{R}.$$

The latter equivalence may be used to show

$$\vdash \mathbf{P} \,\&\, (\mathbf{Q} \,\&\, \mathbf{R}) \Leftrightarrow (\mathbf{P} \,\&\, \mathbf{Q}) \,\&\, \mathbf{R},$$

a result allowing one to write unambiguously '$\mathbf{P} \,\&\, \mathbf{Q} \,\&\, \mathbf{R}$'. Axiom Scheme 3 and 2.16 say

$$\vdash \mathbf{P} \vee \mathbf{Q} \Leftrightarrow \mathbf{Q} \vee \mathbf{P}$$

$$\vdash \mathbf{P} \,\&\, \mathbf{Q} \Leftrightarrow \mathbf{Q} \,\&\, \mathbf{P},$$

so that the order in which the components $\mathbf{P}$, $\mathbf{Q}$ of a disjunction or conjunction are given is immaterial.

2.25. PROOF RULE

If $\mathbf{P} \Leftrightarrow \mathbf{Q}$ is deducible from $\mathcal{A}$, then so are:

$$\begin{gathered}
(\neg\mathbf{P}) \Leftrightarrow (\neg\mathbf{Q}) \\
(\mathbf{R} \vee \mathbf{P}) \Leftrightarrow (\mathbf{R} \vee \mathbf{Q}) \\
(\mathbf{R} \,\&\, \mathbf{P}) \Leftrightarrow (\mathbf{R} \,\&\, \mathbf{Q}) \\
(\mathbf{R} \Rightarrow \mathbf{P}) \Leftrightarrow (\mathbf{R} \Rightarrow \mathbf{Q}) \\
(\mathbf{P} \Rightarrow \mathbf{R}) \Leftrightarrow (\mathbf{Q} \Rightarrow \mathbf{R}) \\
(\mathbf{P} \Leftrightarrow \mathbf{R}) \Leftrightarrow (\mathbf{Q} \Leftrightarrow \mathbf{R}).
\end{gathered}$$

Roughly speaking, then, substitution for a part $\mathbf{P}$ of a formula $\mathbf{A}$ an equivalent formula $\mathbf{Q}$ yields a formula equivalent to $\mathbf{A}$.

The next metatheorem, which is easily established, is a technical result needed later.

2.26. METATHEOREM

$$\vdash \mathbf{P} \Rightarrow (\mathbf{Q} \Rightarrow \mathbf{R}) \Leftrightarrow \mathbf{P} \,\&\, \mathbf{Q} \Rightarrow \mathbf{R}.$$

The following equivalences are all easy to prove.

2.27. METATHEOREM

(1) $\vdash \neg(\mathbf{P} \Rightarrow \mathbf{Q}) \Leftrightarrow \mathbf{P} \mathbin{\&} \neg \mathbf{Q}$,
(2) $\vdash \neg(\mathbf{P} \vee \mathbf{Q}) \Leftrightarrow (\neg \mathbf{P}) \mathbin{\&} (\neg \mathbf{Q})$,
(3) $\vdash \neg(\mathbf{P} \mathbin{\&} \mathbf{Q}) \Leftrightarrow (\neg \mathbf{P}) \vee (\neg \mathbf{Q})$,
(4) $\vdash \mathbf{P} \vee \mathbf{Q} \Leftrightarrow \neg \mathbf{P} \Rightarrow \mathbf{Q}$.

Equivalence (1) is used to show that certain implications are false. Consider the problem of demonstrating the falsity of 'If f is a continuous function, then f is a differentiable function'. Of course, one need only show the falsity of this implication for one particular f (see Chapter 3). A convenient choice for f is the absolute value function. One proves, 'f is a continuous function and f is not a differentiable function'.

Equivalence (4) is likewise useful to simplify proofs. Let n be a prime natural number (that is, n has no divisors other than 1 and n). To prove 'either n is odd or $n = 2$', one proves 'if n is even (that is, not odd), then $n = 2$'.

Equivalences (2) and (3) allow one to express the negation of a disjunction (conjunction) as a conjunction (disjunction). Thus, if X is a set of real numbers, to deny 'either X has no upper bound or X has a least upper bound' is to affirm 'X has an upper bound and X does not have a least upper bound'. To deny '$-1 < x < 1$' is to affirm '$x \leq -1$ or $x \geq 1$'.

Equivalence (2) permits us to present at last the truth table for disjunction.

P	**Q**	**P ∨ Q**
true	true	true
true	false	true
false	true	true
false	false	false

Earlier discussion supplies the rationale for the first three lines of this table. To explain the fourth, suppose **P** and **Q** are both false. Then $\neg\mathbf{P}$ and $\neg\mathbf{Q}$ are both true, so $\neg\mathbf{P} \mathbin{\&} \neg\mathbf{Q}$ is true. By 2.27(2), $\neg(\mathbf{P} \vee \mathbf{Q})$ is true, that is, $\mathbf{P} \vee \mathbf{Q}$ is false.

The truth table shows clearly that '**P** or **Q**' is used mathematically in the inclusive sense '**P** and/or **Q**' rather than in the exclusive sense 'either **P** or **Q** but not both'. In fact, the symbol '$\vee$' derives from the initial letter of the latin *vel* meaning 'and/or', the Latin word for 'either ... or ... but not both' being *aut*. The English language requires more complicated locutions to distinguish the inclusive from the exclusive meaning of 'or'.

The truth table for equivalence may be constructed as follows using the information in the truth tables for conjunction and implication.

$\mathbf{P}$	$\mathbf{Q}$	$\mathbf{P} \Rightarrow \mathbf{Q}$	$\mathbf{Q} \Rightarrow \mathbf{P}$	$(\mathbf{P} \Rightarrow \mathbf{Q})\ \&\ (\mathbf{Q} \Rightarrow \mathbf{P})$
true	true	true	true	true
true	false	false	true	false
false	true	true	false	false
false	false	true	true	true

Here the third and fourth columns are obtained from the first two, and the fifth column is obtained from the third and fourth. Recalling the definition of '$\mathbf{P} \Leftrightarrow \mathbf{Q}$', we merely delete the intermediate third and fourth columns to get the truth table for equivalence.

$\mathbf{P}$	$\mathbf{Q}$	$\mathbf{P} \Leftrightarrow \mathbf{Q}$
true	true	true
true	false	false
false	true	false
false	false	true

The same technique used to construct the truth table of '$(\mathbf{P} \Rightarrow \mathbf{Q})\ \&\ (\mathbf{Q} \Rightarrow \mathbf{P})$' may be used to construct truth tables for more complicated expressions. Here is an example.

$\mathbf{P}$	$\mathbf{Q}$	$\neg\mathbf{P}$	$\mathbf{P} \vee \mathbf{Q}$	$\neg\mathbf{P}\ \&\ (\mathbf{P} \vee \mathbf{Q})$	$\neg\mathbf{P}\ \&\ (\mathbf{P} \vee \mathbf{Q}) \Rightarrow \mathbf{Q}$
true	true	false	true	false	true
true	false	false	true	false	true
false	true	true	true	true	true
false	false	true	false	false	true

Notice that the entries in the last column here are all 'true'. An expression such as '$\neg\mathbf{P}\ \&\ (\mathbf{P} \vee \mathbf{Q}) \Rightarrow \mathbf{Q}$' is called a *tautology* if the entries in the last column of its truth table are all 'true'. Given any combination of the truth-values 'true' and 'false' for the components **P** and **Q** from which the expression is built by means of logical connectives, the whole expression takes the truth-value 'true'. It can be shown that for any expression **E**, **E** is a tautology precisely when **E** is a metatheorem; see Kleene [19].

THE DEDUCTION CRITERION

The most frequently used mode of mathematical reasoning concerns implications. To prove $\mathbf{P} \Rightarrow \mathbf{Q}$, one often proceeds by "assuming **P**," that is, by treating **P** as if it were another axiom, and then one deduces **Q** from **P** together with the genuine axioms.

Here is a fairly typical instance of this procedure.

THEOREM

If f is a constant function, then $f'(x) = 0$ for all x.

Proof

Assume f is a constant function. Then for every $h, f(x+h) = f(x)$, so

$$\lim_{h\to 0} \frac{f(x+h) - f(x)}{h} = \lim_{h\to 0} \frac{0}{h} = 0. \quad \square$$

Pity the poor calculus student who demands a justification for this procedure of assuming the hypothesis. Too often his instructor justifies it by the "method of intimidation": "it works because I say it works." The more sympathetic instructor may explain that the hypothesis is either true or false, the implication is automatically true when its hypothesis is false, and so one need only prove it true when the hypothesis is true. This pseudo-explanation seriously confuses mathematics and metamathematics and grossly misinterprets the law of the excluded middle. The real explanation is the following proof rule.

2.28. PROOF RULE (HERBRAND-TARSKI DEDUCTION CRITERION)

Let $\mathcal{A}$, $\mathbf{P}$ denote the list obtained by adjoining $\mathbf{P}$ to $\mathcal{A}$. Suppose $\mathbf{Q}$ is deducible from $\mathcal{A}$, $\mathbf{P}$. Then $\mathbf{P} \Rightarrow \mathbf{Q}$ is deducible from $\mathcal{A}$ alone.

In particular, if $\mathbf{Q}$ is deducible from $\mathbf{P}$, then $\mathbf{P} \Rightarrow \mathbf{Q}$ is true.

Justification

Consider a deduction $\mathfrak{D}$ of $\mathbf{Q}$ whose premises are entries in $\mathcal{A}$, $\mathbf{P}$ and axioms. Call the entries in $\mathfrak{D}$, in order, $\mathbf{D}_1, \mathbf{D}_2, \ldots, \mathbf{D}_n$. We shall show how to construct from $\mathfrak{D}$ a deduction of $\mathbf{P} \Rightarrow \mathbf{Q}$ whose only premises are entries in $\mathcal{A}$ and axioms.

The last entry $\mathbf{D}_n$ in $\mathfrak{D}$ is $\mathbf{Q}$, so we must show $\mathcal{A} \vdash \mathbf{P} \Rightarrow \mathbf{D}_n$. To do this, we describe a systematic procedure for showing step-by-step

$$\mathcal{A} \vdash \mathbf{P} \Rightarrow \mathbf{D}_1, \mathcal{A} \vdash \mathbf{P} \Rightarrow \mathbf{D}_2, \ldots, \mathcal{A} \vdash \mathbf{P} \Rightarrow \mathbf{D}_n.$$

Let $1 \leq i \leq n$, and suppose we have already shown

$$\mathcal{A} \vdash \mathbf{P} \Rightarrow \mathbf{D}_j$$

for each $j < i$ (when $i = 1$, there is nothing to suppose). We show the next step

$$\mathcal{A} \vdash \mathbf{P} \Rightarrow \mathbf{D}_i.$$

We consider separately the two possibilities accounting for the appearance of $\mathbf{D}_i$ in $\mathfrak{D}$.

The first possibility is that $\mathbf{D}_i$ is a premise of $\mathfrak{D}$, that is, $\mathbf{D}_i$ is $\mathbf{P}$, or $\mathbf{D}_i$ is an axiom, or $\mathbf{D}_i$ is one of the entries in $\mathfrak{A}$. If $\mathbf{D}_i$ is $\mathbf{P}$, then $\mathbf{P} \Rightarrow \mathbf{D}_i$ is true and so is deducible from $\mathfrak{A}$. If $\mathbf{D}_i$ is an axiom, then $\mathbf{P} \Rightarrow \mathbf{D}_i$ is true by 2.18(1) and so is deducible from $\mathfrak{A}$. If $\mathbf{D}_i$ is an entry in $\mathfrak{A}$, then certainly it is deducible from $\mathfrak{A}$, and by 2.18(1) again $\mathbf{P} \Rightarrow \mathbf{D}_i$ is deducible from $\mathfrak{A}$.

The second possibility is that for some $j < i$ and $k < i$, $\mathbf{D}_k$ is

$$\mathbf{D}_j \Rightarrow \mathbf{D}_i$$

(this cannot happen when $i = 1$). By our supposition we have already shown that

$$\mathfrak{A} \vdash \mathbf{P} \Rightarrow \mathbf{D}_j$$

$$\mathfrak{A} \vdash \mathbf{P} \Rightarrow (\mathbf{D}_j \Rightarrow \mathbf{D}_i).$$

Since $\mathbf{P} \Rightarrow \mathbf{P}$ is true, 2.20 says

$$\mathfrak{A} \vdash \mathbf{P} \Rightarrow \mathbf{P} \,\&\, \mathbf{D}_j.$$

By 2.26,

$$\mathfrak{A} \vdash \mathbf{P} \,\&\, \mathbf{D}_j \Rightarrow \mathbf{D}_i.$$

Hence $\mathbf{P} \Rightarrow \mathbf{D}_i$ is in fact deducible from $\mathfrak{A}$. $\square$

Here is a simple application of the deduction criterion to the propositional calculus itself.

2.29. METATHEOREM

$$\vdash \neg\mathbf{P} \,\&\, (\mathbf{P} \vee \mathbf{Q}) \Rightarrow \mathbf{Q}.$$

Metaproof

Denote $\neg\mathbf{P} \,\&\, (\mathbf{P} \vee \mathbf{Q})$ by $\mathbf{A}$. We shall show $\mathbf{A} \vdash \mathbf{Q}$. We have

$$\vdash \mathbf{A} \Rightarrow \neg\mathbf{P},$$

so

(1) $\mathbf{A} \vdash \mathbf{A} \Rightarrow \neg\mathbf{P}$.

Since

(2) $\mathbf{A} \vdash \mathbf{A}$,

we conclude from (1) and (2)

(3) $\mathbf{A} \vdash \neg\mathbf{P}$.

Similarly,

(4) $\mathbf{A} \vdash \mathbf{P} \vee \mathbf{Q}$.

Now by 2.27(4),

$$\vdash \mathbf{P} \vee \mathbf{Q} \Rightarrow (\neg\mathbf{P} \Rightarrow \mathbf{Q})$$

so

$$(5) \quad \mathbf{A} \vdash \mathbf{P} \vee \mathbf{Q} \Rightarrow (\neg \mathbf{P} \Rightarrow \mathbf{Q}).$$

Hence by (4) and (5),

$$(6) \quad \mathbf{A} \vdash \neg \mathbf{P} \Rightarrow \mathbf{Q}.$$

By (3) and (6),

$$\mathbf{A} \vdash \mathbf{Q}. \quad \square$$

The first part of 2.30 will be proved by means of the deduction criterion. Without the deduction criterion its proof is lengthy and complicated indeed.

2.30. METATHEOREM

(1) $\vdash \mathbf{P} \vee (\mathbf{Q} \,\&\, \mathbf{R}) \Leftrightarrow (\mathbf{P} \vee \mathbf{Q}) \,\&\, (\mathbf{P} \vee \mathbf{R})$
(2) $\vdash \mathbf{P} \,\&\, (\mathbf{Q} \vee \mathbf{R}) \Leftrightarrow (\mathbf{P} \,\&\, \mathbf{Q}) \vee (\mathbf{P} \,\&\, \mathbf{R})$.

Metaproof

(1) First we show

$$(*) \qquad \vdash \mathbf{P} \vee (\mathbf{Q} \,\&\, \mathbf{R}) \Rightarrow (\mathbf{P} \vee \mathbf{Q}) \,\&\, (\mathbf{P} \vee \mathbf{R}).$$

Assume $\mathbf{P} \vee (\mathbf{Q} \,\&\, \mathbf{R})$. We use the method of cases to deduce $(\mathbf{P} \vee \mathbf{Q}) \,\&\, (\mathbf{P} \vee \mathbf{R})$. On the one hand,

$$\vdash \mathbf{P} \Rightarrow \mathbf{P} \vee \mathbf{Q}, \qquad \vdash \mathbf{P} \Rightarrow \mathbf{P} \vee \mathbf{R},$$

so

$$\vdash \mathbf{P} \Rightarrow (\mathbf{P} \vee \mathbf{Q}) \,\&\, (\mathbf{P} \vee \mathbf{R}).$$

On the other hand,

$$\vdash \mathbf{Q} \,\&\, \mathbf{R} \Rightarrow \mathbf{Q}, \qquad \vdash \mathbf{Q} \Rightarrow \mathbf{P} \vee \mathbf{Q},$$

$$\vdash \mathbf{Q} \,\&\, \mathbf{R} \Rightarrow \mathbf{R}, \qquad \vdash \mathbf{R} \Rightarrow \mathbf{P} \vee \mathbf{R},$$

so

$$\vdash \mathbf{Q} \,\&\, \mathbf{R} \Rightarrow (\mathbf{P} \vee \mathbf{Q}) \,\&\, (\mathbf{P} \vee \mathbf{R}).$$

Next we show the converse

$$\vdash (\mathbf{P} \vee \mathbf{Q}) \,\&\, (\mathbf{P} \vee \mathbf{R}) \Rightarrow \mathbf{P} \vee (\mathbf{Q} \,\&\, \mathbf{R})$$

of (*). Assume $(\mathbf{P} \vee \mathbf{Q}) \,\&\, (\mathbf{P} \vee \mathbf{R})$. In view of 2.27(4) it suffices to deduce $\neg \mathbf{P} \Rightarrow (\mathbf{Q} \,\&\, \mathbf{R})$. Assume $\neg \mathbf{P}$. From the first assumption we deduce $\mathbf{P} \vee \mathbf{Q}$. By 2.29, we deduce $\mathbf{Q}$. Similarly we can deduce $\mathbf{R}$. Thus,

$$(\mathbf{P} \vee \mathbf{Q}) \,\&\, (\mathbf{P} \vee \mathbf{R}), \neg \mathbf{P} \vdash \mathbf{Q} \,\&\, \mathbf{R},$$

so that

$$(\mathbf{P} \vee \mathbf{Q}) \,\&\, (\mathbf{P} \vee \mathbf{R}) \vdash \neg \mathbf{P} \Rightarrow \mathbf{Q} \,\&\, \mathbf{R}.$$

(2) follows from (1). $\square$

EXERCISES

A. Symbolize:
   (a) either **P** or **Q** but not both.
   (b) neither **P** nor **Q**.
   (c) **P** but not **Q**.
   (d) **P** yet not **Q**.
   (e) at least one of the two conditions **P** and **Q**.
   (f) **P** only when **Q**.
   (g) **P** just in case **Q**.
   (h) **P** unless **Q**.
   (i) **P** provided that **Q**.
   (j) **P** implies **Q**, and conversely.

B. Find expressions equivalent to the negations of the following:
   (a) x is rational and $x > 3$ and $x \neq 10/3$.
   (b) If $\sum_n a_n$ is convergent, then it is absolutely convergent.
   (c) V has a finite basis or V contains an infinite number of linearly independent vectors.

C. Explain the logic of the following arguments concerning real numbers:
   (a) $-1 < 0$, for otherwise $0 = -1 + 1 \geq 1$.
   (b) $x^2 \neq -1$ since $x^2 > 0 > -1$.
   (c) Suppose $x > 0$ and $(x + 1)(x - 2) = 0$. Then $x = 2$.
   (d) $x^2 + y^2 = 0$ if and only if $x = y = 0$. In fact, if $x = y = 0$, then clearly $x^2 + y^2 = 0$. Conversely, if $x \neq 0$ or $y \neq 0$, say $x \neq 0$, then $x^2 + y^2 > x^2 > 0$ so $x^2 + y^2 \neq 0$.
   (e) $\sqrt{x + 1} = \sqrt{x + 2}$ has no solutions, for from $\sqrt{x + 1} = \sqrt{x + 2}$ we deduce $x + 1 = x + 2$ whence $1 = 2$.

D. Let c be a positive real number and x be a real number. Recall the fact

$$(*) \qquad |x| < c \Leftrightarrow -c < x < c.$$

   (a) Explain how to solve the inequality $|x| \geq c$ on the basis of (*) and the results of this chapter.
   (b) Write out a proof of (*) (copy one from a textbook if you must) and justify in detail the logic of your proof.

E. Find three applications of the deduction criterion in any mathematics book.

F. The *inverse* of an implication is the converse of its contrapositive.
   (a) What is the inverse of $\mathbf{P} \Rightarrow \mathbf{Q}$?
   (b) Discuss the truth of the inverse of a true implication.

G. Express $\mathbf{P} \Rightarrow \mathbf{Q}$ solely in terms of **P**, **Q**, $\neg$, and &.

H. (For truth table buffs.) Show that the expressions in Axiom Schemes 1–4 are tautologies. Note that the truth table for an expression involving '**P**', '**Q**', '**R**' will have $8 = 2^3$ lines.

I. Justify: If $\neg \mathbf{P} \Rightarrow \mathbf{P}$ is true, then $\mathbf{P}$ is true.

J. Prove:

(a) $\mathbf{P} \,\&\, \mathbf{P} \Leftrightarrow \mathbf{P}$.

(b) $(\mathbf{P} \Rightarrow \mathbf{Q}) \,\&\, (\mathbf{R} \Rightarrow \mathbf{S}) \Rightarrow (\mathbf{P} \,\&\, \mathbf{R} \Rightarrow \mathbf{Q} \,\&\, \mathbf{S})$.

(c) $\mathbf{P} \,\&\, (\mathbf{P} \vee \mathbf{Q}) \Leftrightarrow \mathbf{P}$.

(d) $\mathbf{P} \,\&\, (\mathbf{P} \Rightarrow \mathbf{Q}) \Rightarrow \mathbf{Q}$.

K. Justify the following proof rules:

(a) If $\mathcal{A} \vdash \mathbf{P} \Leftrightarrow \mathbf{Q}$, then $\mathcal{A} \vdash \mathbf{Q} \Leftrightarrow \mathbf{P}$.

(b) If $\mathcal{A} \vdash \mathbf{P} \Leftrightarrow \mathbf{Q}$ and $\mathcal{A} \vdash \mathbf{Q} \Leftrightarrow \mathbf{R}$, then $\mathcal{A} \vdash \mathbf{P} \Leftrightarrow \mathbf{R}$.

L. Prove each of the following first by using the deduction criterion and then without using it:

(a) $(\mathbf{P} \Rightarrow \mathbf{Q}) \Rightarrow (\mathbf{P} \Rightarrow (\mathbf{P} \Rightarrow \mathbf{Q}))$.

(b) $\mathbf{P} \Rightarrow (\mathbf{Q} \Rightarrow \mathbf{R}) \Leftrightarrow \mathbf{Q} \Rightarrow (\mathbf{P} \Rightarrow \mathbf{R})$.

M. (a) Write a deduction of $\neg \mathbf{Q} \Rightarrow \neg \mathbf{P}$ from $\mathbf{P} \Rightarrow \mathbf{Q}$ and axioms.

(b) By applying to your deduction the *method* used to justify 2.28, construct a proof of

$$(\mathbf{P} \Rightarrow \mathbf{Q}) \Rightarrow (\neg \mathbf{Q} \Rightarrow \neg \mathbf{P}).$$

N. (a) Justify the following converse of the deduction criterion: Suppose $\mathcal{A} \vdash \mathbf{P} \Rightarrow \mathbf{Q}$. Then $\mathcal{A}, \mathbf{P} \vdash \mathbf{Q}$. In particular, if $\vdash \mathbf{P} \Rightarrow \mathbf{Q}$, then $\mathbf{P} \vdash \mathbf{Q}$.

(b) Use this new rule to shorten the metaproof of 2.29.

P. The *Scheffer stroke*, $|$, is defined as follows: $\mathbf{P} \mid \mathbf{Q}$ abbreviates $(\neg \mathbf{P}) \vee (\neg \mathbf{Q})$.

(a) Write the truth table for $\mathbf{P} \mid \mathbf{Q}$.

(b) Prove: $\neg \mathbf{P} \Leftrightarrow (\mathbf{P} \mid \mathbf{P})$,

$$(\neg \mathbf{P}) \mid (\neg \mathbf{Q}) \Leftrightarrow (\mathbf{P} \mid \mathbf{P}) \mid (\mathbf{Q} \mid \mathbf{Q}).$$

(c) Use (b) to find expressions involving solely $\mathbf{P}$, $\mathbf{Q}$, and $|$ which are equivalent to

$$\mathbf{P} \vee \mathbf{Q}, \qquad \mathbf{P} \,\&\, \mathbf{Q}, \qquad \mathbf{P} \Rightarrow \mathbf{Q}, \qquad \mathbf{P} \Leftrightarrow \mathbf{Q}.$$

(This shows that all of the logical connectives can be expressed in terms of the Scheffer stroke.)

3. Quantification

In this chapter we present three axiom schemes governing the meaning of the universal quantifier, define the existential quantifier, and formulate proof rules concerning these quantifiers. These rules constitute the (first-order) *predicate calculus.*

THE UNIVERSAL QUANTIFIER

Consider the statements:

(1) For every x, if x is a natural number, then $x^2 \geq 0$.
(2) For every y, if y is a natural number, then $y^2 \geq 0$.

The two statements are certainly different, for the letter 'x' appears in the first but not in the second. Nevertheless, the two statements have the same intuitive content: The square of every natural number is nonnegative. Axiom Scheme 5, below, provides a guarantee that they are in fact equivalent.

Some notation will be convenient. If **P** is a string, **x** is a letter, and **S** is a string, then

$$[\mathbf{S} \mid \mathbf{x}]\mathbf{P}$$

stands for the string obtained by replacing **x** each time it appears in **P** by **S**. For example, if **P** is

$$(\forall z)(x = z),$$

$\mathbf{y}$ is 'y', and $\mathbf{x}$ is 'x', then $[\mathbf{y} \mid \mathbf{x}]\mathbf{P}$ is

$$(\forall z)(y = z)$$

and if $\mathbf{w}$ is 'w' and $\mathbf{z}$ is 'z', then $[(\iota\mathbf{w})(\mathbf{w} = \mathbf{z}) \mid \mathbf{x}]\mathbf{P}$ is

$$(\forall z)((\iota w)(w = z) = z).$$

If $\mathbf{P}$ is statement (1) above, $\mathbf{y}$ is 'y', and $\mathbf{x}$ is 'x', then $[\mathbf{y} \mid \mathbf{x}]\mathbf{P}$ is statement (2). Of course, $[\mathbf{S} \mid \mathbf{x}]\mathbf{P}$ is $\mathbf{P}$ itself in case $\mathbf{x}$ does not appear in $\mathbf{P}$.

AXIOM SCHEME 5

Let $\mathbf{P}$ be a formula in which the letter $\mathbf{x}$ is free and in which the letter $\mathbf{y}$ does not appear. Then

$$(\forall \mathbf{x})(\mathbf{P}) \Leftrightarrow (\forall \mathbf{y})([\mathbf{y} \mid \mathbf{x}]\mathbf{P})$$

is an axiom.

By this scheme, the formula

$$(\forall x)(x = x) \Leftrightarrow (\forall y)(y = y)$$

is an axiom. Notice that the scheme does *not* say that

$$(\forall x)(x = y) \Leftrightarrow (\forall y)(y = y)$$

is an axiom, for 'y' appears in '$x = y$'.

Although Axiom Scheme 5 will prove to be redundant (see Chapter 3, Exercise N), it has been included for convenience.

The next axiom scheme expresses the idea that if a property $\mathbf{P}$ holds for every class $\mathbf{x}$, then $\mathbf{P}$ holds for any particular class $\mathbf{X}$.

AXIOM SCHEME 6

Let $\mathbf{P}$ be a formula in which the letter $\mathbf{x}$ is free, and let $\mathbf{X}$ be a term. Then

$$(\forall \mathbf{x})(\mathbf{P}) \Rightarrow [\mathbf{X} \mid \mathbf{x}]\mathbf{P}$$

is an axiom.

The intuitive content of the next axiom scheme is roughly as follows. Suppose a property $\mathbf{P}$ concerning an arbitrary but unspecified class $\mathbf{x}$ is implied by a statement $\mathbf{C}$ which does not concern $\mathbf{x}$. Then $\mathbf{C}$ implies that $\mathbf{P}$ holds for every class $\mathbf{x}$.

AXIOM SCHEME 7

Let $\mathbf{C}$ be a formula in which the letter $\mathbf{x}$ does not appear, and let $\mathbf{P}$ be a formula in which $\mathbf{x}$ is free. Then

$$(\mathbf{C} \Rightarrow \mathbf{P}) \Rightarrow (\mathbf{C} \Rightarrow (\forall \mathbf{x})(\mathbf{P}))$$

is an axiom.

In developing the predicate calculus, we use the following conventions for the remainder of this chapter. *The letters* '**P**', '**Q**', *and* '**C**' *always stand for formulas,* '**x**', '**y**', '**z**', *and* '**w**' *for letters, and* '$\mathcal{A}$' *for a list of formulas.*

Axiom Scheme 6 justifies the following two proof rules.

3.1. PROOF RULE (UNIVERSAL SPECIALIZATION—US)

Suppose **x** is free in **P** and **X** is a term. If $(\forall \mathbf{x})(\mathbf{P})$ is deducible from $\mathcal{A}$, so is $[\mathbf{X} \mid \mathbf{x}]\mathbf{P}$.

For example, consider the theorem 'Every differentiable function f is continuous', that is, 'For every f, if f is a differentiable function, then f is a continuous function'. By 3.1 one obtains the particular result 'If the sine function is a differentiable function, then the sine function is a continuous function'.

3.2. PROOF RULE

Suppose **x** is free in **P** and **X** is a term. If $[\mathbf{X} \mid \mathbf{x}]\mathbf{P}$ is false, then $(\forall \mathbf{x})(\mathbf{P})$ is false.

Justification

The contrapositive

$$\neg [\mathbf{X} \mid \mathbf{x}]\mathbf{P} \Rightarrow \neg (\forall \mathbf{x})(\mathbf{P})$$

of the implication in Axiom Scheme 6 is true. □

Thus, one way to "disprove" $(\forall \mathbf{x})(\mathbf{P})$, that is, to show that $(\forall \mathbf{x})(\mathbf{P})$ is false, is to exhibit a term **X** for which $[\mathbf{X} \mid \mathbf{x}]\mathbf{P}$ is false. Such a term is called a *counterexample* to $(\forall \mathbf{x})(\mathbf{P})$. The number '$-1$' is a counterexample to 'For every x, if x is a real number, then $\sqrt{x^2} = x$'.

3.3. PROOF RULE (UNIVERSAL GENERALIZATION—UG)

Let **x** be free in **P**. Suppose **P** is deducible from $\mathcal{A}$. Then

$$(\forall \mathbf{x})(\mathbf{P})$$

is deducible from $\mathcal{A}$.

Justification

Let **y** be a letter different from **x**, and let **C** be the formula

$$\neg (\mathbf{y} = \mathbf{y}) \vee (\mathbf{y} = \mathbf{y})$$

in which **x** does not appear. Since $\mathcal{A} \vdash \mathbf{P}$,

$$\mathcal{A} \vdash \mathbf{C} \Rightarrow \mathbf{P}.$$

By Axiom Scheme 7,

$$\mathcal{A} \vdash \mathbf{C} \Rightarrow (\forall \mathbf{x})(\mathbf{P}).$$

But $\vdash \mathbf{C}$, so $\mathcal{A} \vdash (\forall \mathbf{x})(\mathbf{P})$. □

Recall that no letter which is bound in the conclusion of a deduction is permitted to be free in any of its premises. This restriction must be observed scrupulously when 3.3 is used if contradiction is to be avoided (see Chapter 3, Exercise L).

Let **x** be free in **P**. According to 3.1 and 3.3, $(\forall \mathbf{x})(\mathbf{P})$ is true when **P** is true, and vice versa. Hence one often "proves" a theorem of the form $(\forall \mathbf{x})(\mathbf{P})$ by giving instead a proof of **P**, and then one sometimes adds at the end of the proof of **P** such a phrase as, "But **x** was arbitrary, so $(\forall \mathbf{x})(\mathbf{P})$ is true."

The next three metatheorems establish relationships between the universal quantifier and the logical connectives.

3.4. METATHEOREM

Let **x** be free in **P** and **Q**. Then

$$\vdash (\forall \mathbf{x})(\mathbf{P} \Rightarrow \mathbf{Q}) \Rightarrow ((\forall \mathbf{x})(\mathbf{P}) \Rightarrow (\forall \mathbf{x})(\mathbf{Q})).$$

Metaproof

Let **y** be a letter different from **x** and not appearing in **P** or in **Q**. By Axiom Scheme 6,

$$\vdash (\forall \mathbf{x})(\mathbf{P} \Rightarrow \mathbf{Q}) \Rightarrow ([\mathbf{y} \mid \mathbf{x}])\mathbf{P} \Rightarrow ([\mathbf{y} \mid \mathbf{x}]\mathbf{Q}),$$

so

$$\vdash [\mathbf{y} \mid \mathbf{x}]\mathbf{P} \Rightarrow ((\forall \mathbf{x})(\mathbf{P} \Rightarrow \mathbf{Q}) \Rightarrow [\mathbf{y} \mid \mathbf{x}]\mathbf{Q}).$$

Now

$$\vdash (\forall \mathbf{x})\mathbf{P} \Rightarrow [\mathbf{y} \mid \mathbf{x}]\mathbf{P},$$

so

$$\vdash (\forall \mathbf{x})(\mathbf{P}) \Rightarrow ((\forall \mathbf{x})(\mathbf{P} \Rightarrow \mathbf{Q}) \Rightarrow [\mathbf{y} \mid \mathbf{x}]\mathbf{Q}).$$

Hence

$$\vdash (\forall \mathbf{x})(\mathbf{P}) \;\&\; (\forall \mathbf{x})(\mathbf{P} \Rightarrow \mathbf{Q}) \Rightarrow [\mathbf{y} \mid \mathbf{x}]\mathbf{Q}.$$

By Axiom Scheme 7,

$$\vdash (\forall \mathbf{x})(\mathbf{P}) \;\&\; (\forall \mathbf{x})(\mathbf{P} \Rightarrow \mathbf{Q}) \Rightarrow (\forall \mathbf{y})([\mathbf{y} \mid \mathbf{x}]\mathbf{Q}).$$

By Axiom Scheme 5,

$$\vdash (\forall \mathbf{y})([\mathbf{y} \mid \mathbf{x}]\mathbf{Q}) \Leftrightarrow (\forall \mathbf{x})([\mathbf{x} \mid \mathbf{y}][\mathbf{y} \mid \mathbf{x}]\mathbf{Q}).$$

Now $[\mathbf{x} \mid \mathbf{y}][\mathbf{y} \mid \mathbf{x}]\mathbf{Q}$ is just **Q**, so

$$\vdash (\forall \mathbf{x})(\mathbf{P}) \;\&\; (\forall \mathbf{x})(\mathbf{P} \Rightarrow \mathbf{Q}) \Rightarrow (\forall \mathbf{x})(\mathbf{Q}).$$

This last formula is equivalent to the asserted theorem. □

3.5. METATHEOREM

Let $\mathbf{x}$ be free in $\mathbf{P}$ and $\mathbf{Q}$. Then

$$\vdash (\forall \mathbf{x})(\mathbf{P} \,\&\, \mathbf{Q}) \Leftrightarrow (\forall \mathbf{x})(\mathbf{P}) \,\&\, (\forall \mathbf{x})(\mathbf{Q}).$$

Metaproof

Let $\mathbf{y}$ be a letter different from $\mathbf{x}$ and not appearing in $\mathbf{P}$ or in $\mathbf{Q}$. By Axiom Scheme 6,

$$\vdash (\forall \mathbf{x})(\mathbf{P} \,\&\, \mathbf{Q}) \Rightarrow [\mathbf{y} \mid \mathbf{x}]\mathbf{P} \,\&\, [\mathbf{y} \mid \mathbf{x}]\mathbf{Q},$$

so

$$\vdash (\forall \mathbf{x})(\mathbf{P} \,\&\, \mathbf{Q}) \Rightarrow [\mathbf{y} \mid \mathbf{x}]\mathbf{P},$$

$$\vdash (\forall \mathbf{x})(\mathbf{P} \,\&\, \mathbf{Q}) \Rightarrow [\mathbf{y} \mid \mathbf{x}]\mathbf{Q}.$$

Using Axiom Schemes 7 and 5, we see

$$\vdash (\forall \mathbf{x})(\mathbf{P} \,\&\, \mathbf{Q}) \Rightarrow (\forall \mathbf{x})(\mathbf{P}) \,\&\, (\forall \mathbf{x})(\mathbf{Q}).$$

We show next the truth of the converse of the preceding formula. Since

$$\vdash (\forall \mathbf{x})(\mathbf{P}) \Rightarrow [\mathbf{y} \mid \mathbf{x}]\mathbf{P},$$

$$\vdash (\forall \mathbf{x})(\mathbf{Q}) \Rightarrow [\mathbf{y} \mid \mathbf{x}]\mathbf{Q},$$

we have

$$\vdash (\forall \mathbf{x})(\mathbf{P}) \,\&\, (\forall \mathbf{x})(\mathbf{Q}) \Rightarrow [\mathbf{y} \mid \mathbf{x}]\mathbf{P} \,\&\, [\mathbf{y} \mid \mathbf{x}]\mathbf{Q}.$$

Now $[\mathbf{y} \mid \mathbf{x}]\mathbf{P} \,\&\, [\mathbf{y} \mid \mathbf{x}]\mathbf{Q}$ is just $[\mathbf{y} \mid \mathbf{x}](\mathbf{P} \,\&\, \mathbf{Q})$, so from Axiom Schemes 7 and 5 it follows that

$$\vdash (\forall \mathbf{x})(\mathbf{P}) \,\&\, (\forall \mathbf{x})(\mathbf{Q}) \Rightarrow (\forall \mathbf{x})(\mathbf{P} \,\&\, \mathbf{Q}). \quad \square$$

To illustrate both 3.5 and 3.6, below, suppose we have proved

$$(\forall x)(x \in X \Rightarrow x \in Y), \qquad (\forall x)(x \in Y \Rightarrow x \in X).$$

Then we can conclude

$$\vdash (\forall x)(x \in X \Leftrightarrow x \in Y).$$

3.6. METATHEOREM

Let $\mathbf{x}$ be free in $\mathbf{P}$ and $\mathbf{Q}$. Then

$$\vdash (\forall \mathbf{x})(\mathbf{P} \Leftrightarrow \mathbf{Q}) \Rightarrow ((\forall \mathbf{x})(\mathbf{P}) \Leftrightarrow (\forall \mathbf{x})(\mathbf{Q})).$$

Metaproof

Use 3.4 and 3.5. $\square$

Interchanging the order of two occurrences of the universal quantifier in a statement does not change its truth or falsity:

3.7. METATHEOREM

Let $\mathbf{x}$ and $\mathbf{y}$ be free in $\mathbf{P}$. Then

$$\vdash (\forall \mathbf{x})(\forall \mathbf{y})(\mathbf{P}) \Leftrightarrow (\forall \mathbf{y})(\forall \mathbf{x})(\mathbf{P}).$$

Metaproof

We prove only the implication

$$(\forall \mathbf{x})(\forall \mathbf{y})(\mathbf{P}) \Rightarrow (\forall \mathbf{y})(\forall \mathbf{x})(\mathbf{P}),$$

the proof of the converse being similar.

Let $\mathbf{z}$ and $\mathbf{w}$ be distinct letters not appearing in $\mathbf{P}$ and different from both $\mathbf{x}$ and $\mathbf{y}$. By Axiom Scheme 6,

$$\vdash (\forall \mathbf{z})(\forall \mathbf{w})([\mathbf{z} \mid \mathbf{x}][\mathbf{w} \mid \mathbf{y}]\mathbf{P}) \Rightarrow (\forall \mathbf{w})([\mathbf{w} \mid \mathbf{y}]\mathbf{P}),$$

$$\vdash (\forall \mathbf{w})([\mathbf{w} \mid \mathbf{y}]\mathbf{P}) \Rightarrow \mathbf{P}$$

so

$$\vdash (\forall \mathbf{z})(\forall \mathbf{w})([\mathbf{z} \mid \mathbf{x}][\mathbf{w} \mid \mathbf{y}]\mathbf{P}) \Rightarrow \mathbf{P}.$$

By double application of Axiom Scheme 7,

$$(*) \qquad \vdash (\forall \mathbf{z})(\forall \mathbf{w})([\mathbf{z} \mid \mathbf{x}][\mathbf{w} \mid \mathbf{y}]\mathbf{P}) \Rightarrow (\forall \mathbf{y})(\forall \mathbf{x})(\mathbf{P}).$$

By Axiom Scheme 5,

$$\vdash (\forall \mathbf{w})([\mathbf{z} \mid \mathbf{x}][\mathbf{w} \mid \mathbf{y}]\mathbf{P}) \Leftrightarrow (\forall \mathbf{y})([\mathbf{z} \mid \mathbf{x}]\mathbf{P}).$$

By 3.6,

$$(**) \qquad \vdash (\forall \mathbf{z})(\forall \mathbf{w})([\mathbf{z} \mid \mathbf{x}][\mathbf{w} \mid \mathbf{y}]\mathbf{P}) \Leftrightarrow (\forall \mathbf{z})(\forall \mathbf{y})([\mathbf{z} \mid \mathbf{x}]\mathbf{P}).$$

By Axiom Scheme 5,

$$(***) \qquad \vdash (\forall \mathbf{z})(\forall \mathbf{y})([\mathbf{z} \mid \mathbf{x}]\mathbf{P}) \Leftrightarrow (\forall \mathbf{x})(\forall \mathbf{y})(\mathbf{P}).$$

From (*), (**), and (***) we obtain the truth of the desired implication. □

3.8. METATHEOREM

Suppose $\mathbf{x}$ does not appear in $\mathbf{C}$ and is free in $\mathbf{P}$. Then:

(1) $\vdash (\forall \mathbf{x})(\mathbf{C} \lor \mathbf{P}) \Leftrightarrow \mathbf{C} \lor (\forall \mathbf{x})(\mathbf{P})$.
(2) $\vdash (\forall \mathbf{x})(\mathbf{C} \Rightarrow \mathbf{P}) \Leftrightarrow \mathbf{C} \Rightarrow (\forall \mathbf{x})(\mathbf{P})$.
(3) $\vdash (\forall \mathbf{x})(\mathbf{C} \,\&\, \mathbf{P}) \Leftrightarrow \mathbf{C} \,\&\, (\forall \mathbf{x})(\mathbf{P})$.

THE EXISTENTIAL QUANTIFIER

3.9. DEFINITION

Let **P** be a formula in which the letter **x** is free. Then

$$(\exists \mathbf{x})(\mathbf{P})$$

is defined to be the formula

$$\neg(\forall \mathbf{x})(\neg \mathbf{P})$$

and is read as

there exists an **x** such that **P**
there is some **x** for which **P**
for some **x**, **P**
there is at least one **x** such that **P**

and so on. The symbol '$\exists$' is the *existential quantifier*.

The colloquial meaning of 'there is' is more restricted than its mathematical meaning just defined. The reporter who claims, "There is an elephant standing on top of the Washington Monument," is presumably prepared to support his claim by pointing to a particular pachyderm on top of that monument. The mathematician who claims, citing the fundamental theorem of algebra, "There is a root x of this polynomial p of degree 317," is saying that the denial "No x is a root of p" of his claim is absurd. *Just because* $(\exists \mathbf{x})(\mathbf{P})$ *is true, one need not be able to construct a specific term* **X** *satisfying* **P**, that is, for which $[\mathbf{X} \mid \mathbf{x}]\mathbf{P}$ is true. However, one way to prove $(\exists \mathbf{x})(\mathbf{P})$ is to construct such an **X**: To establish the existence of a root of the quadratic polynomial $ax^2 + bx + c$, the mathematician need only exhibit the particular root $(-b + (b^2 - 4ac)^{1/2})/2a$. This procedure is justified by:

3.10. PROOF RULE (EXISTENTIAL GENERALIZATION—EG)

Suppose **x** is free in **P** and **X** is a term. If $[\mathbf{X} \mid \mathbf{x}]\mathbf{P}$ is deducible from $\mathcal{A}$, so is $(\exists \mathbf{x})(\mathbf{P})$.

Rule 3.10 is justified by:

3.11. METATHEOREM

Let **x** be free in **P**, and let **X** be a term. Then

$$\vdash [\mathbf{X} \mid \mathbf{x}]\mathbf{P} \Rightarrow (\exists \mathbf{x})(\mathbf{P}).$$

Metaproof

Axiom Scheme 6 furnishes the axiom

$$(\forall \mathbf{x})(\neg \mathbf{P}) \Rightarrow [\mathbf{X} \mid \mathbf{x}](\neg \mathbf{P}).$$

Now $[\mathbf{X} \mid \mathbf{x}](\neg \mathbf{P})$ is $\neg [\mathbf{X} \mid \mathbf{x}]\mathbf{P}$, and so

$$\vdash \neg \neg [\mathbf{X} \mid \mathbf{x}]\mathbf{P} \Rightarrow \neg (\forall \mathbf{x})(\neg \mathbf{P}). \quad \square$$

Taking the term **X** to be, in particular, the letter **x**, we obtain from 3.11

$$\vdash \mathbf{P} \Rightarrow (\exists \mathbf{x})(\mathbf{P}).$$

The next metatheorem tells how to negate a quantified statement.

3.12. METATHEOREM

Let **x** be free in **P**. Then:

(1) $\vdash \neg (\forall \mathbf{x})(\mathbf{P}) \Leftrightarrow (\exists \mathbf{x})(\neg \mathbf{P})$,
(2) $\vdash \neg (\exists \mathbf{x})(\mathbf{P}) \Leftrightarrow (\forall \mathbf{x})(\neg \mathbf{P})$.

Metatheorem 3.12 may be used to transform the negation of a multiply quantified statement into a more tractable equivalent statement. For example,

$$\vdash \neg (\forall \mathbf{x})(\exists \mathbf{y})(\mathbf{P}) \Leftrightarrow (\exists \mathbf{x})(\forall \mathbf{y})(\neg \mathbf{P}).$$

The rule for writing such equivalences is: Change '$\forall$' to '$\exists$' and '$\exists$' to '$\forall$', and place '$\neg$' after instead of before the quantifiers.

Metatheorem 3.12 together with the definition of '$\exists$' may be used to derive from results concerning the universal quantifier corresponding results concerning the existential quantifier. For example, in the notation of Axiom Scheme 5, we have

$$\vdash (\exists \mathbf{x})(\mathbf{P}) \Leftrightarrow (\exists \mathbf{y})([\mathbf{y} \mid \mathbf{x}]\mathbf{P}).$$

We shall frequently use such derived results even though they have not been explicitly formulated. A few of these are given below, as are additional relations between the quantifiers and other connectives.

3.13. METATHEOREM

Suppose **x** and **y** are free in **P** and **Q**, and **x** does not appear in **C**. Then the following formulas are true.

(1) $(\forall \mathbf{x})(\mathbf{P}) \vee (\forall \mathbf{x})(\mathbf{Q}) \Rightarrow (\forall \mathbf{x})(\mathbf{P} \vee \mathbf{Q})$.
(2) $(\exists \mathbf{x})(\mathbf{P} \,\&\, \mathbf{Q}) \Rightarrow (\exists \mathbf{x})(\mathbf{P}) \,\&\, (\exists \mathbf{x})(\mathbf{Q})$.
(3) $(\exists \mathbf{x})(\mathbf{P} \vee \mathbf{Q}) \Leftrightarrow (\exists \mathbf{x})(\mathbf{P}) \vee (\exists \mathbf{x})(\mathbf{Q})$.

(4) $(\exists \mathbf{x})(\mathbf{P} \Rightarrow \mathbf{Q}) \Leftrightarrow (\forall \mathbf{x})(\mathbf{P}) \Rightarrow (\exists \mathbf{x})(\mathbf{Q})$.
(5) $(\exists \mathbf{x})(\exists \mathbf{y})(\mathbf{P}) \Leftrightarrow (\exists \mathbf{y})(\exists \mathbf{x})(\mathbf{P})$.
(6) $(\exists \mathbf{x})(\mathbf{C} \vee \mathbf{P}) \Leftrightarrow \mathbf{C} \vee (\exists \mathbf{x})(\mathbf{P})$.
(7) $(\exists \mathbf{x})(\mathbf{C} \,\&\, \mathbf{P}) \Leftrightarrow \mathbf{C} \,\&\, (\exists \mathbf{x})(\mathbf{P})$.
(8) $(\exists \mathbf{x})(\mathbf{C} \Rightarrow \mathbf{P}) \Leftrightarrow \mathbf{C} \Rightarrow (\exists \mathbf{x})(\mathbf{P})$.
(9) $(\exists \mathbf{x})(\mathbf{P} \Rightarrow \mathbf{C}) \Leftrightarrow (\forall \mathbf{x})(\mathbf{P}) \Rightarrow \mathbf{C}$.
(10) $(\forall \mathbf{x})(\mathbf{P} \Rightarrow \mathbf{C}) \Leftrightarrow (\exists \mathbf{x})(\mathbf{P}) \Rightarrow \mathbf{C}$.

From Axiom Scheme 7 we obtain the following result.

3.14. METATHEOREM

Suppose $\mathbf{x}$ is free in $\mathbf{P}$ and does not appear in $\mathbf{C}$. Then

$$\vdash (\mathbf{P} \Rightarrow \mathbf{C}) \Rightarrow ((\exists \mathbf{x})(\mathbf{P}) \Rightarrow \mathbf{C}).$$

Consider the two formulas:

(1) $(\exists \mathbf{x})(\forall \mathbf{y})(\mathbf{P})$,
(2) $(\forall \mathbf{y})(\exists \mathbf{x})(\mathbf{P})$.

Formula (2) says, roughly, that for each $\mathbf{y}$ there is some $\mathbf{x}$, $\mathbf{x}$ depending on the particular $\mathbf{y}$, such that $\mathbf{P}$ holds. In contrast, (1) says that there is a single $\mathbf{x}$ such that, whatever $\mathbf{y}$ may be, $\mathbf{P}$ holds. The two formulas certainly have different intuitive content. We show by example that *the implication*

(3) $(\forall \mathbf{y})(\exists \mathbf{x})(\mathbf{P}) \Rightarrow (\exists \mathbf{x})(\forall \mathbf{y})(\mathbf{P})$

may be false.

Take for $\mathbf{P}$ the formula '$x = y$'. From axioms given later we shall prove '$y = y$'. Since '$[y \mid x](x = y)$' is '$y = y$', '$(\exists x)(x = y)$' is true. Hence,

$$\vdash (\forall y)(\exists x)(x = y).$$

Now

$$\vdash \neg(\exists x)(\forall y)(x = y) \Leftrightarrow (\forall x)(\exists y)(\neg x = y).$$

We shall see later that

$$\vdash (\forall x)(\exists y)(\neg x = y).$$

Then

$$(\exists x)(\forall y)(x = y)$$

is false. Hence the implication (3) for the particular $\mathbf{P}$ here is false.

The converse of (3) is true, however. It is the reason, for example, that a uniformly continuous function is continuous (see Chapter 3, Exercise C).

3.15. METATHEOREM

Let $\mathbf{x}$ and $\mathbf{y}$ be free in $\mathbf{P}$. Then

$$\vdash (\exists \mathbf{x})(\forall \mathbf{y})(\mathbf{P}) \Rightarrow (\forall \mathbf{y})(\exists \mathbf{x})(\mathbf{P}).$$

Metaproof

Let $\mathbf{z}$ be a letter different from $\mathbf{x}$ and $\mathbf{y}$ not appearing in $\mathbf{P}$. By 3.11,

$$\vdash \mathbf{P} \Rightarrow (\exists \mathbf{z})([\mathbf{z} \mid \mathbf{x}]\mathbf{P}),$$

so by 3.4 and 3.3,

$$\vdash (\forall \mathbf{y})(\mathbf{P}) \Rightarrow (\forall \mathbf{y})(\exists \mathbf{z})([\mathbf{z} \mid \mathbf{x}]\mathbf{P}).$$

By 3.14,

$$\vdash (\exists \mathbf{x})(\forall \mathbf{y})(\mathbf{P}) \Rightarrow (\forall \mathbf{y})(\exists \mathbf{z})([\mathbf{z} \mid \mathbf{x}]\mathbf{P}).$$

But

$$\vdash (\forall \mathbf{y})(\exists \mathbf{z})([\mathbf{z} \mid \mathbf{x}]\mathbf{P}) \Leftrightarrow (\forall \mathbf{y})(\exists \mathbf{x})(\mathbf{P}). \quad \square$$

To motivate the next proof rule, consider an example from analysis. We are given two sequences $(a_n \mid n = 1, 2, \ldots)$ and $(b_n \mid n = 1, 2, \ldots)$ of real numbers and a positive real number ϵ. In addition we are given the information

$$(1) \quad (\exists N_1)(N_1 \text{ is a natural number} \\ \& \ (n > N_1 \Rightarrow |a_n| < \epsilon/2))$$

$$(2) \quad (\exists N_2)(N_2 \text{ is a natural number} \\ \& \ (n > N_2 \Rightarrow |b_n| < \epsilon/2)).$$

We wish to prove

$$(3) \quad (\exists N)(N \text{ is a natural number} \\ \& \ (n > N \Rightarrow |a_n + b_n| < \epsilon)).$$

To do this "choose" a natural number N_1 such that

$$n > N_1 \Rightarrow |a_n| < \epsilon/2,$$

and "choose" a natural number N_2 such that

$$n > N_2 \Rightarrow |b_n| < \epsilon/2;$$

such N_1 and N_2 exist by (1) and (2). Then $N_1 + N_2$ is a natural number,

$$n > N_1 + N_2 \Rightarrow |a_n| < \epsilon/2 \ \& \ |b_n| < \epsilon/2,$$

and since

$$|a_n| < \epsilon/2 \ \& \ |b_n| < \epsilon/2 \Rightarrow |a_n + b_n| < \epsilon,$$

(4) $N_1 + N_2$ is a natural number

$$\& \ (n > N_1 + N_2 \Rightarrow |\, a_n + b_n \,| < \epsilon).$$

From (4) we deduce (3) by existential generalization.

What does it mean to "choose" N_1 and N_2? We did not have any particular N_1 and N_2 given to us by (1) and (2), which are simply existential statements. Let us indicate more carefully the format of the supposed proof of (3). Let **A** and **B** be the respective statements

$$N_1 \text{ is a natural number } \& \ (n > N_1 \Rightarrow |\, a_n \,| < \epsilon/2),$$
$$N_2 \text{ is a natural number } \& \ (n > N_2 \Rightarrow |\, b_n \,| < \epsilon/2).$$

Then (1) and (2) give the truth of

$$(\exists \mathbf{x})(\mathbf{P}),$$

where **P** is

$$\mathbf{x} \text{ is a pair } (N_1, N_2) \ \& \ \mathbf{A} \ \& \ \mathbf{B}.$$

Let **C** be statement (4). We showed that

$$\mathbf{P} \Rightarrow \mathbf{C}$$

is true. We then concluded that **C** is true. The correctness of this argument is assured by the next proof rule.

3.16. PROOF RULE (METHOD OF CHOICE)

Let **x** be free in **P** and not appear in **C**. Suppose

$$(\exists \mathbf{x})(\mathbf{P}), \qquad \mathbf{P} \Rightarrow \mathbf{C}$$

are deducible from $\mathcal{A}$. Then **C** is deducible from $\mathcal{A}$.

Justification

First of all,

$$\mathcal{A} \vdash \neg\mathbf{C} \Rightarrow \neg\mathbf{P}.$$

By Axiom Scheme 7,

$$\mathcal{A} \vdash \neg\mathbf{C} \Rightarrow (\forall \mathbf{x})(\neg\mathbf{P}),$$

hence

$$\mathcal{A} \vdash \neg(\forall \mathbf{x})(\neg\mathbf{P}) \Rightarrow \neg\neg\mathbf{C}.$$

Then

$$\mathcal{A} \vdash (\exists \mathbf{x})(\mathbf{P}) \Rightarrow \mathbf{C}.$$

But $\mathcal{A} \vdash (\exists \mathbf{x})(\mathbf{P})$. □

A typical use of the method of choice to prove **C** true might run: "Choose an **x** such that **P**; such an **x** exists because .... Thus from the assertion that **P** holds for our chosen **x**, we conclude that **C** also holds. Hence **C** is true."

In employing this analog of the intuitive "choice" of an **x** such that **P**, the Herbrand-Tarski deduction criterion is often used to show that $\mathbf{P} \Rightarrow \mathbf{C}$ is deducible from $\mathcal{A}$. Moreover, existential generalization is often used to show that $(\exists \mathbf{x})(\mathbf{P})$ is deducible from $\mathcal{A}$.

UNIQUE EXISTENCE

The existential and universal quantifiers may be combined to express the concept of unique existence.

3.17. DEFINITION

Let **x** be free in **P**, and let the letter $\mathbf{x}'$ not appear in **P**. Then

there exists at most one **x** such that **P**

is defined to mean

$$(\forall \mathbf{x})(\forall \mathbf{x}')(\mathbf{P} \,\&\, [\mathbf{x}' \mid \mathbf{x}]\mathbf{P} \Rightarrow \mathbf{x} = \mathbf{x}').$$

Moreover,

$$(\exists! \, \mathbf{x})(\mathbf{P})$$

is defined to mean

$(\exists \mathbf{x})(\mathbf{P})$ & (there exists at most one **x** such that **P**)

and is read

there exists a unique **x** such that **P**
there is one and only one **x** for which **P**
for exactly one **x**, **P**

and the like.

According to its very definition, $(\exists! \, \mathbf{x})(\mathbf{P})$ may be proved by showing separately *existence*—$(\exists \mathbf{x})(\mathbf{P})$—and *uniqueness*—there exists at most one **x** such that **P**. For example, consider

(*) There exists a unique x such that $x + 2 = 3$.

Existence follows by existential generalization from '$1 + 2 = 3$'. Uniqueness follows from

$$x + 2 = 3 \,\&\, x' + 2 = 3 \Rightarrow x + 2 = x' + 2 \Rightarrow x = x'.$$

The method just used to prove (*) has the disadvantage that one must first know a class **X** for which $\mathbf{X} + 2 = 3$. This disadvantage is overcome by a different approach which consists essentially of proving uniqueness first: "If $x + 2 = 3$, then $x = 1$; thus the only possible choice for a class X to satisfy $X + 2 = 3$ is the class 1. And in fact, $1 + 2 = 3$." That this approach proves (*) follows from 3.18, whose complete justification depends on results of Chapter 4.

3.18. PROOF RULE

Let $\mathbf{x}$ be free in $\mathbf{P}$, and let $\mathbf{X}$ be a term. Suppose

$$\mathbf{P} \Rightarrow \mathbf{x} = \mathbf{X}, \qquad [\mathbf{X} \mid \mathbf{x}]\mathbf{P}$$

are deducible from $\mathcal{A}$. Then $(\exists! \, \mathbf{x})(\mathbf{P})$ is deducible from $\mathcal{A}$.

Justification

Since $\mathcal{A} \vdash [\mathbf{X} \mid \mathbf{x}]\mathbf{P}$,

$$\mathcal{A} \vdash (\exists \mathbf{x})(\mathbf{P})$$

by existential generalization. Since

$$\mathcal{A} \vdash \mathbf{P} \Rightarrow \mathbf{x} = \mathbf{X},$$

also

$$\mathcal{A} \vdash [\mathbf{x}' \mid \mathbf{x}]\mathbf{P} \Rightarrow \mathbf{x}' = \mathbf{X}.$$

Hence

$$\mathcal{A} \vdash \mathbf{P} \,\&\, [\mathbf{x}' \mid \mathbf{x}]\mathbf{P} \Rightarrow \mathbf{x} = \mathbf{X} \,\&\, \mathbf{x}' = \mathbf{X}.$$

By Theorem 4.2,

$$\mathbf{x} = \mathbf{X} \,\&\, \mathbf{x}' = \mathbf{X} \Rightarrow \mathbf{x} = \mathbf{x}'.$$

Hence

$$\mathcal{A} \vdash \mathbf{P} \,\&\, [\mathbf{x}' \mid \mathbf{x}]\mathbf{P} \Rightarrow \mathbf{x} = \mathbf{x}'.$$

By universal generalization,

$\mathcal{A} \vdash$ there exists at most one $\mathbf{x}$ such that $\mathbf{P}$. □

RESTRICTED QUANTIFICATION

Many a theorem is not an unqualified universal statement 'for every x, $x = x$' but rather a restricted, qualified statement such as 'for every element x of X, $x \neq y$' or 'for every real number $x < 1$, $x^2 < 1$' in which only those x's satisfying an auxiliary condition are of interest. Similarly, in 'there exists a real number $x < 1$ such that $x^2 < \frac{1}{2}$' the additional restriction that x be a real number less than 1 is imposed.

3.19. DEFINITION

Let $\mathbf{x}$ be free in $\mathbf{C}$ and in $\mathbf{P}$. Then

$$(\forall \mathbf{x}, \mathbf{C})(\mathbf{P})$$

is defined to mean

$$(\forall \mathbf{x})(\mathbf{C} \Rightarrow \mathbf{P}),$$

and

$$(\exists \mathbf{x}, \mathbf{C})(\mathbf{P})$$

is defined to mean

$$(\exists \mathbf{x})(\mathbf{C} \,\&\, \mathbf{P}).$$

For example, $(\forall \mathbf{x}, \mathbf{x} \in \mathbf{X})(\mathbf{P})$ means $(\forall \mathbf{x})(\mathbf{x} \in \mathbf{X} \Rightarrow \mathbf{P})$. More specifically, 'for every $x \in \mathbb{R}$, $x^2 \geq 0$' means 'for every x, if $x \in \mathbb{R}$, then $x^2 \geq 0$'; here $\mathbb{R}$ is the set of all real numbers.

A statement of the form $(\forall \mathbf{x}, \mathbf{C})(\mathbf{P})$ or $(\exists \mathbf{x}, \mathbf{C})(\mathbf{P})$ is said to be obtained by "restricted quantification," and $\mathbf{C}$ is called the *restriction* (*on* $\mathbf{x}$).

One often symbolizes $(\forall \mathbf{x}, \mathbf{C})(\mathbf{P})$ by displaying $\mathbf{P}$ on a separate line with the restriction $\mathbf{C}$ enclosed in parentheses to the extreme right. For example,

$$x^2 \geq 0 \qquad\qquad (x \in \mathbb{R})$$

stands for '$(\forall x, x \in \mathbb{R})(x^2 \geq 0)$'.

3.20. METATHEOREM

Let $\mathbf{x}$ be free in $\mathbf{C}$ and in $\mathbf{P}$. Then:

(1) $\vdash \neg(\forall \mathbf{x}, \mathbf{C})(\mathbf{P}) \Leftrightarrow (\exists \mathbf{x}, \mathbf{C})(\neg \mathbf{P})$.
(2) $\vdash \neg(\exists \mathbf{x}, \mathbf{C})(\mathbf{P}) \Leftrightarrow (\forall \mathbf{x}, \mathbf{C})(\neg \mathbf{P})$.

Metatheorem 3.20 allows us to negate formulas having repeated restricted quantification in the same way as for ordinary quantification. For example, if $\mathbf{x}$ and $\mathbf{y}$ are free in $\mathbf{C}$, $\mathbf{D}$, and $\mathbf{P}$, then

$$\neg(\forall \mathbf{x}, \mathbf{C})(\exists \mathbf{y}, \mathbf{D})(\mathbf{P}) \Leftrightarrow (\exists \mathbf{x}, \mathbf{C})(\forall \mathbf{y}, \mathbf{D})(\neg \mathbf{P}).$$

More specifically, to deny

for each $x \in \mathbb{R}$ there exists some $y \in \mathbb{R}$
for which $y > x$

is to affirm

there exists an $x \in \mathbb{R}$ such that for each
$y \in \mathbb{R}$, $y \not> x$.

3.21. PROOF RULE

Let $\mathbf{x}$ be free in $\mathbf{C}$ and in $\mathbf{P}$.

(1) If $(\exists \mathbf{x}, \mathbf{C})(\mathbf{P})$ is true, so is $(\exists \mathbf{x})(\mathbf{C})$.
(2) If $\mathbf{C}$ is false, then $(\forall \mathbf{x}, \mathbf{C})(\mathbf{P})$ is true.

Justification

(2) Suppose $\mathbf{C}$ is false. Then the formula $\mathbf{C} \Rightarrow \mathbf{P}$ is true, so $(\forall \mathbf{x})(\mathbf{C} \Rightarrow \mathbf{P})$ is also true. □

Rule 3.21 says, for example, that if a class $\mathbf{X}$ has no elements, then the statement 'For every element $\mathbf{x}$ of $\mathbf{X}$, $\mathbf{P}$' is true.

Most of our earlier results concerning ordinary quantification yield useful analogous properties of restricted quantification. The reader should have no difficulty in formulating these.

EXERCISES

A. Symbolize:

(a) For no $\mathbf{x}$, $\mathbf{P}$.

(b) For all $\mathbf{x}$, not $\mathbf{P}$.

(c) For not all $\mathbf{x}$ is $\mathbf{P}$.

(d) $\mathbf{P}$ for some but not all $\mathbf{x}$.

(e) For arbitrary $\mathbf{x} \in \mathbf{X}$, $\mathbf{P}$.

(f) There are an $\mathbf{x}$ and a $\mathbf{y}$ such that $\mathbf{P}$.

(g) For each $\mathbf{x}$ and each $\mathbf{y}$, $\mathbf{P}$.

(h) There exists an $\mathbf{x} \in \mathbf{X}$.

(i) There are at least two $\mathbf{x}$'s such that $\mathbf{P}$.

(j) If any $\mathbf{x} = \mathbf{x}$, then $1 = 1$.

(k) If any $\mathbf{x} \neq \mathbf{x}$, then $1 + 1 = 5$.

B. Find a statement with no negation sign preceding the quantifiers equivalent to:

(a) $(\exists \mathbf{x})(\mathbf{P} \,\&\, (\forall \mathbf{y})(\mathbf{Q}))$.

(b) $(\exists ! \mathbf{x})(\mathbf{P})$.

(c) $x \in X \Leftrightarrow x > 1 \qquad (x \in \mathbb{R})$.

(d) The standard ϵ, δ-definition of

$$\lim_{x \to a} f(x) = b.$$

C. Let f be a real-valued function defined on the entire real line. Using the results of this chapter:

(a) Symbolize the usual ϵ, δ-definitions of 'f is uniformly continuous', 'f is continuous'.

(b) Prove that uniform continuity of f implies continuity of f.

(c) Negate the definitions of 'f is uniformly continuous', 'f is continuous'.

(d) Prove that continuity of f does not necessarily imply uniform continuity of f.

D. Explain the conventional meanings of:

(a) $\tan x$ exists for $-\pi/2 < x < \pi/2$ but not for $x = -\pi/2$, $\pi/2$.

(b) $\lim_{x \to 0} f(x)$ exists.

(c) If f is continuous on $[a,b]$, then

$$\int_a^b f(x)\,dx$$

exists.

E. One statement of the fundamental theorem of the calculus is: If f is a continuous function on $[a,b]$, then f has an antiderivative on $[a,b]$. To prove it one shows

$$\frac{d}{dx}\int_a^x f(t)\,dt = f(x) \qquad (x \in [a,b]).$$

Why does this constitute a proof?

F. Let **P** be the statement

x is a real number and y is a real number and $x < y$.

Explain the intuitive meanings of $(\forall x)(\exists y)(\mathbf{P})$ and $(\exists y)(\forall x)(\mathbf{P})$ and discuss the logical relationships between these two statements.

G. Make explicit the implicit use of proof rules in the following proofs:

(a) $\lim_{x\to a} 2 = 2$. *Proof.* Let $\epsilon > 0$. We want to find a $\delta > 0$ such that $|2 - 2| < \epsilon$ for $|x - a| < \delta$. However, $|2 - 2| = 0 < \epsilon$, so any $\delta > 0$ will do.

(b) $\lim_{x\to a} x = a$. *Proof.* Let $\epsilon > 0$ be given. We shall find a δ depending on ϵ with the property that $|x - a| < \delta$ implies $|x - a| < \epsilon$. We may take $\delta = \epsilon$.

(c) $\lim_{x\to a} 2x = 2a$. *Proof.* Let $\epsilon > 0$. Since $\lim_{x\to a} x = a$, there exists $\delta > 0$ such that $|x - a| < \epsilon/2$ when $|x - a| < \delta$. For this δ, $|x - a| < \delta$ yields $|2x - 2a| = 2|x - a| < 2(\epsilon/2) = \epsilon$.

H. Symbolize:

(a) The differential equation $f'(x) = 0$ has a unique solution satisfying the initial condition $f(0) = 2$.

(b) Every polynomial with complex coefficients has a complex root.

(c) $\int x\,dx = x^2/2 + C$.

I. Are the converses of 3.13(1) and (2) always true?

J. Prove: $(\forall \mathbf{x})(\mathbf{P}) \Rightarrow (\exists \mathbf{x})(\mathbf{P})$. [*Hint*: A letter **y** different from **x** and not appearing in **P** is a term.]

K. Prove:

(a) $(\forall \mathbf{x})(\forall \mathbf{y})(\forall \mathbf{z})(\mathbf{P}) \Leftrightarrow (\forall \mathbf{y})(\forall \mathbf{z})(\forall \mathbf{x})(\mathbf{P})$.

(b) The analog of (a) for '$\exists$'.

L. (a) Prove: $(\forall \mathbf{x})(\mathbf{P}) \Rightarrow \mathbf{P}$.

(b) Show that

(*) $$\mathbf{P} \Rightarrow (\forall \mathbf{x})(\mathbf{P})$$

is not necessarily true. [*Hint*: Take **P** to be '$x = 0$'.]

(c) Discover the flaw in the following purported proof of (*): Assume **P**. Now $\mathbf{P} \Rightarrow \mathbf{P}$ is true, so **P** is deducible from **P**. By 3.3, $(\forall \mathbf{x})(\mathbf{P})$ is also deducible from **P**. Thus (*) is true by the Herbrand-Tarski deduction criterion.

M. Find the flaw in the following "proof" of the converse of 3.15: Assume

$$(\forall \mathbf{y})(\exists \mathbf{x})(\mathbf{P}).$$

Then

$$(\exists \mathbf{x})(\mathbf{P})$$

follows by universal specialization. Choose some **x** for which **P** holds. By 3.3, $(\forall \mathbf{y})(\mathbf{P})$ holds. Hence

$$(\exists \mathbf{x})(\forall \mathbf{y})(\mathbf{P})$$

follows by existential generalization.

N. Show that Axiom Scheme 5 is redundant by deducing it from Axiom Schemes 6 and 7.

4. Classes and Sets

The study of set theory proper is begun in this chapter. The connection between equality and membership is presented. The totality of all classes is split into two kinds—sets and proper classes. The classifier, a device for constructing specific classes, is defined in terms of the descriptor.

In this chapter we move closer to the informal style of exposition used by working mathematicians.

EQUALITY AND MEMBERSHIP

For a formula **P** in which **x** is free and for a term **z**, to assert $[\mathbf{z} \mid \mathbf{x}]\mathbf{P}$ is to say that **z** has the property expressed by **P**. Hence Axiom Scheme 8 says that if two objects are equal, then they have precisely the same properties.

AXIOM SCHEME 8

Let **P** be a formula in which the letter **x** is free, and let **X** and **Y** be terms. Then

$$\mathbf{X} = \mathbf{Y} \Rightarrow ([\mathbf{X} \mid \mathbf{x}]\mathbf{P} \Leftrightarrow [\mathbf{Y} \mid \mathbf{x}]\mathbf{P})$$

is an axiom.

For example, take **P** to be '$x = z$'. Then Axiom Scheme 8 gives the implicit axiom

$$X = Y \Rightarrow (X = z \Leftrightarrow Y = z).$$

We remind the reader that he may meaningfully call two classes equal, that is, write '$\mathbf{X} = \mathbf{Y}$' for terms **X** and **Y**, and may call two statements equivalent, that is, write '$\mathbf{P} \Leftrightarrow \mathbf{Q}$' for formulas **P** and **Q**, but not vice versa. Although we say, in the metalanguage, that a statement named or abbreviated **P** *is* a statement named or abbreviated **Q**, the expression '**P** equals **Q**' for formulas **P** and **Q** has no meaning either in the metalanguage or in the formal language.

Our first explicit axiom says that two classes are equal if they have the same elements.

AXIOM 1 (AXIOM OF EXTENSION)

$$(\forall x)(x \in X \Leftrightarrow x \in Y) \Rightarrow (X = Y).$$

Actually, two classes are equal if and only if they have the same elements.

4.1. THEOREM

Let X and Y be classes. Then

$$X = Y \Leftrightarrow (\forall x)(x \in X \Leftrightarrow x \in Y).$$

Proof

In Axiom Scheme 8 take **x** to be 'y', **P** to be '$x \in y$', **X** to be 'X', and **Y** to be 'Y' to obtain the axiom

$$X = Y \Rightarrow ([X \mid y]x \in y \Leftrightarrow [Y \mid y]x \in y),$$

that is,

$$X = Y \Rightarrow (x \in X \Leftrightarrow x \in Y).$$

Then

$$X = Y \Rightarrow (\forall x)(x \in X \Leftrightarrow x \in Y).$$

The converse of this implication is just Axiom 1. ☐

The peculiar admixture of informal, formal, and metalanguage in the formulations of this theorem and its proof deserves comment. The second sentence after 'Theorem' says 'Then **P**', where **P** is the (abbreviated) formula

$$X = Y \Leftrightarrow (\forall x)(x \in X \Leftrightarrow x \in Y)$$

of the formal language. The sentence's intended meaning is, of course, 'Then **P** is true'. The first sentence, 'Let **X** and **Y** be classes', is entirely superfluous and is intended solely to guide our understanding. In the technical sense of 'theorem', the theorem asserted is simply **P**, or what amounts to the same thing, $(\forall X)(\forall Y)(\mathbf{P})$.

The three sentences following 'Proof' might be called an *informal proof*; by their (implicit) use of our proof rules they only suggest how to construct a formal proof in the sense of Chapter 1.

Abbreviate

$$x \in X \Leftrightarrow x \in Y$$

by **A**. Then the following formulas are the principal steps in a formal proof of 4.1:

(1) $X = Y \Rightarrow \mathbf{A}$.
(2) $(X = Y \Rightarrow \mathbf{A}) \Rightarrow (X = Y \Rightarrow (\forall x)(\mathbf{A}))$.
(3) $X = Y \Rightarrow (\forall x)(\mathbf{A})$.
(4) $(\forall x)(\mathbf{A}) \Rightarrow X = Y$.
(5) $X = Y \Leftrightarrow (\forall x)(\mathbf{A})$.

Here (1) and (2) are implicit axioms obtained from Axiom Schemes 8 and 7 respectively. Line (3) appears below (1) and (2) in accordance with condition (D2) in the definition of 'deduction'. Line (4) is just Axiom 1. Proof Rule 2.19 informally justifies writing (5) after (3) and (4); in an actual formal proof many additional steps would be inserted between (4) and (5), these steps repeating the deduction of (5) from (3) and (4) suggested by the justification of 2.19.

4.2. THEOREM

Let X, Y, and Z be classes. Then:

(1) $X = X$.
(2) If $X = Y$, then $Y = X$.
(3) If $X = Y$ and $Y = Z$, then $X = Z$.

Proof

(1) The idea of the proof is to replace 'Y' by 'X' in Axiom 1 to obtain

$$(\forall x)(x \in X \Leftrightarrow x \in X) \Rightarrow X = X$$

and then to note that '$(\forall x)(x \in X \Leftrightarrow x \in X)$' is true. Formally:

$$(\forall x)(x \in X \Leftrightarrow x \in Y) \Rightarrow X = Y$$
$$(\forall Y)((\forall x)(x \in X \Leftrightarrow x \in Y) \Rightarrow X = Y)$$
$$(\forall Y)((\forall x)(x \in X \Leftrightarrow x \in Y) \Rightarrow X = Y) \Rightarrow$$
$$((\forall x)(x \in X \Leftrightarrow x \in X) \Rightarrow X = X)$$
$$(\forall x)(x \in X \Leftrightarrow x \in X) \Rightarrow X = X$$
$$x \in X \Leftrightarrow x \in X$$
$$(\forall x)(x \in X \Leftrightarrow x \in X)$$
$$X = X.$$

Again, this is not a complete formal proof. For example, many steps between the first and second lines have been omitted.

(2) By Axiom Scheme 8,

$$X = Y \Rightarrow (x \in X \Leftrightarrow x \in Y).$$

Since

$$(x \in X \Leftrightarrow x \in Y) \Rightarrow (x \in Y \Leftrightarrow x \in X),$$

$$X = Y \Rightarrow (x \in Y \Leftrightarrow x \in X).$$

By Axiom Scheme 7,

$$X = Y \Rightarrow (\forall x)(x \in Y \Leftrightarrow x \in X).$$

By Axiom 1,

$$(\forall x)(x \in Y \Leftrightarrow x \in X) \Rightarrow Y = X.$$

Hence

$$X = Y \Rightarrow Y = X.$$

(3) follows from Axiom Scheme 8, Axiom 1, and the implication

$$(x \in X \Leftrightarrow x \in Y) \;\&\; (x \in Y \Leftrightarrow x \in Z)$$
$$\Rightarrow (x \in X \Leftrightarrow x \in Z). \quad \square$$

The "transitivity" of equality expressed by (3) suggests the notation '$X = Y = Z$' for '$X = Y \;\&\; Y = Z$'.

Notice the use of the expression 'If . . ., then . . .' in (2) and (3) to replace the implication sign '$\Rightarrow$'. We shall more and more be replacing symbols in the formal language, and their abbreviations, by their renderings into ordinary language.

4.3. DEFINITION

If **X** and **Y** are terms, then

$$\mathbf{X} \neq \mathbf{Y}$$

is defined to mean $\neg(\mathbf{X} = \mathbf{Y})$ and is read "**X** is not equal to **Y**;" one also says that **X** is *distinct from* **Y** to mean $\mathbf{X} \neq \mathbf{Y}$.

If **x** and **X** are terms, then

$$\mathbf{x} \notin \mathbf{X}$$

is defined to mean $\neg(\mathbf{x} \in \mathbf{X})$ and is read "**x** is not an element of **X**," "**x** does not belong to **X**," and the like.

Theorem 4.1 provides a useful test for the inequality of two classes X and Y:

$$X \neq Y \Leftrightarrow (\exists x \in X)(x \notin Y) \vee (\exists y \in Y)(y \notin X).$$

SUBCLASSES

4.4. DEFINITION

Let $\mathbf{X}$ and $\mathbf{Y}$ be terms. One says that $\mathbf{X}$ is *contained in* $\mathbf{Y}$ (or is *included in* $\mathbf{Y}$) and calls $\mathbf{X}$ a *subclass of* $\mathbf{Y}$ and writes

$$\mathbf{X} \subset \mathbf{Y}$$

to mean each element of $\mathbf{X}$ is an element of $\mathbf{Y}$, that is,

$$(\forall \mathbf{x})(\mathbf{x} \in \mathbf{X} \Rightarrow \mathbf{x} \in \mathbf{Y}),$$

where $\mathbf{x}$ is a letter appearing neither in $\mathbf{X}$ nor in $\mathbf{Y}$. Instead of $\mathbf{X} \subset \mathbf{Y}$, one also writes

$$\mathbf{Y} \supset \mathbf{X}$$

and says that $\mathbf{Y}$ *contains* $\mathbf{X}$. The negations of $\mathbf{X} \subset \mathbf{Y}$ and $\mathbf{Y} \supset \mathbf{X}$ are abbreviated by $\mathbf{X} \not\subset \mathbf{Y}$ and $\mathbf{Y} \not\supset \mathbf{X}$. Finally, one says that $\mathbf{X}$ is *strictly contained in* $\mathbf{Y}$ and calls $\mathbf{X}$ a *strict subclass of* $\mathbf{Y}$ and writes

$$\mathbf{X} \subsetneq \mathbf{Y}$$

to mean $\mathbf{X} \subset \mathbf{Y} \;\&\; \mathbf{X} \neq \mathbf{Y}$ (we avoid the more common terms 'properly' and 'proper' for 'strictly' and 'strict' because we introduce 'proper' in another context below).

The distinction between $\mathbf{X}$ being contained in $\mathbf{Y}$, $\mathbf{X} \subset \mathbf{Y}$, and $\mathbf{X}$ being an element of $\mathbf{Y}$, $\mathbf{X} \in \mathbf{Y}$, must be observed scrupulously. For example, the set of all positive integers is contained in, but is not an element of, the set of all integers. However, the assertions $\mathbf{X} \subset \mathbf{Y}$ and $\mathbf{X} \in \mathbf{Y}$ are by no means inconsistent, for our definition of natural numbers will demand that $m \subset n \Leftrightarrow m \in n$ for any two distinct natural numbers m and n.

4.5. THEOREM

Let X, Y, and Z be classes. Then:

(1) $X \subset X$.
(2) If $X \subset Y$ and $Y \subset Z$, then $X \subset Z$.
(3) $X = Y$ if and only if $X \subset Y$ and $Y \subset X$.

Proof

(3) By 4.1,

$$X = Y \Leftrightarrow (\forall x)(x \in X \Leftrightarrow x \in Y).$$

By 3.5,

$$(\forall x)(x \in X \Leftrightarrow x \in Y) \Leftrightarrow$$
$$(\forall x)(x \in X \Rightarrow x \in Y) \;\&\; (\forall x)(x \in Y \Rightarrow x \in X),$$

that is,

$$(\forall x)(x \in X \Leftrightarrow x \in Y) \Leftrightarrow X \subset Y \;\&\; Y \subset X. \quad \square$$

In view of (1), the truth of $X \subset Y$ does not preclude the possibility that $X = Y$. Statement (2) suggests the notation $X \subset Y \subset Z$ for $X \subset Y$ & $Y \subset Z$. Statement (3) is just a restatement of 4.1 and says that two classes are equal precisely when each contains the other.

THE DESCRIPTOR

Axiom Scheme 9 is technical in nature. It says that the letter **x** in $(\iota\mathbf{x})(\mathbf{P})$ is a "dummy variable" and, subject to certain restrictions, may be replaced by any other letter. The situation is analogous to that in calculus, where one has

$$\int_a^b f(x)\,dx = \int_a^b f(y)\,dy;$$

of course one would not replace the dummy variable 'x' in the integral by 'f'.

AXIOM SCHEME 9

Let **P** be a formula in which the letter **x** is free and in which the letter **y** does not appear. Then

$$(\iota\mathbf{x})(\mathbf{P}) = (\iota\mathbf{y})([\mathbf{y} \mid \mathbf{x}]\mathbf{P})$$

is an axiom.

The next scheme says that if two properties are equivalent, then the object having the first property is equal to the object having the second (if there are such objects).

AXIOM SCHEME 10

Let **P** and **Q** be formulas in which the letter **x** is free. Then

$$(\forall\mathbf{x})(\mathbf{P} \Leftrightarrow \mathbf{Q}) \Rightarrow (\iota\mathbf{x})(\mathbf{P}) = (\iota\mathbf{x})(\mathbf{Q})$$

is an axiom.

Axiom Scheme 11 furnishes the interpretation of $(\iota\mathbf{x})(\mathbf{P})$ as the one, unique object having property $\mathbf{P}$: If there exists a unique $\mathbf{x}$ satisfying $\mathbf{P}$, then $(\iota\mathbf{x})(\mathbf{P})$ is equal to any object having property $\mathbf{P}$.

AXIOM SCHEME 11

Let $\mathbf{P}$ be a formula in which the letter $\mathbf{x}$ is free and in which the letter $\mathbf{y}$ does not appear. Then

$$(\exists!\mathbf{x})(\mathbf{P}) \Rightarrow (\forall\mathbf{y})([\mathbf{y} \mid \mathbf{x}]\mathbf{P} \Leftrightarrow \mathbf{y} = (\iota\mathbf{x})(\mathbf{P}))$$

is an axiom.

To illustrate the interpretation of this scheme, let s be the sequence $(1, \frac{1}{2}, \frac{1}{4}, \frac{1}{8}, \ldots, 2^{-n}, \ldots)$ and take for $\mathbf{P}$ the statement 's converges to x'. Now s converges to 0, so $(\exists x)(\mathbf{P})$ is true; a general theorem says that if s converges to x and also to x', then $x = x'$. Hence $(\exists! x)(\mathbf{P})$ is true. Thus

$$s \text{ converges to } y \Leftrightarrow y = \text{the } x \text{ such that } s \text{ converges to } x.$$

Since s converges to 0,

$$0 = \text{the } x \text{ such that } s \text{ converges to } x.$$

Suppose there does not exist a unique $\mathbf{x}$ such that $\mathbf{P}$. Suppose, in other words, either there is no $\mathbf{x}$ such that $\mathbf{P}$, or there exist two distinct objects satisfying $\mathbf{P}$. In this event Axiom Scheme 11 gives us no information, and we can say nothing of interest about $(\iota\mathbf{x})(\mathbf{P})$. In practice, we do not construct $(\iota\mathbf{x})(\mathbf{P})$ unless we know $(\exists!\mathbf{x})(\mathbf{P})$ is true.

We illustrate the use of Axiom Scheme 11 in proving the next result.

4.6. PROPOSITION

Let Y be a class. Then

$$Y = (\iota X)(X = Y).$$

Proof

We first show there is a unique class X such that $X = Y$. We know $Y = Y$. Since '$[Y \mid X](X = Y)$' is just '$Y = Y$', it follows that '$(\exists X)(X = Y)$' is true. This proves existence of X. To prove uniqueness, we note that if $X = Y$ and $X' = Y$, then $X = X'$.

By what we have just proved and Axiom Scheme 11,

$$(\forall z)(z = Y \Leftrightarrow z = (\iota X)(X = Y));$$

we have taken 'z' for $\mathbf{y}$, 'X' for $\mathbf{x}$, and '$X = Y$' for $\mathbf{P}$ in Axiom Scheme 11. Then by universal specialization,

$$Y = Y \Leftrightarrow Y = (\iota X)(X = Y).$$

But $Y = Y$, so we obtain the desired result. $\square$

SETS AND THE CLASSIFIER

4.7. DEFINITION

We write

$$\mathbf{X} \text{ is a set}$$

to mean

$$(\exists \mathbf{Y})(\mathbf{X} \in \mathbf{Y}).$$

Thus we call a class that is an element of some class a *set*. Intuitively, a set is a collection small enough to belong to some collection.

Notice that the expression '**X** is a set' is (an abbreviation for) a formula in the formal language. It is therefore quite meaningful to say 'There exists a set'. At present we cannot establish the existence of any sets; a later axiom is designed to allow us to do this. Many familiar objects are in fact sets: the collection of all real numbers, the collection of all real-valued functions of a real variable, the collection of all lines in a plane.

In place of saying '**X** is a set and **X** is contained in **Y**', we may also say '**X** is a *subset* of **Y**'.

Our final axiom scheme says that, given a property **P**, there is a class whose elements are precisely those *sets* having property **P**.

AXIOM SCHEME 12 (AXIOM SCHEME OF ABSTRACTION)

Let **P** be a formula in which the letter **x** is free and in which the letter **X** does not appear. Then

$$(\exists \mathbf{X})(\forall \mathbf{x})(\mathbf{x} \in \mathbf{X} \Leftrightarrow \mathbf{x} \text{ is a set } \& \ \mathbf{P})$$

is an axiom.

Axiom Scheme 12, due to A. P. Morse and W. V. O. Quine, is a stronger form of the axiom scheme of abstraction than is standard in the von Neumann-Bernays-Gödel system. Usually the additional condition is imposed that all bound letters in **P** be restricted to sets, and then the weaker form is equivalent to a finite number of explicit axioms (see Section 14.2 of Rubin [34]).

Something more than Axiom Scheme 12 is true:

4.8. METATHEOREM

Let **P** be a formula in which the letter **x** is free and in which the letter **X** does not appear. Then

$$(\exists! \mathbf{X})(\forall \mathbf{x})(\mathbf{x} \in \mathbf{X} \Leftrightarrow \mathbf{x} \text{ is a set } \& \ \mathbf{P}).$$

Metaproof

Existence of $\mathbf{X}$ follows from Axiom Scheme 12. We show uniqueness. Assume

$$(\forall \mathbf{x})(\mathbf{x} \in \mathbf{X} \Leftrightarrow \mathbf{x} \text{ is a set } \& \mathbf{P})$$

$$(\forall \mathbf{x})(\mathbf{x} \in \mathbf{X}' \Leftrightarrow \mathbf{x} \text{ is a set } \& \mathbf{P}).$$

Then

$$(\forall \mathbf{x})(\mathbf{x} \in \mathbf{X} \Leftrightarrow \mathbf{x} \in \mathbf{X}'),$$

so $\mathbf{X} = \mathbf{X}'$. □

4.9. DEFINITION

Let $\mathbf{P}$, $\mathbf{x}$, and $\mathbf{X}$ be as in 4.8. Then

$$\{\mathbf{x} \mid \mathbf{P}\}$$

is defined to be the term

$$(\iota \mathbf{X})(\forall \mathbf{x})(\mathbf{x} \in \mathbf{X} \Leftrightarrow \mathbf{x} \text{ is a set } \& \mathbf{P}).$$

One calls $\{\mathbf{x} \mid \mathbf{P}\}$ *the class of all (sets)* $\mathbf{x}$ *such that* $\mathbf{P}$. The symbol '{ | }' is the *classifier*.

Let $\mathbf{Q}$ be the formula

$$(\forall \mathbf{x})(\mathbf{x} \in \mathbf{X} \Leftrightarrow \mathbf{x} \text{ is a set } \& \mathbf{P}).$$

Metatheorem 4.8 says $(\exists! \mathbf{X})(\mathbf{Q})$ is true. Now Axiom Scheme 11 says

$$(\exists! \mathbf{X})(\mathbf{Q}) \Rightarrow (\forall \mathbf{Y})([\mathbf{Y} \mid \mathbf{X}]\mathbf{Q} \Leftrightarrow \mathbf{Y} = (\iota \mathbf{X})(\mathbf{Q})),$$

where $\mathbf{Y}$ is a letter not appearing in $\mathbf{P}$. Hence

$$[\mathbf{Y} \mid \mathbf{X}]\mathbf{Q} \Leftrightarrow \mathbf{Y} = (\iota \mathbf{X})(\mathbf{Q}).$$

According to the preceding definition, $(\iota \mathbf{X})(\mathbf{Q})$ is $\{\mathbf{x} \mid \mathbf{P}\}$, and $[\mathbf{Y} \mid \mathbf{X}]\mathbf{Q}$ is

$$(\forall \mathbf{x})(\mathbf{x} \in \mathbf{Y} \Leftrightarrow \mathbf{x} \text{ is a set } \& \mathbf{P}).$$

This proves:

4.10. METATHEOREM

Let $\mathbf{P}$ be a formula in which the letter $\mathbf{x}$ is free and in which the letter $\mathbf{Y}$ does not appear. Then

$$\mathbf{Y} = \{\mathbf{x} \mid \mathbf{P}\}$$

if and only if

$$(\forall \mathbf{x})(\mathbf{x} \in \mathbf{Y} \Leftrightarrow \mathbf{x} \text{ is a set } \& \mathbf{P}).$$

Since $\{\mathbf{x} \mid \mathbf{P}\} = \{\mathbf{x} \mid \mathbf{P}\}$, we obtain from 4.10 the crucial property

$$(*) \qquad \mathbf{x} \in \{\mathbf{x} \mid \mathbf{P}\} \Leftrightarrow \mathbf{x} \text{ is a set } \& \mathbf{P}.$$

Thus *the elements of* $\{\mathbf{x} \mid \mathbf{P}\}$ *are exactly those sets satisfying* $\mathbf{P}$. For example,

$$x \in \{x \mid x \in \mathbb{R} \,\&\, 0 < x < 1\} \Leftrightarrow x \in \mathbb{R} \,\&\, 0 < x < 1,$$
$$x \in \{x \mid x = 0\} \Leftrightarrow x = 0,$$
$$x \in \{x \mid x = 0 \vee x = 1\} \Leftrightarrow x = 0 \vee x = 1$$

(here we anticipate the fact that 0 and 1 are sets).

According to Axiom Scheme 9,

$$(^{**}) \qquad \{\mathbf{x} \mid \mathbf{P}\} = \{\mathbf{y} \mid [\mathbf{y} \mid \mathbf{x}]\mathbf{P}\},$$

where $\mathbf{y}$ is a letter not appearing in $\mathbf{P}$. For example,

$$\{x \mid x \in \mathbb{R} \,\&\, 0 < x < 1\} = \{y \mid y \in \mathbb{R} \,\&\, 0 < y < 1\}.$$

According to Axiom Scheme 10,

$$(^{***}) \qquad (\forall \mathbf{x})\,(\mathbf{P} \Leftrightarrow \mathbf{Q}) \Rightarrow \{\mathbf{x} \mid \mathbf{P}\} = \{\mathbf{x} \mid \mathbf{Q}\},$$

where $\mathbf{Q}$ is a formula in which $\mathbf{x}$ is free. For example,

$$\{x \mid x \in \mathbb{R} \,\&\, x(x - 1) = 0\} = \{x \mid x = 0 \vee x = 1\}$$

because

$$x \in \mathbb{R} \,\&\, x(x - 1) = 0 \Leftrightarrow x = 0 \vee x = 1.$$

Axiom Schemes 9–12 were designed almost exclusively to make (*), (**), (***) true. No great harm will be done by now forgetting the definition of $\{\mathbf{x} \mid \mathbf{P}\}$. What really matters is that the classifier has the properties (*), (**), (***).

4.11. PROPOSITION

Let X be a class. Then

$$X = \{x \mid x \in X\}.$$

Proof

If $x \in X$, then x is a set. Hence

$$x \in X \Leftrightarrow x \text{ is a set } \& \, x \in X.$$

Now use 4.10, taking 'X' for $\mathbf{Y}$ and '$x \in X$' for $\mathbf{P}$ there. □

SOME PARTICULAR CLASSES

We next use the classifier to construct some specific classes of special importance. The first is suggested by our earlier discussion of Russell's paradox.

4.12. DEFINITION

The *Russell class*, denoted by $\mathcal{R}$, is defined to be the term

$$\{x \mid x \notin x\}.$$

Thus

$$x \in \mathcal{R} \Leftrightarrow x \text{ is a set } \& \ x \notin x.$$

4.13. THEOREM

The Russell class $\mathcal{R}$ is not a set.

Proof

Just suppose $\mathcal{R}$ is a set. Either $\mathcal{R} \notin \mathcal{R}$ or $\mathcal{R} \in \mathcal{R}$. If $\mathcal{R} \notin \mathcal{R}$, then $\mathcal{R}$ is a set and $\mathcal{R} \notin \mathcal{R}$, so $\mathcal{R} \in \mathcal{R}$ by definition of $\mathcal{R}$. If $\mathcal{R} \in \mathcal{R}$, then since no element of $\mathcal{R}$ belongs to itself, $\mathcal{R} \notin \mathcal{R}$. Thus $\mathcal{R} \in \mathcal{R}$ if and only if $\mathcal{R} \notin \mathcal{R}$, a contradiction. Hence $\mathcal{R}$ is not a set. □

Theorem 4.13 shows how our theory eliminates Russell's paradox.

4.14. DEFINITION

Let X be a class. One calls X a *proper class* when X is not a set.

In this terminology, 4.13 says that $\mathcal{R}$ is a proper class.

4.15. DEFINITION

The *universe*, denoted by $\mathcal{U}$, is defined to be the term

$$\{x \mid x = x\}.$$

The universe $\mathcal{U}$ is the class of all sets:

4.16. THEOREM

$x \in \mathcal{U}$ if and only if x is a set.

Proof

By definition of $\mathcal{U}$,

$$x \in \mathcal{U} \Leftrightarrow x \text{ is a set } \& \ x = x.$$

Now $x = x$, so

$$x \text{ is a set } \& \ x = x \Leftrightarrow x \text{ is a set.} \quad \square$$

Notice that $\mathcal{U}$ is not "the class of all classes." We shall prove later that $\mathcal{U}$ is a proper class, in fact, $\mathcal{U} = \mathcal{R}$.

The universe is the largest class:

4.17. THEOREM

Let X be a class. Then

$$X \subset \mathfrak{U}.$$

Proof

If $x \in X$, then x is a set and so $x \in \mathfrak{U}$ by 4.16. ▯

We may think of $\mathfrak{U}$ as a large container holding every set. Then we may conceive of a container holding nothing, and hence of a class having no elements whatsoever.

4.18. DEFINITION

The *empty class*, denoted by $\varnothing$, is defined to be the term

$$\{x \mid x \neq x\}.$$

One calls a class X *empty* (or *null* or *vacuous*) if $X = \varnothing$. The negation of "empty" is "nonempty." (The symbol $\varnothing$ is not the Greek letter phi, but rather a letter of the Danish and Norwegian alphabets. The symbols □ and Λ also appear in the literature for $\varnothing$.)

The empty class $\varnothing$ is really the class having no elements:

4.19. THEOREM

For every x, $x \notin \varnothing$.

Proof

We have

$$x \in \varnothing \Leftrightarrow x \text{ is a set } \& \ x \neq x.$$

Now '$x = x$' is true, so '$x \neq x$' is false. Then '$x \in \varnothing$' is false. ▯

The empty class is the smallest class:

4.20. THEOREM

Let X be a class. Then

$$\varnothing \subset X.$$

Proof

Since '$x \notin \varnothing$' is true, so is

$$x \notin X \Rightarrow x \notin \varnothing.$$

Hence '$x \in \varnothing$ implies $x \in X$' is true. ▯

Although we have demonstrated by example the existence of at least one proper class, we cannot prove the existence of a single set without additional axioms.

AXIOM 0

$\varnothing$ is a set.

Because of Axiom 0, we also call the empty class $\varnothing$ the *empty set.*

We have formulated Axiom 0 here purely as a convenient source of examples. It will be replaced later by a much stronger axiom—the axiom of infinity—and so it is purely temporary. The theory developed prior to introduction of the axiom of infinity does not depend on Axiom 0, although a number of examples do.

EXERCISES

A. Prove all unproved theorems, complete all incomplete proofs, and supply all missing details in this chapter. (This is, of course, a standing exercise for every chapter, indeed for any mathematical work.)

B. Prove the implication

$$x = y \Rightarrow (\forall X)(x \in X \Leftrightarrow y \in X)$$

and express in words its intuitive content.

C. (a) Show that

$$X \subsetneq Y \Leftrightarrow (\forall x \in X)(x \in Y) \,\&\, (\exists y \in Y)(y \notin X).$$

(b) Use (a) to express $\neg(X \subsetneq Y)$ in terms of elements of X and Y.

D. Assuming $X \neq Y$ and $Y \neq Z$, what can be said about the truth of $X \neq Z$?

E. (a) Is it possible to have both $X \subsetneq Y$ and $Y \subsetneq X$?

(b) If $X \subsetneq Y$ and $Y \subsetneq Z$, must $X \subsetneq Z$?

F. Write in the formal language the formula abbreviated by '$X \subset Y$'.

G. Symbolize in terms of the descriptor:

(a) $\lim_{x \to 1} f(x)$.

(b) The real solution of the equation

$$z^2 - (1 + i)z + i = 0.$$

(c) $\int_0^1 x dx$.

H. Which of the following sets are equal to and which are contained in the others?

$$\{x \mid x = 0 \vee x = 1\}, \quad \{x \mid x = 0 \,\&\, x = 1\}, \quad \{x \mid x = 0\},$$

$$\{y \mid y = 0 \vee y = 1\}, \quad \{z \mid z = 1\}, \quad \{x \mid x = 1 \vee. x = 0\}.$$

I. Prove or disprove:
(a) $\{x \mid x \in X \,\&\, (x \in Y \vee x \in Z)\}$
$\subset \{x \mid (x \in X \,\&\, x \in Y) \vee (x \in X \,\&\, x \in Z)\}$.
(b) $\{f \mid f$ is a continuous function$\}$
$= \{f \mid f$ is a differentiable function$\}$.
("Function" here means "real-valued function defined on the set of all real numbers.")

J. Let X and Y be sets. Prove:
(a) If $X = Y$, then $\{x \mid x \in X\} = \{y \mid y \in Y\}$.
(b) If $X = Y$, then $\{x \mid X \in x\} = \{x \mid Y \in x\}$.

K. Prove or disprove: $\mathcal{U} = \varnothing$.

L. Let X be $\{x \mid x = \varnothing\}$. Prove that $X \neq \varnothing$. Is $X \in X$?

M. Is $\{x \mid x = \mathcal{R}\}$ empty?

N. Let $\mathbf{P}$ be a formula in which the letter $\mathbf{x}$ is free. If

$$(\forall \mathbf{x})\,(\mathbf{x} \in \mathcal{U} \Rightarrow \neg \mathbf{P}),$$

show that $\{\mathbf{x} \mid \mathbf{P}\} = \varnothing$. Deduce in particular that $\{\mathbf{x} \mid \mathbf{P}\} = \varnothing$ in case $\mathbf{P}$ is false.

P. Formulate and prove a result concerning $\{\mathbf{x} \mid \mathbf{P}\} = \mathcal{U}$ analogous to Exercise N.

Q. Prove the converse of the implication in Axiom Scheme 11.

5. The Calculus of Classes

Thus far only two specific classes, the empty set and the universe, have been constructed. In order that others can be constructed, elementary operations upon classes are now introduced which yield new classes from old.

COMPLEMENTATION

5.1. DEFINITION

For a class X, the *complement of* X, denoted by X^c, is the class $\{x \mid x \notin X\}$ consisting of all sets which are not members of X. (The notations $\mathcal{C}X$ and X' also appear in the literature.)

We may visualize X^c by means of a "Venn diagram" as shown in Figure 5.1. The region inside the rectangle represents $\mathcal{U}$, the unshaded region represents X, and the shaded region represents X^c.

5.2. THEOREM

(1) $\varnothing^c = \mathcal{U}$ and $\mathcal{U}^c = \varnothing$.
(2) For any class X, $(X^c)^c = X$.
(3) For any classes X and Y, $X \subset Y \Leftrightarrow Y^c \subset X^c$.
(4) For any classes X and Y, $X = Y \Leftrightarrow X^c = Y^c$.

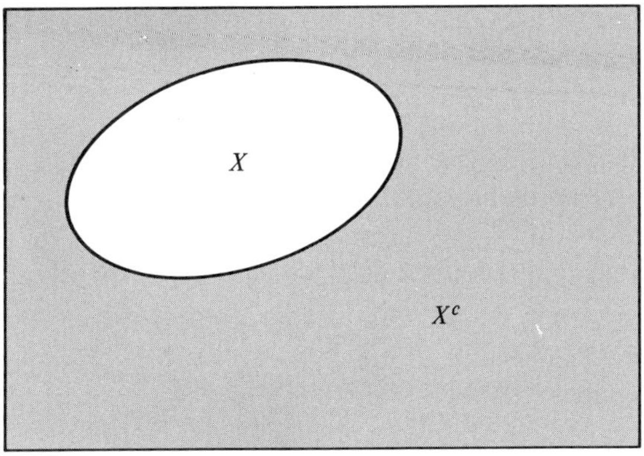

Figure 5.1

Proof

(1) If x is a set, then

$$x \in \varnothing^c \Leftrightarrow x \notin \varnothing \Leftrightarrow \neg(x \in \varnothing) \Leftrightarrow \neg(x \neq x)$$
$$\Leftrightarrow x = x \Leftrightarrow x \in \mathcal{U}.$$

This proves $\varnothing^c = \mathcal{U}$. Similarly, $\mathcal{U}^c = \varnothing$.

(3) We first show

$$(*) \qquad X \subset Y \Rightarrow Y^c \subset X^c.$$

Assume $X \subset Y$. If $z \in Y^c$, then $z \notin Y$. Now $z \in X \Rightarrow z \in Y$, so $z \notin Y \Rightarrow z \notin X$. Then $z \in Y^c$ implies $z \notin X$, that is, $z \in X^c$. Hence $Y^c \subset X^c$.

To prove the converse of (*), we replace X by Y^c and Y by X^c in (*) to obtain

$$Y^c \subset X^c \Rightarrow (X^c)^c \subset (Y^c)^c.$$

Now $(X^c)^c = X$ and $(Y^c)^c = Y$, so

$$Y^c \subset X^c \Rightarrow X \subset Y.$$

(4) We know

$$X = Y \Leftrightarrow X \subset Y \,\&\, Y \subset X,$$
$$X^c = Y^c \Leftrightarrow X^c \subset Y^c \,\&\, Y^c \subset X^c.$$

Now apply (3). □

In view of (1) above, the *complement of a set need not be a set.*

5.3. DEFINITION

Let X and Y be classes. The *(relative) complement of Y in X*, denoted by $X \setminus Y$, is the class

$$\{x \mid x \in X \,\&\, x \notin Y\}$$

consisting of all elements of X which are not elements of Y. (Other notations are $X - Y$, $X \sim Y$, and $\mathcal{C}_X Y$.)

A Venn diagram displaying $X \setminus Y$ appears in Figure 5.2. The entire shaded region represents X, and the heavily shaded region represents $X \setminus Y$.

Notice that $X \setminus Y$ is defined even when $Y \not\subset X$. Clearly,

$$X \setminus Y \subset X.$$

Also,

$$X \setminus \varnothing = X.$$

Dually,

$$\mathfrak{U} \setminus X = X^c,$$

so that relative complementation generalizes complementation as defined in 5.1.

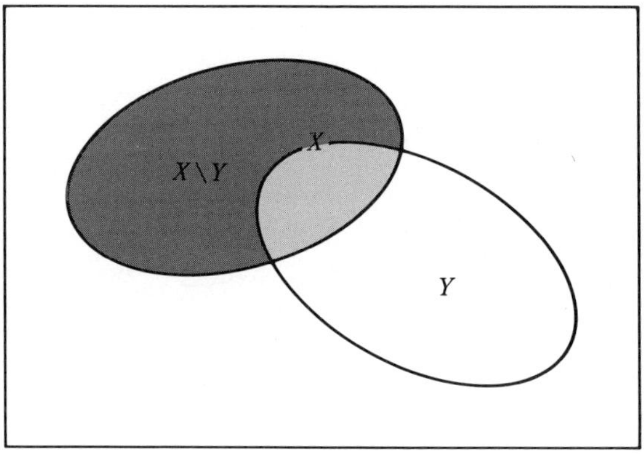

Figure 5.2

UNION AND INTERSECTION OF TWO CLASSES

Just as negation is used to construct the complement of a given class, so disjunction and conjunction may be used to construct new classes from two given classes.

5.4. DEFINITION

Let X and Y be classes. The *union of X and Y*, denoted $X \cup Y$, is the class

$$\{x \mid x \in X \vee x \in Y\}$$

consisting of all sets belonging to at least one of the classes X and Y. The *intersection of* X *and* Y, denoted $X \cap Y$, is the class

$$\{x \mid x \in X \,\&\, x \in Y\}$$

consisting of all sets belonging both to X and to Y.

A Venn diagram depicting $X \cup Y$ and $X \cap Y$ appears in Figure 5.3. The entire shaded region represents $X \cup Y$, and the heavily shaded region represents $X \cap Y$.

One says that X is *disjoint from* Y to mean that $X \cap Y = \varnothing$, in other words, that X and Y have no elements in common.

The "algebra" of classes is given by the following rules for computing intersections, unions, and complements.

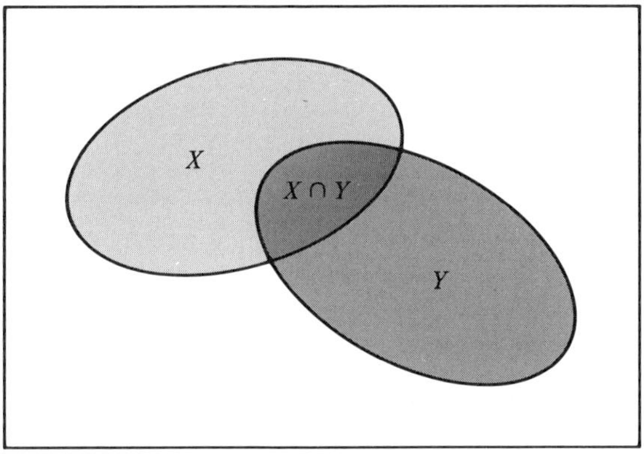

Figure 5.3

5.5. THEOREM

Let X, Y, and Z be classes. Then:

(1) $(X \cup Y)^c = X^c \cap Y^c$.
(1′) $(X \cap Y)^c = X^c \cup Y^c$. (De Morgan's laws)
(2) $(X \cup Y) \cup Z = X \cup (Y \cup Z)$.
(2′) $(X \cap Y) \cap Z = X \cap (Y \cap Z)$. (associative laws)
(3) $X \cup Y = Y \cup X$.
(3′) $X \cap Y = Y \cap X$. (commutative laws)
(4) $X \cup (Y \cap Z) = (X \cup Y) \cap (X \cup Z)$.
(4′) $X \cap (Y \cup Z) = (X \cap Y) \cup (X \cap Z)$. (distributive laws)
(5) $X \cup X = X$.
(5′) $X \cap X = X$. (idempotency laws)
(6) $X \cup \varnothing = X$.
(6′) $X \cap \mathfrak{U} = X$.
(7) $X \cup \mathfrak{U} = \mathfrak{U}$.

(7′) $X \cap \varnothing = \varnothing$.
(8) $X \cup X^c = \mathfrak{U}$.
(8′) $X \cap X^c = \varnothing$.
(9) $X \subset X \cup Y$.
(9′) $X \cap Y \subset X$.
(10) $X \subset Y \Leftrightarrow X \cup Y = Y$.
(10′) $Y \subset X \Leftrightarrow X \cap Y = Y$.

Proof

(1) If x is a set, then

$$\begin{aligned} x \in (X \cup Y)^c &\Leftrightarrow x \notin X \cup Y \\ &\Leftrightarrow \neg(x \in X \vee x \in Y) \\ &\Leftrightarrow x \notin X \,\&\, x \notin Y \\ &\Leftrightarrow x \in X^c \,\&\, x \in Y^c \\ &\Leftrightarrow x \in X^c \cap Y^c. \end{aligned}$$

Aside from the definitions involved, the only thing used here is the logical law 2.27(2). One proves (1′) by a similar use of 2.27(3). □

The various parts of Theorem 5.5 occur in pairs. Statement (n') is obtained from statement (n) by replacing $\cup$ by $\cap$, $\cap$ by $\cup$, $\varnothing$ by $\mathfrak{U}$, $\mathfrak{U}$ by $\varnothing$, and $\subset$ by $\supset$, and statement (n) is obtained from (n') by the same formal replacements. Moreover, for $n > 1$ the truth of statement (n') can be deduced from that of (n), and vice versa, by means of the De Morgan laws (1) and (1′). Let us illustrate.

Suppose we have already proved (3). Since in (3) X and Y are arbitrary classes, we also have

$$X^c \cup Y^c = Y^c \cup X^c.$$

By De Morgan's laws,

$$X^c \cup Y^c = (X \cap Y)^c, \qquad Y^c \cup X^c = (Y \cap X)^c.$$

Then

$$(X \cap Y)^c = (Y \cap X)^c.$$

Using 5.2(4) we conclude

$$X \cap Y = Y \cap X,$$

which is just (3′).

If this procedure is used to deduce (3′) from (3), then of course (3) must be proved by some method other than by using De Morgan's laws to deduce (3) from (3′). A good method is a straightforward application of propositional calculus together with the definitions of the terms involved.

The technique just illustrated effectively halves the work of proving many theorems. Moreover, the method described to transform statement

(n) into its *dual* (n') permits one to invent new theorems. The entire phenomenon is known as *duality*. (Actually, things are more complicated than they seem; see Gottschalk [13].)

POWER CLASS OF A CLASS

Our second explicit axiom says: Given a *set* X, there is some *set* Z of which every subclass of X is a member.

AXIOM 2 (AXIOM OF SUBSETS)

$$(\forall X \in \mathcal{U})(\exists Z \in \mathcal{U})(\forall Y)(Y \subset X \Rightarrow Y \in Z).$$

Axiom 2 guarantees that certain classes are sets once it is known that certain other classes are sets.

5.6. THEOREM

Let X be a class. If X is contained in some set, then X is itself a set.

Proof

Assume there exists some $Y \in \mathcal{U}$ such that $X \subset Y$. Let Y be such a set. By Axiom 2, there exists a set Z such that $E \subset Y \Rightarrow E \in Z$. Choose such a Z. Since $X \subset Y$, $X \in Z$. Hence X is a set. □

Thus a subclass of a set Y is a subset of Y. The contrapositive of 5.6 says that if a class Y contains some proper class, then Y is itself a proper class.

5.7. DEFINITION

Let X be a class. The *power class of* X, denoted $\mathcal{P}(X)$, is defined to be the class

$$\{Y \mid Y \subset X\}$$

consisting of all subsets of X.

5.8. THEOREM

Let X be a set. Then $\mathcal{P}(X)$ is a set, and

$$Y \in \mathcal{P}(X) \Leftrightarrow Y \subset X.$$

Proof

By definition,

$$Y \in \mathcal{P}(X) \Leftrightarrow Y \in \mathcal{U} \;\&\; Y \subset X.$$

But $Y \subset X$ implies $Y \in \mathcal{U}$ by 5.6, so

$$Y \in \mathcal{U} \ \& \ Y \subset X \Leftrightarrow Y \subset X.$$

In view of Axiom 2, we may choose a set Z such that $Y \subset X \Rightarrow Y \in Z$. By the preceding paragraph, $Y \in \mathcal{P}(X) \Rightarrow Y \in Z$. Hence $\mathcal{P}(X) \subset Z$. Now Z is a set, so $\mathcal{P}(X)$ is also a set by 5.6. ▯

For any class X,

$$\varnothing \in \mathcal{P}(X)$$

and

$$X \in \mathcal{P}(X) \text{ if } X \text{ is a set.}$$

However, $X \notin \mathcal{P}(X)$ if X is a proper class, for only sets can be elements of $\mathcal{P}(X)$.

5.9. THEOREM

The universe $\mathcal{U}$ is a proper class. Moreover, $\mathcal{P}(\mathcal{U}) = \mathcal{U}$.

Proof

Since $\mathcal{R} \subset \mathcal{U}$ and $\mathcal{R}$ is a proper class, so is $\mathcal{U}$.

We already know $\mathcal{P}(\mathcal{U}) \subset \mathcal{U}$. To prove the opposite inclusion $\mathcal{U} \subset \mathcal{P}(\mathcal{U})$, let $X \in \mathcal{U}$. Since $X \subset \mathcal{U}$, X is a subset of $\mathcal{U}$. Hence $X \in \mathcal{P}(\mathcal{U})$. ▯

DOUBLETONS AND SINGLETONS

Starting with a class X, we may construct a class having no elements other than X itself.

5.10. DEFINITION

Let X be a class. Then *singleton* X, denoted $\{X\}$, is defined to be the class

$$\{x \mid x = X\}.$$

We have

$$x \in \{X\} \Leftrightarrow x \text{ is a set} \ \& \ x = X.$$

Hence *if X is a set, then*

$$x \in \{X\} \Leftrightarrow x = X,$$

so that $\{X\}$ is the class having X as its sole element. However, *if X is a proper class, then* $\{X\} = \varnothing$, for $x \in \{X\} \Rightarrow x = X$, and only sets can be members of $\{X\}$. Thus it is of no practical interest to construct $\{X\}$ when X is a proper class.

If X is a class and z is a set, then

$$X \setminus \{z\} = \{x \mid x \in X \,\&\, x \neq z\}.$$

For example, $\mathcal{U} \setminus \{\varnothing\}$ is the class of all nonempty sets. By way of contrast, $\mathcal{U} \setminus \varnothing = \mathcal{U}$.

One must not confuse $\{X\}$ with X itself. For example, $\{\varnothing\} \neq \varnothing$ since $\varnothing \in \{\varnothing\}$ and $\varnothing \notin \varnothing$. The distinction between X and $\{X\}$ is just as real as that between a lion and a cage holding the lion!

Let us now cage two lions at once.

5.11. DEFINITION

Let X and Y be classes. Then *doubleton* X, Y, denoted $\{X, Y\}$, is the class

$$\{x \mid x = X \vee x = Y\}.$$

Clearly

$$\{X, Y\} = \{Y, X\},$$

and accordingly $\{X, Y\}$ is also called the *unordered pair* X, Y.

Doubletons are related to singletons as follows:

$$\{X\} = \{X, X\},$$

$$\{X, Y\} = \{X\} \cup \{Y\}.$$

The latter equation and earlier remarks about singletons show that *if X and Y are sets, then*

$$x \in \{X, Y\} \Leftrightarrow x = X \vee x = Y.$$

As an application of doubletons, we compute $\mathcal{P}(\{\varnothing\})$. If $E \subset \{\varnothing\}$, then $x \in E$ implies $x = \varnothing$, so $E = \varnothing$ or else $E = \{\varnothing\}$. Hence $\mathcal{P}(\{\varnothing\}) = \{\varnothing, \{\varnothing\}\}$. In contrast, $\mathcal{P}(\varnothing) = \{\varnothing\}$.

5.12. THEOREM

Let x, y, a, and b be sets. Then

$$\{x, y\} = \{a, b\}$$

if and only if

$$(x = a \,\&\, y = b) \text{ or } (x = b \,\&\, y = a).$$

Proof

Assume $\{x, y\} = \{a, b\}$. Just suppose it is not the case that $x = a \,\&\, y = b$. Then $x \neq a$ or $y \neq b$. We distinguish three cases.

Case (1)

$x \neq a$ and $y = b$. Since $x \in \{x, y\} = \{a, b\}$, $x = a$ or $x = b$. Then $x = b = y$. But $a \in \{a, b\} = \{x, y\}$, so $a = x$ or $a = y$. If $a = y$, then $a = x$. Hence $a = x$, contradicting the assumption in this case that $x \neq a$. Thus case (1) is impossible.

Case (2)

$x = a \,\&\, y \neq b$. An argument similar to the one just given shows that this case is also impossible. Hence we are left with the only remaining possibility:

Case (3)

$x \neq a$ and $y \neq b$. Since $x \in \{x, y\} = \{a, b\}$, $x = a$ or $x = b$. Hence $x = b$. Similarly $y = a$ since $y \in \{x, y\} = \{a, b\}$.

It follows that

$$\{x, y\} = \{a, b\} \Rightarrow (x = a \,\&\, y = b) \vee (x = b \,\&\, y = a).$$

The proof of the converse is trivial. □

5.13. COROLLARY

If x and a are sets, then

$$\{x\} = \{a\} \Leftrightarrow x = a.$$

It should be clear how to define the notions of *tripleton*, *quadrupleton*, and so on. For example, $\{X, Y, Z\}$ is defined to be

$$\{x \mid x = X \vee x = Y \vee x = Z\}.$$

To guarantee that a set is obtained when a doubleton is constructed from two sets, we need a new axiom.

AXIOM 3 (AXIOM OF PAIRING)

If X and Y are sets, then $\{X, Y\}$ is a set.

In particular, $\{X\}$ is a set when X is a set.

UNION AND INTERSECTION OF A CLASS OF SETS

In what follows, we sometimes speak of a "class of sets." This language is redundant, for the elements of a class are automatically sets, but it is useful for its suggestion of a hierarchy of objects. At one level there are elements x (points in the plane, for example), at the next level sets X of such elements (lines consisting of points), and at the third level classes $\mathcal{A}$ of such sets (collections of lines). Of course, what is at the top of the hierarchy in one context may be at the bottom in another: The molecule is to the chemist what the star is to the astronomer.

5.14. DEFINITION

Let $\mathcal{A}$ be a class of sets. The *union of* $\mathcal{A}$, denoted $\bigcup\mathcal{A}$, is defined to be the class

$$\{x \mid (\exists A \in \mathcal{A})(x \in A)\}$$

consisting of those sets belonging to at least one member of $\mathcal{A}$. The *intersection of* $\mathcal{A}$, denoted $\bigcap\mathcal{A}$, is the class

$$\{x \mid (\forall A \in \mathcal{A})(x \in A)\}$$

consisting of those sets belonging to every member of $\mathcal{A}$.

The union (intersection) of two sets as defined earlier can be expressed as the union (intersection) of a class as just defined.

5.15. THEOREM

Let X and Y be sets. Then

$$X \cup Y = \bigcup\{X, Y\}, \qquad X \cap Y = \bigcap\{X, Y\}.$$

Proof

We verify only the first equation. We have

$$\begin{aligned} x \in \bigcup\{X, Y\} &\Leftrightarrow (\exists Z)(Z \in \{X, Y\} \,\&\, x \in Z) \\ &\Leftrightarrow (\exists Z)((Z = X \vee Z = Y) \,\&\, x \in Z) \\ &\Leftrightarrow (\exists Z)((Z = X \,\&\, x \in Z) \vee (Z = Y \,\&\, x \in Z)) \\ &\Leftrightarrow (\exists Z)(Z = X \,\&\, x \in Z) \vee (\exists Z)(Z = Y \,\&\, x \in Z) \\ &\Leftrightarrow x \in X \vee x \in Y \\ &\Leftrightarrow x \in X \cup Y. \quad \square \end{aligned}$$

5.16. COROLLARY

If X is a set, then

$$\bigcup\{X\} = X = \bigcap\{X\}.$$

The restriction in 5.15 that X and Y be sets is essential. For example,

$$\mathcal{U} \cup \mathcal{U} = \mathcal{U} \neq \varnothing = \bigcup\{\varnothing\} = \bigcup\{\mathcal{U}, \mathcal{U}\}.$$

Thus, Definition 5.14 is only a partial generalization of Definition 5.4.

Although the results in the next theorem may seem peculiar at first sight, they are necessary consequences of our definitions.

5.17. THEOREM

(1) $\bigcup\varnothing = \varnothing$.
(2) $\bigcap\varnothing = \mathcal{U}$.
(3) $\bigcup\mathcal{U} = \mathcal{U}$.
(4) $\bigcap\mathcal{U} = \varnothing$.

Proof

(1) Suppose $x \in \bigcup\varnothing$. Then there exists $X \in \varnothing$ such that $x \in X$. But $X \in \varnothing$ is false, so

$$(\exists X \in \varnothing)(x \in X)$$

is false. Hence $x \notin \bigcup\varnothing$.

(2) We have

$$x \in \bigcap\varnothing \Leftrightarrow x \in \mathcal{U} \;\&\; (\forall X)(X \in \varnothing \Rightarrow x \in X).$$

But $X \in \varnothing \Rightarrow x \in X$ is true since $X \in \varnothing$ is false. Hence $x \in \bigcap\varnothing \Leftrightarrow x \in \mathcal{U}$.

(3) Let $x \in \mathcal{U}$. Then $\{x\} \in \mathcal{U}$ and $x \in \{x\}$, so $x \in \bigcup\mathcal{U}$. Hence $\mathcal{U} \subset \bigcup\mathcal{U}$. We already know $\bigcup\mathcal{U} \subset \mathcal{U}$.

(4) Suppose $x \in \bigcap\mathcal{U}$. Then x is a set. Since $\varnothing \subset x$, $\varnothing$ is a set, that is, $\varnothing \in \mathcal{U}$. Then $x \in \varnothing$, an impossibility. ☐

Let $\mathcal{A}$ be a class of sets. One easily proves

$$X \in \mathcal{A} \Rightarrow \bigcap\mathcal{A} \subset X \subset \bigcup\mathcal{A}.$$

If $\mathcal{B}$ is another class of sets, then

$$\mathcal{A} \subset \mathcal{B} \Rightarrow \bigcup\mathcal{A} \subset \bigcup\mathcal{B} \;\&\; \bigcap\mathcal{B} \subset \bigcap\mathcal{A}.$$

5.18. THEOREM

Let $\mathcal{A}$ be a class of sets. Then $\bigcap\mathcal{A}$ is a set if and only if $\mathcal{A}$ is nonempty.

Proof

Assume $\mathcal{A}$ is nonempty. Choose $X \in \mathcal{A}$. Then $\bigcap\mathcal{A} \subset X$. Now X is a set, so $\bigcap\mathcal{A}$ is a set by 5.6.

Conversely, assume $\mathcal{A}$ is empty. Then $\bigcap\mathcal{A} = \bigcap\varnothing = \mathcal{U}$, and $\mathcal{U}$ is a proper class. ☐

In contrast to 5.18, the union of a nonempty class of sets need not be a set. For example, $\mathcal{U} \neq \varnothing$ and $\bigcup\mathcal{U} = \mathcal{U}$. A new axiom is now appropriate.

AXIOM 4 (AXIOM OF UNION)

If $\mathcal{A}$ is a set, then $\bigcup\mathcal{A}$ is a set.

As an immediate consequence of Axioms 3 and 4 and Theorem 5.15, we have:

5.19. THEOREM

If X and Y are sets, then $X \cup Y$ is a set.

EXERCISES

A. Prove Theorem 5.5 completely. Better yet, see Chapter 4, Exercise A.

B. Generalize the De Morgan laws 5.5(1), (1′) to relative complements.

C. Prove:
   (a) $X \setminus Y = X \cap Y^c$.
   (b) $(X \setminus Y) \cup Y = X \cup Y$.
   (c) $X \setminus (X \cap Y) = X \setminus Y = (X \cup Y) \setminus Y$.

D. Prove the "absorption law" $X \cap (X \cup Y) = X$. State and prove a dual result.

E. If $X = Y$, then of course $Z \cap X = Z \cap Y$ and $Z \cup X = Z \cup Y$. Is the converse true?

F. Let

$$A = \{\varnothing\}, \qquad B = \{\varnothing, A\}, \qquad D = \{\varnothing, A, B\}.$$

   (a) Compute the power set of each of these three sets and the union and intersection of each pair of these three sets.
   (b) Considering A, B, and D together with the sets you computed in (a), determine which are elements of others, subsets of others, and equal to others.
   (c) Show that $B = A \cup \{A\}$ and $D = B \cup \{B\}$.

G. Here ω denotes the set of all natural numbers 0, 1, 2, 3, and so on, and $\mathbb{R}$ denotes the set of all real numbers. Moreover, for $a < b$, $]a,b[$ denotes the open interval with endpoints a and b, and $[a,b]$ denotes the closed interval with the same endpoints. Compute

$$\bigcap\{X \mid (\exists n \in \omega \setminus \{0\})(X =]-1/n, 1/n[)\}$$

and

$$\bigcup\{X \mid (\exists n \in \omega \setminus \{0\})(X = [-1/n, 1/n])\}.$$

H. Let

$$A = \{x \mid x \in \mathbb{R} \;\&\; x^2 < 1\}, \qquad B = \{x \mid x \in \mathbb{R} \;\&\; x^2 = 1\}.$$

Express the following sets in terms of intervals (or rays) via set-theoretic operations: A, $\mathbb{R} \setminus A$, $\mathbb{R} \setminus B$, $A \cup B$, $A \cap B$.

I. The *symmetric difference of X and Y*, denoted $X \triangle Y$, is defined to be the class

$$(X \setminus Y) \cup (Y \setminus X).$$

Draw a Venn diagram for $X \triangle Y$. Prove:

(a) $X \triangle (Y \triangle Z) = (X \triangle Y) \triangle Z$.

(b) $X \triangle Y = Y \triangle X$.

(c) $X \triangle \varnothing = X$.

(d) $X \cap (Y \triangle Z) = (X \cap Y) \triangle (X \cap Z)$.

(e) $X = Y \Leftrightarrow X \triangle Y = \varnothing$.

(f) $X \triangle Y = (X \cup Y) \setminus (X \cap Y)$.

J. Let Y be a given subclass of a given class X. Does the equation $Y \triangle Z = X$ have a solution $Z \subset X$? If so, does it have a unique solution $Z \subset X$?

K. Let X be a set. Suppose $X \cup Y = X$ for every set Y. Show that $X = \varnothing$.

L. If X and Y are sets, show that $X \cap Y$, $X \setminus Y$, and $X \triangle Y$ are sets.

M. Is X^c ever a set when X is a set?

N. (a) Compute $\{X, Y\}$ in case at least one of the classes X, Y is not a set.

(b) Show that $\{X, Y\}$ is always a set.

P. (a) State and prove a result for tripletons similar to 5.12.

(b) Show that $\{X, Y, Z\}$ is always a set.

Q. Simplify the proof of 5.17(4) by using Axiom 0.

R. If X is a set, show that $\bigcup \mathcal{P}(X) = X$.

S. Let X and Y be sets.

(a) Is $\mathcal{P}(X \cup Y) = \mathcal{P}(X) \cup \mathcal{P}(Y)$? Is $\mathcal{P}(X \cap Y) = \mathcal{P}(X) \cap \mathcal{P}(Y)$?

(b) Solve the equation

$$\mathcal{P}(X \cup Y) = \mathcal{P}(X) \cup \mathcal{P}(Y) \cup \mathcal{A}$$

for $\mathcal{A}$, obtaining $\mathcal{A}$ explicitly in terms of subsets of X and Y.

(c) Use (b) to compute

$$\mathcal{P}(X \cup \{y\}) \setminus \mathcal{P}(X)$$

when y is a set with $y \notin X$.

T. Assume $X \notin X$ for any set X. Given a set X, find another set disjoint from X.

U. Let X be a class and $\mathcal{A}$ be a class of sets. If

$$\mathcal{B} = \{Y \mid (\exists A \in \mathcal{A})(Y = X \setminus A)\},$$

show that

$$X \setminus \bigcup \mathfrak{A} = \bigcap \mathfrak{B}.$$

Establish a similar result for $X \setminus \bigcap \mathfrak{A}$.

V. Let X be a set. A class $\mathfrak{I}$ of subsets of X is called a *topology* on X if $X \in \mathfrak{I}$, if $\bigcup \mathfrak{A} \in \mathfrak{I}$ for each $\mathfrak{A} \subset \mathfrak{I}$, and if $A \cap B \in \mathfrak{I}$ for all A, $B \in \mathfrak{I}$.

(a) If $\mathfrak{I}$ is a topology on X, show that $\varnothing \in \mathfrak{I}$.

(b) If X has at least two different elements, show that there exist at least two distinct topologies on X.

(c) Determine all topologies on the set $\{0,1,2\}$.

(d) Let $Y \subset X$ and let $\mathfrak{I}$ be a topology on X. Show that the class

$$\{E \mid (\exists A \in \mathfrak{I})(E = A \cap Y)\}$$

is a topology on Y.

(e) Prove that any topology on X is a set and that the class of all topologies on X is a set.

6. The Set of Natural Numbers

Among the most ancient, familiar, and important mathematical objects are the natural numbers, 0, 1, 2, 3, ... From them all the other important number systems—integers, rational numbers, real numbers, complex numbers—can be constructed (the construction of the integers and rational numbers is effected in the Appendix). Until the advent of set theory the natural numbers themselves were regarded as such primitive objects that they were incapable of being analyzed into more basic entities. In point of fact, the natural numbers can be easily constructed as certain sets.

In this chapter the natural numbers are defined as an application of the set-theoretic constructions introduced in Chapter 5. An axiom is stated which guarantees that the class of all natural numbers is a set. The Peano postulates are derived as consequences of the definition. Finally, a property of natural numbers is generalized with the aid of an axiom which implies that no set is an element of itself.

DEFINITION OF NATURAL NUMBERS

How ought natural numbers be defined as sets? It is reasonable to require that the first natural number 0 be a set having no elements, so 0 may be defined as the empty set $\varnothing$. Next, 1 should be a set having one element, so 1 may be defined as the singleton $\{0\}$. Next, 2 should be a set having two distinct elements, so 2 may be defined as the doubleton $\{0,1\}$. Next, 3 may be defined as the tripleton $\{0,1,2\}$.

This procedure is deficient, for in order to define the natural numbers 4 and 5, the notions "quadrupleton" and "quintupleton" must first be defined. Then the problem of defining 6 would arise. Thus, the procedure can only be used to define as many natural numbers as time and paper allow. What is needed is a procedure for obtaining the totality of all the natural numbers at once. Such a procedure is suggested by the preceding tentative definitions.

Observe that the definitions of 0, 1, 2, 3 indicated above require

$$\begin{aligned} 0 &= \varnothing, \\ 1 &= \{0\} = 0 \cup \{0\}, \\ 2 &= \{0,1\} = \{0\} \cup \{1\} = 1 \cup \{1\}, \\ 3 &= 2 \cup \{2\}. \end{aligned}$$

Then the first natural number should be $\varnothing$, and in order to define the natural number m coming next after an already defined natural number n, one should take m to be $n \cup \{n\}$. Moreover, each natural number except the first ought to be constructed in this way. Hence the following definitions.

6.1. DEFINITION

If x is a class, then the *successor of* x, denoted by x^+, is defined to be the class $x \cup \{x\}$.

If x is a set, then $\{x\}$ is a set and hence x^+ is also a set, namely, the set whose elements are x together with the elements of x.

6.2. DEFINITION

A class X is said to be *inductive* if

$$\varnothing \in X \;\&\; (\forall x \in X)(x^+ \in X).$$

The universe is an example of an inductive class. The class of all natural numbers is going to be defined as a certain inductive set. Accordingly, we need a new axiom.

AXIOM 5 (AXIOM OF INFINITY)

There exists an inductive set.

The name of this axiom comes from the fact that an inductive set is "infinite."

Since the empty class is a member of every inductive class, it follows from Axiom 5 that $\varnothing$ is a set. Hence the temporary Axiom 0 can now be dispensed with.

In view of Axiom 5, it is plausible (and in fact true) that many inductive sets exist. Which particular one should be chosen as the set of all natural numbers? In order to insure that each natural number other than 0 is the successor of another, we choose the smallest inductive set possible.

6.3. DEFINITION

The class ω is defined to be

$$\bigcap\{X \mid X \text{ is inductive}\}.$$

By a *natural number* is meant an element of ω.

6.4. THEOREM

(1) The class ω is an inductive set.
(2) If X is an inductive set, then $\omega \subset X$.

Proof

(1) Let

$$\mathcal{A} = \{X \mid X \text{ is inductive}\}.$$

By Axiom 5, $\mathcal{A} \neq \varnothing$. Hence $\omega = \bigcap\mathcal{A}$ is a set.

We show that ω is inductive. First $\varnothing \in \omega$, since $\varnothing \in X$ for each $X \in \mathcal{A}$. Now assume $x \in \omega$; we must show $x^+ \in \omega$. If $X \in \mathcal{A}$, then $x \in X$ and so $x^+ \in X$. Hence $x^+ \in \bigcap\mathcal{A} = \omega$.

(2) If X is an inductive set, then $X \in \mathcal{A}$ and hence $\omega = \bigcap\mathcal{A} \subset X$. □

Part of 6.4 justifies calling ω *the set of all natural numbers*. Of course, the usual names for natural numbers are applied to the members of ω.

6.5. DEFINITION

One defines 0 to be $\varnothing$, 1 to be 0^+, 2 to be 1^+, 3 to be 2^+, 4 to be 3^+, 5 to be 4^+, 6 to be 5^+, 7 to be 6^+, 8 to be 7^+, and 9 to be 8^+.

Later, when addition of natural numbers has been defined, we shall see that $n^+ = n + 1$ for each $n \in \omega$.

VERIFICATION OF THE PEANO POSTULATES

The fundamental properties of the set ω are summarized by:

6.6. PEANO POSTULATES

(1) $0 \in \omega$.
(2) If $n \in \omega$, then $n^+ \in \omega$.
(3) If $n \in \omega$, then $n^+ \neq 0$.

(4) If $n, m \in \omega$ and if $n^+ = m^+$, then $n = m$.
(5) If $X \subset \omega$, if $0 \in X$, and if $n^+ \in X$ whenever $n \in X$, then $X = \omega$.

Essentially these statements were used by G. Peano as axioms for the natural numbers. For us they are consequences of our set-theoretic definition of ω. In fact, (1) and (2) follow from 6.4. To prove (3), note that if $n \in \omega$, then $n \in n \cup \{n\} = n^+$. We prove (4) below. Property (5) is just a trivial consequence of the definition of ω yet is so important as to deserve a separate statement.

6.7. THEOREM (PRINCIPLE OF MATHEMATICAL INDUCTION)

Let $X \subset \omega$. Assume that $0 \in X$ and that $n^+ \in X$ for each $n \in X$. Then $X = \omega$.

Proof

Since $X \subset \omega$, X is a set. By assumption, X is inductive. By 6.4, $\omega \subset X$. Hence $X = \omega$. ▯

Theorem 6.7 justifies the familiar method of proving something true of every natural number by showing that it is true of 0 and that it is true of n^+ whenever it is true of n. More precisely:

6.8. PROOF RULE (METHOD OF INDUCTION)

Let $\mathbf{P}$ be a formula in which the letter $\mathbf{n}$ is free. Suppose

$$\vdash [0 \mid \mathbf{n}]\mathbf{P}$$

and

$$\vdash (\forall \mathbf{n} \in \omega)(\mathbf{P} \Rightarrow [\mathbf{n}^+ \mid \mathbf{n}]\mathbf{P}).$$

Then

$$\vdash (\forall \mathbf{n} \in \omega)(\mathbf{P}).$$

Justification

Let $\mathbf{X}$ denote the subset

$$\{\mathbf{n} \mid \mathbf{n} \in \omega \,\&\, \mathbf{P}\}$$

of ω. Then

$$(\forall \mathbf{k} \in \omega)([\mathbf{k} \mid \mathbf{n}]\mathbf{P} \Leftrightarrow \mathbf{k} \in \mathbf{X}).$$

We need therefore to show that $\mathbf{X} = \omega$ is true. By supposition, $0 \in \mathbf{X}$ and $(\forall \mathbf{n} \in \mathbf{X})(\mathbf{n}^+ \in \mathbf{X})$ are true. Hence $\mathbf{X} = \omega$ is true by 6.7. ▯

Application of this rule to prove $(\forall \mathbf{n} \in \omega)(\mathbf{P})$ is indicated by referring to *induction* (*on* $\mathbf{n}$). In such an induction proof, the formula

$$\mathbf{n} \in \omega \,\&\, \mathbf{P} \Rightarrow [\mathbf{n}^+ \mid \mathbf{n}]\mathbf{P}$$

is frequently proved by use of the deduction criterion, and then the assumed formula **P** is called the *inductive hypothesis*.

The method of induction is illustrated by the proof of the next theorem. First, a definition.

6.9. DEFINITION

A class X is said to be *full* if

$$(\forall x \in X)(x \subset X).$$

Hence X is full if and only if

$$(\forall x)(\forall n)(x \in X \,\&\, n \in x \Rightarrow n \in X).$$

6.10. THEOREM

If $n \in \omega$, then n is full.

Proof

We use induction on n. To be precise, take **P** to be the statement 'n is full'.

We first prove $[0 \mid n]\mathbf{P}$, that is, '0 is full'. Since '$x \in 0$' is false, it is vacuously true that '0 is full'.

We next prove the inductive step

$$(\forall n \in \omega)(\mathbf{P} \Rightarrow [n^+ \mid n]\mathbf{P}),$$

that is,

$$(\forall n \in \omega)(n \text{ full} \Rightarrow n^+ \text{ full}).$$

Assume $n \in \omega$ and n is full; we deduce that n^+ is full. We must deduce

$$(*) \qquad m \in n^+ \Rightarrow m \subset n^+.$$

Assume $m \in n^+$. Since $n^+ = n \cup \{n\}$, either $m \in n$ or $m = n$. If $m \in n$, then $m \subset n$ by the inductive hypothesis, and so $m \subset n^+$ since $n \subset n^+$. On the other hand, if $m = n$, then $m \subset n^+$ by definition of n^+. Thus $m \subset n^+$ in either case. This establishes (*) and completes the inductive step. □

The present interest in 6.10 resides in one of its consequences.

6.11. THEOREM

If $n \in \omega$, then $n \notin n$. If $m, n \in \omega$, then $m \notin n$ or $n \notin m$.

Proof

We prove the first statement by induction on n. Since $0 = \varnothing$, $0 \notin 0$. Let $n \in \omega$, and assume $n \notin n$. We show that $n^+ \notin n^+$. Just suppose $n^+ \in n^+$. Then $n^+ \in n$ or $n^+ = n$. If $n^+ \in n$, then $n^+ \subset n$ since n is full, and since $n \in n^+$, $n \in n$ contradicts the inductive hypothesis. On the other hand, if $n^+ = n$, then $n \in n^+ = n$, again a contradiction. Hence $n^+ \notin n^+$.

To prove the second statement, let $m, n \in \omega$. If $m \in n$ and $n \in m$, then $m \subset n$ and $n \subset m$ since m and n are full, $m = n$, and so $n \in n$, in contradiction to what we proved above. $\square$

6.12. THEOREM

Let $n, m \in \omega$ with $n^+ = m^+$. Then $n = m$.

Proof

Just suppose $n \neq m$. Since $n \in m^+$, $n \in m$ or $n = m$. Then $n \in m$. Since $m \in n^+$, $m \in n$ or $m = n$. Hence $m \in n$ and $n \in m$. This contradicts 6.11. $\square$

The following theorem says that each nonzero natural number n is indeed the successor of another natural number, uniquely determined by n. The uniqueness assertion is a corollary of 6.12, and the existence assertion is proved by induction.

6.13. THEOREM

If $n \in \omega$ and $n \neq 0$, then there exists a *unique* $m \in \omega$ such that $n = m^+$.

ORDERING OF THE NATURAL NUMBERS

Now that all five Peano postulates have been verified, we consider next the definition and properties of the familiar "ordering" of natural numbers by virtue of which some natural numbers are "less than" others. To motivate the definition, we note that if $m \in n$, then $m \subsetneq n$, so m is in a sense smaller than n.

6.14. DEFINITION

Let $m, n \in \omega$. Then both $m < n$ and $n > m$ are defined to mean $m \in n$; the negation of $m < n$ is written $m \nless n$ or $n \ngtr m$. Both $m \leq n$ and $n \geq m$ are defined to mean

$$(m < n) \vee (m = n);$$

the negation of $m \leq n$ is also written $m \nleq n$ or $n \ngeq m$. The symbols $<$, $>$, $\leq$, $\geq$ are read respectively as "less than," "greater than," "less than or equal to," "greater than or equal to."

Clearly $n < n^+$ always. Also, $m < n \Rightarrow m \leq n$. In fact, 6.11 shows

$$m < n \Leftrightarrow m \leq n \,\&\, m \neq n.$$

On the other hand, by definition,

$$m \leq n \Leftrightarrow m < n \text{ or } m = n.$$

Hence properties of $<$ yield properties of $\leq$, and vice versa.

Each part of the next theorem either is a consequence of 6.11 or may be proved by an easy induction.

6.15. THEOREM

Let $m, n, p \in \omega$. Then:

(1) If $m < n$ and $n < p$, then $m < p$.
(2) $n \nless n$.
(3) If $m < n$, then $n \nless m$.
(4) If $m < n$, then $m^+ \leq n$.
(5) If $n \neq 0$, then $0 < n$.

6.16. COROLLARY

Let $m, n, p \in \omega$. Then:

(1) If $m \leq n$ and $n \leq p$, then $m \leq p$.
(2) $n \leq n$.
(3) If $m \leq n$ and $n \leq m$, then $m = n$.
(4) If $m \leq n$, then $m^+ \leq n^+$.
(5) $0 \leq n$.

6.17. THEOREM

If $n \in \omega$ and $m \in \omega$, then $m \leq n$ or $n \leq m$.

Proof

We use induction on n to show

$$m \in \omega \Rightarrow m \leq n \vee n \leq m.$$

First, for each $m \in \omega$, $0 \leq m$.

Let $n \in \omega$. Assume that for each $m \in \omega$, $m \leq n$ or $n \leq m$. Let $k \in \omega$. We must show $k \leq n^+$ or $n^+ \leq k$. If $k = 0$, then $k \leq n^+$. Now suppose $k \neq 0$. By 6.13, $k = m^+$ for some $m \in \omega$. By the inductive hypothesis, $m \leq n$ or $n \leq m$. If $m \leq n$, then $k = m^+ \leq n^+$; if $n \leq m$, then $n^+ \leq m^+ = k$. ☐

6.18. COROLLARY

Let $m, n \in \omega$. Then exactly one of the following alternatives holds:

$$m < n, \qquad m = n, \qquad n < m.$$

Proof

That at most one alternative holds follows from 6.11. That at least one holds follows from 6.17. ▯

If $X \subset \omega$ and $0 \in X$, then $0 \leq m$ for all $m \in X$. Let us generalize.

6.19. DEFINITION

Let $X \subset \omega$. One calls n a *least element of* X if

$$n \in X \;\&\; (\forall m \in X)\,(n \leq m).$$

One says that X *has a least element* to mean there exists n such that n is a least element of X. Synonyms for 'least' are 'first' and 'smallest'.

6.20. THEOREM (WELL-ORDERING PRINCIPLE)

Let X be a nonempty subset of ω. Then X has a least element.

Proof

Just suppose X does not have a least element. We shall use induction to show

$$n \in \omega \Rightarrow (\forall m \in X)\,(n < m),$$

so that X would be empty.

If $m \in X$, then $m \neq 0$, for otherwise 0 is a least element of X. Hence $0 < m$ for each $m \in X$.

Let $n \in \omega$ and assume that $n < m$ for every $m \in X$. Just suppose $n^+ \nless m$ for some $m \in X$. We show that then n^+ is a least element of X. We have $m \leq n^+$. Now $n < m$ by the inductive hypothesis. Hence $n^+ = m \in X$. If $k \in X$, then $n < k$ by the inductive hypothesis, and so $n^+ \leq k$. Thus n^+ is a least element of X, a contradiction. ▯

Let X be a nonempty subset of ω. By 6.20, X has a least element. Now if m and n are both least elements of X, then $m \leq n$ and $n \leq m$, so that $m = n$. Hence X has a unique least element, and we call

$$(\iota n)\,(n \text{ is a least element of } X)$$

simply *the least element of* X.

Theorem 6.20 is frequently used instead of induction, as in the proof of the next theorem.

6.21. THEOREM

The set ω is full.

Proof

Just suppose $n \not\subset \omega$ for some $n \in \omega$. Then the set

$$X = \{n \mid n \in \omega \;\&\; n \not\subset \omega\}$$

of natural numbers is nonempty. Let k be the least element of X. Since $0 \subset \omega$, $k \neq 0$. Then $k = m^+$ for some $m \in \omega$. Since $m < k$, $m \notin X$. Hence $m \subset \omega$, and

$$k = m^+ = m \cup \{m\} \subset \omega \cup \omega = \omega,$$

contradicting the fact that $k \in X$. $\square$

Thus, each natural number is a set of natural numbers. More precisely, each natural number n is the set of all natural numbers less than itself:

$$n = \{i \mid i \in \omega \,\&\, i < n\}.$$

In fact, $i \in n \Rightarrow i \in \omega$ by 6.21, and $n = \{i \mid i \in n\}$.

The fact that ω and each $n \in \omega$ are full will be utilized later to generalize to the notion of ordinal numbers.

THE AXIOM OF FOUNDATION

We proved above that $n \notin n$ for every natural number n, and $m \notin n$ or $n \notin m$ for any natural numbers m and n. We now extend these results to arbitrary classes by means of an axiom due to von Neumann and Bernays.

AXIOM 6 (AXIOM OF FOUNDATION)

If X is nonempty, then there exists $a \in X$ such that $a \cap X = \varnothing$.

6.22. THEOREM

Let X and Y be classes. Then $X \notin Y$ or $Y \notin X$.

Proof

Just suppose $X \in Y$ and $Y \in X$. Then X and Y are both sets, so that $X \in \{X, Y\}$ and $Y \in \{X, Y\}$. By Axiom 6, there exists $a \in \{X, Y\}$ such that $a \cap \{X, Y\} = \varnothing$. Choose such an a. If $a = X$, then $X \cap \{X, Y\} = \varnothing$, so $Y \notin X$. If $a = Y$, then $Y \cap \{X, Y\} = \varnothing$, so $X \notin Y$. $\square$

Taking $Y = X$ in 6.22, we obtain:

6.23. THEOREM

Let X be a class. Then $X \notin X$.

In Chapter 4 the Russell class $\mathcal{R}$ was shown to be a proper class. In Chapter 5 the universe $\mathcal{U}$ was shown to be a proper class. Now $\mathcal{R} \subset \mathcal{U}$, and an immediate consequence of 6.23 is:

6.24. COROLLARY

$$\mathcal{R} = \mathcal{U}.$$

EXERCISES

A. Compute $\bigcup\omega$ and $\bigcap\omega$.

B. Let $n \in \omega$. Show there is no $k \in \omega$ for which $n < k < n^+$.

C. Let $m, n \in \omega$. Show that

$$m^+ < n^+ \Rightarrow m < n.$$

Is the corresponding statement obtained by changing '$<$' to '$\leq$' also true?

D. Let $m, n \in \omega$. Compute $m \cup n$ and $m \cap n$.

E. Let $X \subset \omega$. Suppose $n \subset X \Rightarrow n \in X$ (this condition is the converse of the condition for X to be full). Prove that $X = \omega$.

F. Mathematical induction as described in the text always "starts with 0." Show that it can start with any natural number k, that is, prove:
Let $k \in \omega$ and let

$$X \subset \{n \mid n \in \omega \,\&\, n \geq k\}.$$

If $k \in X$ and if $n^+ \in X$ for each $n \in X$, then

$$X = \{n \mid n \in \omega \,\&\, n \geq k\}.$$

G. Theorem 6.7 is sometimes referred to as the principle of *simple* induction; its hypothesis says that a natural number belongs to X whenever the "greatest" natural number less than it does. Prove the following principle of *complete* induction whose hypothesis says that a natural number belongs to X whenever all natural numbers less than it do:
If $X \subset \omega$ and if

$$(\forall n \in \omega)((\forall i < n)(i \in X) \Rightarrow n \in X),$$

then $X = \omega$.
Explain why no special hypothesis $0 \in X$ is required here.

H. Formulate and justify proof rules based on Exercises F and G analogous to the method of induction 6.8.

I. Let $n \in \omega$ and let $X \subset \omega$ such that $i \leq n$ for each $i \in X$. Suppose

$$i \in X \,\&\, i < n \Rightarrow i^+ \in X.$$

Show that $X = \{i \mid i \in \omega \,\&\, i \leq n\}$.

J. Prove 6.21 by induction.

K. For a subset X of ω, give a reasonable definition of "a greatest element of X." Must a greatest element of X be unique? Explain any relationship between greatest elements of subsets of ω and least elements of their complements in ω.

L. Deduce 6.7 from 6.20 and 6.13 without using the definition or any other special properties of ω.

M. Let X be full. Show that either $X = \varnothing$ or else $\varnothing \in X$.

N. Deduce 6.23 directly from Axiom 6 without using 6.22.

P. Show that ω^+ is full. Do the same for $(\omega^+)^+$. Does a generalization suggest itself?

Q. Without using Axiom 6, show that Axiom 6 is a consequence of the following two assumptions:

(a) If x is a set, then $x \subset y$ for some full set y.

(b) If x is a nonempty *set*, then there exists $z \in x$ such that $z \cap x = \varnothing$.

[*Hint:* Let X be a nonempty class. Choose any $u \in X$. Suppose $u \cap X \neq \varnothing$. Apply (a) to the set $u \cap X$.] [Proposition (a) can be proved without any use of Axiom 6 by a technique introduced later (see Chapter 10, Exercise M). Hence Axiom 6 is actually a consequence of the apparently weaker statement (b).]

7. Relations

The notion of a relation is introduced in this chapter as a certain kind of class, and the elementary algebra of relations is developed. Later, special types of relations—functional relations, equivalence relations, and ordering relations—will be studied that are central to mathematics. As a prerequisite of the definition of relations, ordered pairs and products are discussed.

ORDERED PAIRS

Let x and y be sets. We would like to define an object (x,y) determined by x and y which assigns the order "first x, then y" to the two sets. In other words, $(x,y) = (a,b)$ should imply $x = a$ and $y = b$, but (x,y) should be different from (y,x) if x is distinct from y. One class determined by x and y is the doubleton $\{x,y\}$. Unfortunately $\{x,y\} = \{y,x\}$, so the doubleton does not itself give any preference to x as "first" and y as "second." The way we get around this obstacle is to pick from the doubleton $\{x,y\}$ its subset $\{x\}$ in order to place x first.

7.1. DEFINITION

Let x and y be classes. One defines the *ordered pair* x, y, denoted by (x,y), to be the class

$$\{\{x\}, \{x,y\}\}.$$

(A synonym for 'ordered pair' is 'couple'.)

This definition actually accomplishes our objective.

7.2. THEOREM

Let x, y, a, and b be sets. Then

$$(x,y) = (a,b)$$

if and only if

$$x = a \quad \text{and} \quad y = b.$$

Proof

If $x = a$ and $y = b$, then $\{x\} = \{a\}$ and $\{x,y\} = \{a,b\}$, so $(x,y) = (a,b)$.

Conversely, assume $(x,y) = (a,b)$, that is,

$$\{\{x\}, \{x,y\}\} = \{\{a\}, \{a,b\}\}.$$

Then either

(1) $$\{x\} = \{a\} \text{ and } \{x,y\} = \{a,b\}$$

or

(2) $$\{x\} = \{a,b\} \text{ and } \{x,y\} = \{a\}.$$

Consider first case (1). We have $x = a$. Also,

$$x = a \quad \text{and} \quad y = b, \quad \text{or} \quad x = b \quad \text{and} \quad y = a.$$

If $x = b$ and $y = a$, then $x = a = y = b$, so $x = a$ and $y = b$.

Consider next case (2). Since $\{x\} = \{a,b\}$, $x = a = b$, so we need only show $y = x$. But $x = a = y$ since $\{x,y\} = \{a\}$.

Thus $x = a$ and $y = b$ in either case. ☐

7.3. COROLLARY

If x and y are distinct sets, then $(x,y) \neq (y,x)$.

Note that if x and y are proper classes, then $(x,y) = \{\varnothing\}$. Hence the assumption in 7.2 that x and y be sets is really needed.

7.4. THEOREM

Let x and y be classes. Then (x,y) is a set.

Proof

We consider only the case that x and y are both sets. Then $\{x\}$ and $\{x,y\}$ are sets by the axiom of pairing, so (x,y) is a set by the same axiom. ☐

7.5. DEFINITION

One says that a set z is an *ordered pair* (*of sets*) to mean there exist sets x and y such that $z = (x,y)$.

Let z be an ordered pair. By 7.2,

$$(\exists!\, x \in \mathfrak{U})(\exists\, y \in \mathfrak{U})(z = (x,y)),$$

$$(\exists!\, y \in \mathfrak{U})(\exists\, x \in \mathfrak{U})(z = (x,y)).$$

Hence one may define the *first coordinate of* z, denoted 1st coord z, to be the set x such that $z = (x,y)$ for some set y. Similarly, one may define the *second coordinate of* z, denoted 2nd coord z, to be the set y such that $z = (x,y)$ for some set x. Then

$$z = (\text{1st coord } z, \text{2nd coord } z).$$

If x and y are sets, then

$$\text{1st coord } (x,y) = x,$$

$$\text{2nd coord } (x,y) = y.$$

We can now express 7.4 as follows: Let z and w be ordered pairs. Then $z = w$ if and only if 1st coord z = 1st coord w and 2nd coord z = 2nd coord w.

PRODUCT OF TWO CLASSES

The following notation will be useful below. Let **P** be a formula in which the letters **x** and **y** are free and in which the letter **z** does not appear. Then one writes

$$\{(\mathbf{x},\mathbf{y}) \mid \mathbf{P}\}$$

to denote the class

$$\{\mathbf{z} \mid (\exists \mathbf{x} \in \mathcal{U})(\exists \mathbf{y} \in \mathcal{U})(\mathbf{P} \,\&\, \mathbf{z} = (\mathbf{x},\mathbf{y}))\}$$

of all ordered pairs **z** of sets whose first and second coordinates satisfy **P**; this class is often called the *graph of* **P**. For example,

$$\{(m,n) \mid m \in \omega \,\&\, n \in \omega \,\&\, n = m^+\}$$

is the class of ordered pairs (m,n) whose second coordinate is the successor of its first coordinate.

7.6. DEFINITION

Let X and Y be classes. The (*Cartesian* or *direct*) *product of* X *and* Y, denoted by $X \times Y$, is defined to be the class

$$\{(x,y) \mid x \in X \,\&\, y \in Y\}$$

of all ordered pairs whose first coordinates belong to X and whose second coordinates belong to Y. The symbol $\times$ is read "cross."

The product of X and Y may be represented schematically as follows. Represent X and Y as two perpendicular sides of a rectangle. Then $X \times Y$ consists of all points enclosed by that rectangle. The coordinates of a point $(x,y) \in X \times Y$ are obtained by projecting (x,y) onto the sides X and Y (see Figure 7.1).

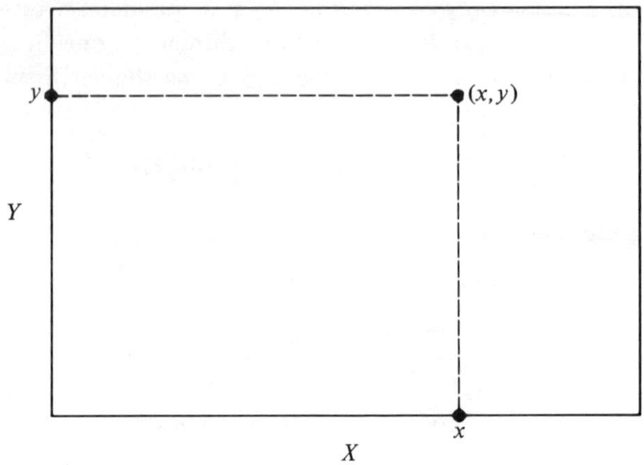

Figure 7.1

The analogy between this schematic representation of a product and Descartes' representation of points in the plane by pairs of real numbers is evident. If, as Descartes essentially said, a line "is" the set $\mathbb{R}$ of all real numbers, then the plane is the product $\mathbb{R} \times \mathbb{R}$ of the set of all real numbers with itself.

Other geometric representations of products can be useful. For example, if X is a circle in the plane and Y is a closed interval in the line, then $X \times Y$ may be represented as a cylindrical surface in space (see Figure 7.2).

Note that there are sets for which $X \times Y \neq Y \times X$. For example, $1 \times 2 \neq 2 \times 1$. On the other hand, $0 \times 2 = 2 \times 0$. More generally,

$$X \times \varnothing = \varnothing = \varnothing \times X$$

for any class X.

7.7. THEOREM

If X and Y are sets, then $X \times Y$ is a set.

Proof

Assume that X and Y are sets. Let $z \in X \times Y$. Then

$$z = \{\{x\}, \{x,y\}\}$$

where $x \in X$ and $y \in Y$. Now

$$\{x\} \subset X \subset X \cup Y, \qquad \{x,y\} \subset X \cup Y,$$

so

$$\{x\} \in \mathcal{P}(X \cup Y), \qquad \{x,y\} \in \mathcal{P}(X \cup Y).$$

Then

$$z \subset \mathcal{P}(X \cup Y),$$

and hence

$$z \in \mathcal{P}(\mathcal{P}(X \cup Y)).$$

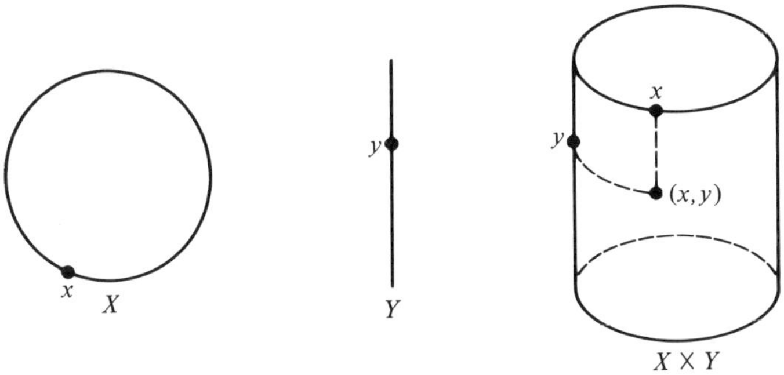

Figure 7.2

We have just proved that $X \times Y \subset \mathcal{P}(\mathcal{P}(X \cup Y))$. Since X and Y are sets, so is $X \cup Y$. Double use of the axiom of subsets shows that $\mathcal{P}(\mathcal{P}(X \cup Y))$ is a set. Hence $X \times Y$ is a set. $\square$

7.8. THEOREM

Let X, Y, and Z be classes. Then:

(1) $X \times (Y \cup Z) = (X \times Y) \cup (X \times Z)$,
(2) $X \times (Y \cap Z) = (X \times Y) \cap (X \times Z)$,
(3) $X \times (Y \setminus Z) = (X \times Y) \setminus (X \times Z)$.

We have similarly, $(Y \cup Z) \times X = (Y \times X) \cup (Z \times X)$, and so on.

7.9. THEOREM

Let X, Y, and Z be classes. If $Y \subset Z$, then

$$X \times Y \subset X \times Z, \qquad Y \times X \subset Z \times X.$$

Conversely, if $X \times Y \subset X \times Z$ and if $X \neq \varnothing$, then $Y \subset Z$.

The hypothesis $X \neq \varnothing$ in the second assertion of 7.9 is surely needed, for $\varnothing \times 1 = \varnothing = \varnothing \times 0$, but $1 \not\subset 0$.

GENERALIZED ORDERED PAIRS

The notion of ordered pair introduced above suffers the defect that a proper class can never be the first or second coordinate of an ordered pair. We give now another method of defining "first X, then Y" applicable to *any* classes X and Y (although it will be of interest primarily only when X and Y are proper classes). The idea is to tag elements of X with 1 and elements of Y with 2.

7.10. DEFINITION

If X and Y are classes, one defines $\langle X,Y\rangle$ to be the class

$$(X \times \{1\}) \cup (Y \times \{2\}).$$

Later, we shall even need an analog of ordered triples (see Chapter 7, Exercise H) for arbitrary classes. If X, Y, and Z are classes, we define $\langle X,Y,Z\rangle$ to be the class

$$(X \times \{1\}) \cup (Y \times \{2\}) \cup (Z \times \{3\}),$$

and we call X (respectively, Y, Z) the *first* (respectively, *second, third*) *argument of* $\langle X,Y,Z\rangle$.

If X and Y are classes, then clearly

$$\langle X,Y\rangle = \langle X,Y,\varnothing\rangle.$$

Hence we limit our discussion below to classes of the form $\langle X,Y,Z\rangle$.
Let $w \in \langle X,Y,Z\rangle$. Then w is an ordered pair,

$$w = (u,v)$$

with $u \in X \cup Y \cup Z$ and $v \in \{1,2,3\}$. We have

$$u \in X \Leftrightarrow v = 1$$

$$u \in Y \Leftrightarrow v = 2$$

$$u \in Z \Leftrightarrow v = 3.$$

Thus, X is the class of all x of the form $x = $ 1st coord w, where $w \in \langle X,Y,Z\rangle$ and 2nd coord $w = 1$. Similarly, for Y and for Z. Then $\langle X,Y,Z\rangle$ is found by tagging elements of X with 1, elements of Y with 2, and elements of Z with 3.

7.11. THEOREM

Let X, Y, Z, A, B, C be classes. Then

$$\langle X,Y,Z\rangle = \langle A,B,C\rangle$$

if and only if

$$X = A \;\&\; Y = B \;\&\; Z = C.$$

Proof

Assume $\langle X,Y,Z\rangle = \langle A,B,C\rangle$. We show that $X = A$. The proofs that $Y = B$ and $Z = C$ are similar.

We show that $X \subset A$. Let $x \in X$. Then $(x,1) \in \langle X,Y,Z\rangle$, so $(x,1) \in \langle A,B,C\rangle$. Now

$$\langle A,B,C\rangle = (A \times \{1\}) \cup (B \times \{2\}) \cup (C \times \{3\}),$$

so $(x,1) \in A \times \{1\}$. It follows that $x \in A$.

By reversing the roles of $\langle X,Y,Z\rangle$ and $\langle A,B,C\rangle$ in the preceding paragraph, we obtain $A \subset X$. Hence $X = A$. $\square$

RELATIONS

Roughly speaking, a "relation" is a rule for assigning to certain sets certain other sets. For example, we may assign to each natural number greater than 2 each natural number greater than it. Now such a rule is completely described by forming the class R of all those ordered pairs (x,y) having the property that x is a set to which the rule assigns the set y. (In the example, the class

$$\{(x,y) \mid x \in \omega \,\&\, x > 2 \,\&\, y \in \omega \,\&\, y > x\}$$

is the class R.) Moreover, any class R of ordered pairs determines such a rule, namely, the rule assigning y to x if and only if $(x,y) \in R$. Hence we may define a relation to be a class of ordered pairs.

7.12. DEFINITION

A class R is called a *relation* if each element of R is an ordered pair of sets. If R is a relation, one often abbreviates $(x,y) \in R$ by

$$xRy$$

and interprets xRy to mean that R assigns y to x; the negation of xRy is written $x\not{R}y$.

Let R be a relation. The *domain of* R, denoted dmn R, is defined to be the class

$$\{x \mid (\exists z \in R)(x = \text{1st coord } z)\}$$

of all sets to which R assigns sets. The *range of* R, denoted rng R, is defined to be the class

$$\{y \mid (\exists z \in R)(y = \text{2nd coord } z)\}$$

of all sets assigned by R to elements of dmn R.

Let X and Y be classes, and let R be a relation. We say R is a relation *in X to Y* if dmn $R \subset X$ and rng $R \subset Y$. If R is a relation in X to X, we say simply that R is a relation *in* X.

If R is a relation, then clearly

$$R \subset \text{dmn } R \times \text{rng } R,$$

but this inclusion may be strict [see 7.13(4)]. If dmn R and rng R are both sets, it follows that R is itself a set.

7.13. EXAMPLES

(1) Let X and Y be classes. Then the empty set $\varnothing$ is a relation in X to Y, and $\operatorname{dmn} \varnothing = \operatorname{rng} \varnothing = \varnothing$. The product $X \times Y$ is also a relation in X to Y, and if $X \times Y \neq \varnothing$, then

$$\operatorname{dmn}(X \times Y) = X, \qquad \operatorname{rng}(X \times Y) = Y.$$

If R is any relation in X to Y, then $\varnothing \subset R \subset X \times Y$.

(2) Let X be a class. The class

$$\{(x, y) \mid x \in X \,\&\, y = x\}$$

is a relation in X, called the *identity relation in* X and denoted Δ_X. We have $\operatorname{dmn} \Delta_X = \operatorname{rng} \Delta_X = X$.

(3) The class

$$\{(x,y) \mid y \in \mathfrak{U} \,\&\, x \in y\}$$

is a relation, called the *elementhood relation*, whose domain is $\mathfrak{U}$ and range is $\mathfrak{U} \setminus \{\varnothing\}$.

(4) The class

$$\{(m,n) \mid m \in \omega \,\&\, n \in \omega \,\&\, m < n\}$$

is a relation, called the (*usual*) *strict order relation in* ω and denoted by $<_\omega$ or simply $<$. We have

$$\operatorname{dmn} < \; = \omega, \qquad \operatorname{rng} < \; = \omega \setminus \{0\},$$

so

$$< \; \subsetneq (\operatorname{dmn} <) \times (\operatorname{rng} <).$$

(5) The class

$$\{(X,Y) \mid X \in \mathfrak{U} \,\&\, Y \in \mathfrak{U} \,\&\, X \subset Y\}$$

is a relation in $\mathfrak{U}$, called the *inclusion relation*. Hence a statement of the form $\mathbf{X} \subset \mathbf{Y}$ is often called an "inclusion relation."

The remainder of this chapter is concerned with the construction of new relations from given relations.

If S is a relation and if $R \subset S$, then R is also a relation, and

$$\operatorname{dmn} R \subset \operatorname{dmn} S, \qquad \operatorname{rng} R \subset \operatorname{rng} S.$$

Hence if R and S are two relations, then $R \cap S$ is a relation, $\operatorname{dmn}(R \cap S) \subset \operatorname{dmn} R \cap \operatorname{dmn} S$, and $\operatorname{rng}(R \cap S) \subset \operatorname{rng} R \cap \operatorname{rng} S$.

COMPOSITE AND INVERSE OF RELATIONS

7.14. DEFINITION

Let R and S be relations. The *composition* (or *composite* or *product*) *of* S *and* R, denoted by $S \circ R$ or simply SR, is defined to be the relation

$$\{(x,z) \mid (\exists y)((x,y) \in R \,\&\, (y,z) \in S)\};$$

one reads "$S \circ R$" as "S composed with R," "S following R," or "S with R." Then

$$x(S \circ R)z \Leftrightarrow (\exists y)(xRy \,\&\, yRz),$$

so that $S \circ R$ assigns z to x exactly when S assigns z to some set assigned by R to x.

Despite our habit of reading from left to right, the definition of $S \circ R$ suggests "first R, then S." Although it might therefore seem natural to denote the composition of S and R by $R \circ S$, the notation adopted here is justified by its traditional usage in the case of relations which are "functional."

7.15. EXAMPLES

(1) Let

$$R = \{(0,0), (0,1), (2,1), (3,3)\},$$

$$S = \{(0,0), (1,2)\}.$$

Then

$$S \circ R = \{(0,0), (0,2), (2,2)\},$$

$$R \circ S = \{(0,0), (0,1), (1,1)\}.$$

Note that $S \circ R \neq R \circ S$. Also

$$\operatorname{dmn}(S \circ R) = \{0,2\} \neq \{0,2,3\} = \operatorname{dmn} R,$$

$$\operatorname{rng}(R \circ S) = \{0,1\} \neq \{0,1,3\} = \operatorname{rng} R.$$

(2) Let X, Y, and Z be classes, and define

$$R = X \times Y, \qquad S = Y \times Z.$$

Then

$$R \circ S = \begin{cases} Y \times Y & \text{if } X \cap Z \neq \varnothing \\ \varnothing & \text{if } X \cap Z = \varnothing, \end{cases}$$

$$S \circ R = \begin{cases} X \times Z & \text{if } Y \neq \varnothing \\ \varnothing & \text{if } Y = \varnothing. \end{cases}$$

7.16. PROPOSITION

Let R and S be relations. Then

$$\operatorname{dmn}(S \circ R) \subset \operatorname{dmn} R,$$

$$\operatorname{rng}(S \circ R) \subset \operatorname{rng} S.$$

Proof

If $x \in \operatorname{dmn}(S \circ R)$, then $(x,z) \in S \circ R$ for some z, $(x,y) \in R$ and $(y,z) \in S$ for some y, $(x,y) \in R$ for some y, and so $x \in \operatorname{dmn} R$. This proves the first inclusion. The second inclusion is proved similarly. ▯

Example 7.15(1) shows that the inclusions in 7.16 may be strict.

7.17. PROPOSITION (ASSOCIATIVE LAW)

Let R, S, and T be relations. Then

$$T \circ (S \circ R) = (T \circ S) \circ R.$$

Proof

Any element of $T \circ (S \circ R)$ or of $(T \circ S) \circ R$ is an ordered pair (x,w) of sets. If x and w are sets, then

$$\begin{aligned}(x,w) \in (T \circ S) \circ R &\Leftrightarrow (\exists y)((x,y) \in R \;\&\; (y,w) \in T \circ S) \\ &\Leftrightarrow (\exists y)((x,y) \in R \;\&\; (\exists z)((y,z) \in S \;\&\; (z,w) \in T)) \\ &\Leftrightarrow (\exists z)((\exists y)((x,y) \in R \;\&\; (y,z) \in S) \;\&\; (z,w) \in T) \\ &\Leftrightarrow (\exists z)((x,z) \in S \circ R \;\&\; (z,w) \in T) \\ &\Leftrightarrow (x,w) \in T \circ (S \circ R). \quad ▯\end{aligned}$$

In view of 7.17, we can unambiguously write $T \circ S \circ R$ for each of the equal relations $T \circ (S \circ R)$ and $(T \circ S) \circ R$.

7.18. PROPOSITION

Let R be a relation in X to Y. Then

$$R \circ \Delta_X = R = \Delta_Y \circ R.$$

Proof

We show only $R \circ \Delta_X = R$. If $(x,y) \in R \circ \Delta_X$, then $(x,z) \in \Delta_X$ and $(z,y) \in R$ for some z, $x = z$, and so $(x,y) \in R$. If $(x,y) \in R$, then $(x,x) \in \Delta_X$ since $x \in \operatorname{dmn} R \subset X$, and so $(x,y) \in R \circ \Delta_X$. ▯

7.19. DEFINITION

Let R be a relation. The *inverse* (or *opposite*, or *reverse*) *of* R, denoted by R^{-1}, is the relation

$$\{(x,y) \mid (y,x) \in R\}.$$

Then

$$xR^{-1}y \Leftrightarrow yRx,$$

so that R^{-1} assigns y to x if and only if R assigns x to y.

7.20. EXAMPLES

(1) We have

$$(<_\omega)^{-1} = \{(n,m) \mid n \in \omega \,\&\, m \in \omega \,\&\, n > m\}.$$

(2) If X is a class, then

$$(\Delta_X)^{-1} = \Delta_X.$$

(3) If X and Y are classes, then

$$(X \times Y)^{-1} = Y \times X.$$

7.21. PROPOSITION

Let R be a relation. Then

$$\text{dmn } R^{-1} = \text{rng } R, \qquad \text{rng } R^{-1} = \text{dmn } R,$$

and

$$(R^{-1})^{-1} = R.$$

The inverse of the composite of two relations is the composite, taken in opposite order, of their inverses:

7.22. PROPOSITION

Let R and S be relations. Then

$$(S \circ R)^{-1} = R^{-1} \circ S^{-1}.$$

Proof

If x and z are sets, then

$$\begin{aligned}(z,x) \in (S \circ R)^{-1} &\Leftrightarrow (x,z) \in S \circ R \\ &\Leftrightarrow (\exists y)((x,y) \in R \,\&\, (y,z) \in S) \\ &\Leftrightarrow (\exists y)((y,x) \in R^{-1} \,\&\, (z,y) \in S^{-1}) \\ &\Leftrightarrow (z,x) \in R^{-1} \circ S^{-1}. \quad \square\end{aligned}$$

7.23. PROPOSITION

Let R be a relation. Then:

$$\Delta_{\text{dmn } R} \subset R^{-1} \circ R$$

$$\Delta_{\text{rng } R} \subset R \circ R^{-1}.$$

Proof

If $z \in \Delta_{\text{dmn } R}$, then $z = (x,x)$ for some $x \in \text{dmn } R$, $(x,y) \in R$ for some y, $(y,x) \in R^{-1}$ for such a y, and so $z = (x,x) \in R^{-1} \circ R$. Hence $\Delta_{\text{dmn } R} \subset R^{-1} \circ R$.

By 7.21 and what we have just proved,

$$\Delta_{\text{rng } R} = \Delta_{\text{dmn } R^{-1}} \subset (R^{-1})^{-1} \circ R^{-1} = R \circ R^{-1}. \quad \square$$

IMAGES UNDER A RELATION

7.24. DEFINITION

Let R be a relation. If A is a class, then the *(direct) image of* A *under* R, denoted by $R\langle A\rangle$, is defined to be the class

$$\{y \mid (\exists x \in A)(xRy)\}$$

of all those sets assigned by R to elements of A.

If B is a class, then the *inverse image of* B *under* R, denoted by $R^{-1}\langle B\rangle$, is defined to be the class

$$\{x \mid (\exists y \in B)(xRy)\}$$

of all those sets to which R assigns elements of B. The notation $R^{-1}\langle B\rangle$ does not conflict with that introduced in the preceding paragraph, since the inverse image of B under R is the direct image of B under R^{-1}.

7.25. EXAMPLES

(1) If X and A are classes, then

$$\Delta_X\langle A\rangle = A \cap X = \Delta_X{}^{-1}\langle A\rangle.$$

(2) If $R = <_\omega$, then

$$R\langle\{0\}\rangle = \omega \setminus \{0\},$$
$$R^{-1}\langle\{0\}\rangle = \varnothing.$$

(3) If R is any relation, then

$$R\langle \operatorname{dmn} R\rangle = \operatorname{rng} R, \qquad R^{-1}\langle \operatorname{rng} R\rangle = \operatorname{dmn} R.$$

If $A \cap \operatorname{dmn} R = \varnothing$, then $R\langle A\rangle = \varnothing$; in particular, $R\langle\varnothing\rangle = \varnothing$. If $B \cap \operatorname{rng} R = \varnothing$, then $R^{-1}\langle B\rangle = \varnothing$; in particular, $R^{-1}\langle\varnothing\rangle = \varnothing$.

If R is a relation and if $A \subset B$, then $R\langle A\rangle \subset R\langle B\rangle$. This result may be used to prove the following proposition.

7.26. PROPOSITION

Let R be a relation, and let A and B be classes. Then:

(1) $R\langle A \cup B\rangle = R\langle A\rangle \cup R\langle B\rangle$.
(2) $R\langle A \cap B\rangle \subset R\langle A\rangle \cap R\langle B\rangle$.
(3) $R\langle A \setminus B\rangle \supset R\langle A\rangle \setminus R\langle B\rangle$.

The inclusions in (2) and (3) may be strict. Moreover, (1) and (2) can be generalized to the case of the union or intersection of any class of sets.

Since the inverse image of a class under R is just the direct image of that class under R^{-1}, 7.26 gives results concerning inverse images:

$$R^{-1}\langle A \cup B\rangle = R^{-1}\langle A\rangle \cup R^{-1}\langle B\rangle,$$

$$R^{-1}\langle A \cap B\rangle \subset R^{-1}\langle A\rangle \cap R^{-1}\langle B\rangle.$$

7.27. PROPOSITION

Let R and S be relations, and let A be a class. Then

$$(S \circ R)\langle A\rangle = S\langle R\langle A\rangle\rangle.$$

Proof

We have

$$\begin{aligned} z \in (S \circ R)\langle A\rangle &\Leftrightarrow (\exists x \in A)((x,z) \in S \circ R) \\ &\Leftrightarrow (\exists x \in A)(\exists y)(xRy \;\&\; ySz) \\ &\Leftrightarrow (\exists y)((\exists x \in A)(xRy) \;\&\; ySz) \\ &\Leftrightarrow (\exists y)(y \in R\langle A\rangle \;\&\; ySz) \\ &\Leftrightarrow z \in S\langle R\langle A\rangle\rangle. \quad \square \end{aligned}$$

EXERCISES

A. Is $\{1,2\}$ an ordered pair?

B. If x and y are sets, show that $(x,y) \neq \{x,y\}$.

C. Compute (x,y) in each of the following cases:

(a) x is a set and y is a proper class.

(b) x is a proper class and y is a set.

D. Let z be an ordered pair of sets. Verify the formulas

$$\text{1st coord } z = \bigcap\bigcap z,$$

$$\text{2nd coord } z = (\bigcap\bigcup z) \cup ((\bigcup\bigcup z) \setminus (\bigcup\bigcap z)).$$

E. If X is a circle in the plane, draw a picture of $X \times X$.

F. When is $X \times Y$ empty?

G. When does $X \times Y = Y \times X$?

H. If x, y, and z are sets, one may define the *ordered triple* (x,y,z) to be the ordered pair $((x,y), z)$.

(a) Represent (x,y,z) as an unordered pair of unordered pairs.

(b) State and prove an analog of 7.2 for ordered triples.

I. Let $\mathcal{A}$ be a class of sets and X be a class. Express $X \times (\bigcup \mathcal{A})$ as the union of a class of products.

J. If X, Y, and Z are sets, is $\langle X,Y,Z\rangle$ a set?

K. Show that

$$\langle X,Y,Z\rangle \subset (X \cup Y \cup Z) \times \{1,2,3\}.$$

L. (a) Construct all relations in the set $\{0,1\}$.

(b) Construct all relations in $\{0,1,2\}$ to $\{0,1\}$.
(c) Construct all relations in $\{0,1\}$ to $\{0,1,2\}$.

M. Let R and S be relations. Show that $R \cup S$ is a relation and compute $\text{dmn}(R \cup S)$ and $\text{rng}(R \cup S)$.

N. Construct relations R and S for which $\text{rng}(R \cap S) \neq \text{rng}\, R \cap \text{rng}\, S$.

P. Given relations R and S, express $(R \cup S)^{-1}$ and $(R \cap S)^{-1}$ in terms of R^{-1} and S^{-1}.

Q. If R, S, T are relations, does $(T \cup S) \circ R = (T \circ R) \cup (S \circ R)$? What happens if union is replaced by intersection?

R. Let $X = \{x \mid x \in \mathbb{R} \;\&\; 0 \leq x \leq 1\}$. Let

$$R = \{(x,y) \mid (x,y) \in X \times X \;\&\; x < y\}.$$

Draw $X \times X$, R, Δ_X, and $R \cup \Delta_X$.

S. Regard the plane as $\mathbb{R} \times \mathbb{R}$. If **P** is a formula in which the letters 'x' and 'y' are free, then the usual "locus" L of **P** is just the relation

$$\{(x,y) \mid x \in \mathbb{R} \;\&\; y \in \mathbb{R} \;\&\; \mathbf{P}\}.$$

Explain the geometric meaning of the condition $L^{-1} = L$.

T. Construct a relation R and sets A and B such that

$$R\langle A \cap B\rangle \neq R\langle A\rangle \cap R\langle B\rangle.$$

U. Let R be a relation. Show that

$$A \subset \text{dmn}\, R \Rightarrow A \subset R^{-1}\langle R\langle A\rangle\rangle.$$

Does $B \subset \text{rng}\, R \Rightarrow B \subset R\langle R^{-1}\langle B\rangle\rangle$?

V. If X is a nonempty set, then a nonempty class $\mathcal{A}$ whose elements are relations in X is called a *uniformity on* X if the following properties hold:

(1) $R \in \mathcal{A} \Rightarrow \Delta_X \subset R$.
(2) $R \in \mathcal{A} \;\&\; S \in \mathcal{A} \Rightarrow R \cap S \in \mathcal{A}$.
(3) $R \subset X \times X \;\&\; (\exists\, S \in \mathcal{A})(R \supset S) \Rightarrow R \in \mathcal{A}$.
(4) $R \in \mathcal{A} \Rightarrow R^{-1} \in \mathcal{A}$.
(5) $R \in \mathcal{A} \;\&\; S \in \mathcal{A} \Rightarrow S \circ R \in \mathcal{A}$.

(a) For each real number $r > 0$ let

$$R_r = \{(x,y) \mid (x,y) \in \mathbb{R} \times \mathbb{R} \;\&\; |x - y| < r\},$$

and let

$$\mathcal{A} = \{S \mid S \subset \mathbb{R} \times \mathbb{R} \;\&\; (\exists\, r \in \mathbb{R})(r > 0 \;\&\; S \supset R_r\}.$$

Verify that $\mathcal{A}$ is a uniformity on $\mathbb{R}$.

(b) Show that a uniformity $\mathcal{A}$ on a nonempty set X is necessarily a set.

8. Maps

Of at least equal importance as the concept of a class is the concept of a "map." A map is a rule which assigns to each element x of a class X an element uniquely determined by x belonging to a class Y. Such a rule is defined below as a certain class determined by X, Y, and a relation describing the assignment of elements of Y to elements of X. Practical methods for constructing particular maps are presented. Composition of maps is defined and related to composition of relations. The highly useful tool of commutative diagrams and the family notation for maps are explained.

FUNCTIONAL RELATIONS

A relation may very well assign to a given element of its domain many elements of its range. This possibility is excluded for a relation which is to define a map.

8.1. DEFINITION

A relation f is said to be *functional* in case

$$(\forall x \in \operatorname{dmn} f)(\exists! y)((x,y) \in f).$$

The relation

$$f = \{(m,n) \mid m \in \omega \,\&\, n = 2\}$$

is functional, but the relation

$$<_\omega = \{(m,n) \mid m \in \omega \,\&\, n \in \omega \,\&\, m < n\}$$

is not.

A functional relation is one which assigns to each element of its domain exactly one element of its range. Now by the definition of the domain of a relation f,

$$(\forall x \in \operatorname{dmn} f)(\exists y)((x,y) \in f).$$

Hence *a relation f is functional if and only if*

$$(x,y) \in f \,\&\, (x,z) \in f \Rightarrow y = z,$$

in other words, f assigns at most one y to any given set x (of course, f assigns something to x only when $x \in \operatorname{dmn} f$). Thus, a set x can appear as the first coordinate of at most one member of a functional relation f. By way of contrast, a set y may appear as the second coordinate of many members of a functional relation; see the example in the preceding paragraph.

Special language is used to describe the unique y a functional relation assigns to a given element of its domain.

8.2. DEFINITION

Let f be a functional relation and let $x \in \operatorname{dmn} f$. The set

$$(\iota y)((x,y) \in f)$$

is denoted by $f(x)$ and is called the *value of f at x* or the *image of x under f*. Thus

$$y = f(x) \Leftrightarrow (x,y) \in f.$$

If $y = f(x)$, one says that f *maps* (or *sends*) x *to* y and writes

$$f\colon x \mapsto y.$$

Suppose f is a relation with $\operatorname{dmn} f = X$ and $\operatorname{rng} f \subset Y$. Represent $X \times Y$ as usual as a rectangle with base X and side Y, so that f is represented by a collection of points in this rectangle. Then the fact that f is a functional relation with domain X may be construed to say geometrically that the vertical line through each point x in X intersects f in exactly one point, whose projection on Y is $f(x)$ (see Figure 8.1).

Each element of the range of a functional relation f is assigned to one or more elements of the domain of f, so $\operatorname{rng} f$ is in a sense no bigger than $\operatorname{dmn} f$. Hence the next statement, our penultimate axiom, is quite reasonable.

AXIOM 7 (AXIOM OF REPLACEMENT)

If f is a functional relation and if $\operatorname{dmn} f$ is a set, then $\operatorname{rng} f$ is a set.

The corresponding statement for arbitrary relations is false. As the most spectacular example, the relation $\{0\} \times \mathcal{U}$ has the set $\{0\}$ as its domain and the proper class $\mathcal{U}$ as its range.

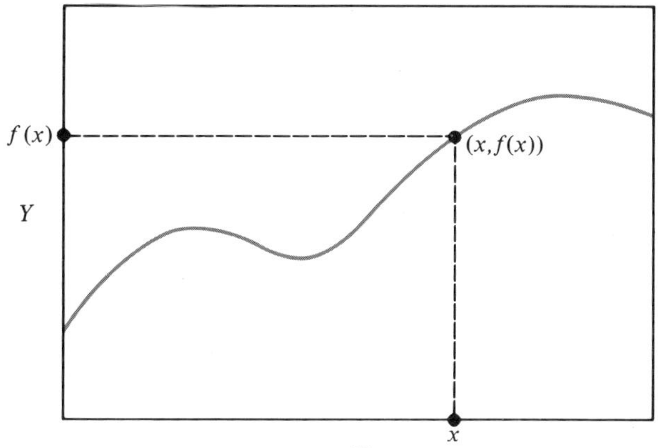

Figure 8.1

8.3. PROPOSITION

Let f be a functional relation. Then $\operatorname{dmn} f$ is a set if and only if f is a set.

Proof

Assume $\operatorname{dmn} f$ is a set. By Axiom 7, $\operatorname{rng} f$ is a set, so $\operatorname{dmn} f \times \operatorname{rng} f$ is a set. Now

$$f \subset \operatorname{dmn} f \times \operatorname{rng} f,$$

so f is a set.

Conversely, assume f is a set. The relation g defined to be

$$\{(z,x) \mid z \in f \;\&\; x = \text{1st coord } z\}$$

is clearly functional. Moreover,

$$\operatorname{dmn} g = f, \qquad \operatorname{rng} g = \operatorname{dmn} f.$$

By assumption, $\operatorname{dmn} g$ is a set. Hence $\operatorname{dmn} f$ is a set by Axiom 7. ∎

THE CONCEPT OF A MAP

One gets a map by putting together three things: (i) a class X to each of whose elements x a set $f(x)$ is going to be assigned, (ii) a functional relation f making such an assignment, and (iii) a class Y of which the value $f(x)$ assigned by f to each $x \in X$ is a member.

8.4. DEFINITION

A class φ is said to be a *map* or *mapping* or *function* if $\varphi = \langle X, f, Y \rangle$ for some class X, some class Y, and some functional relation f such that

$$\operatorname{dmn} f = X, \qquad \operatorname{rng} f \subset Y.$$

One calls X the *domain*, f the *graph*, and Y the *codomain* of a map $\varphi = \langle X, f, Y \rangle$ and denotes these by dmn φ, graph φ, and codmn φ respectively. A map $\langle X, f, Y \rangle$ is denoted by

$$f\colon X \to Y$$

or

$$X \xrightarrow{f} Y$$

and is said to be *from* (or *on*) X *to* (or *into*) Y.

Synonyms for 'map' in special contexts are 'operator' and 'transformation'.

When it makes sense, we apply, with the obvious meaning, terminology concerning a functional relation f to a map φ whose graph is f. Thus we call rng f the *range of* φ and denote it by rng φ, and for $x \in X$ we let $\varphi(x)$ mean $f(x)$.

We often denote a map $\varphi = \langle X, f, Y \rangle$ whose domain is X and codomain is Y by

$$\varphi\colon X \to Y.$$

Of course we make this abuse of notation only in contexts where no ambiguity results, for φ is not the same as f. A map and the graph of that map are different creatures, as the following discussion will make clear.

8.5. EXAMPLES

(1) Let X be a class. Then $\langle X, \Delta_X, X \rangle$ is a map, called the *identity map of* X and denoted by 1_X. We have

$$\text{dmn } 1_X = X = \text{codmn } 1_X, \qquad \text{graph } 1_X = \Delta_X,$$

$$1_X(x) = \Delta_X(x) = x \qquad (x \in X).$$

Other notations for 1_X are

$$\Delta_X\colon X \to X$$

and

$$1_X\colon X \to X.$$

(2) Let X and Y be classes with $X \subset Y$. Then $\langle X, \Delta_X, Y \rangle$ is a map, called the *inclusion* (or *insertion*) *map of* X *into* Y. Denoting this map by j, we have

$$\text{dmn } j = X, \qquad \text{codmn } j = Y, \qquad \text{graph } j = \Delta_X.$$

However, $j = 1_X$ only when codmn j = codmn 1_X, that is, when $X = Y$. Moreover, rng $j = X \neq Y =$ codmn j if $X \neq Y$.

(3) Let Y be any class. Since $\varnothing$ is vacuously a functional relation, $\langle \varnothing, \varnothing, Y \rangle$ is a map whose domain is $\varnothing$ and whose codomain is Y. Moreover, this is the unique map whose domain is $\varnothing$ and whose codomain is Y. By

way of contrast, there does not exist any map whose domain is a nonempty class and whose codomain is $\varnothing$.

(4) The class

$$f = \{(x,y) \mid y = x^+\}$$

is a functional relation which is the graph of the map $\langle \mathcal{U}, f, \mathcal{U} \rangle$, called the *successor map*. If this map is denoted by φ, then $\varnothing \in \mathcal{U}$ and $\varnothing \notin \operatorname{rng} \varphi = \operatorname{rng} f$, so

$$\operatorname{codmn} \varphi \neq \operatorname{rng} \varphi.$$

The connection between the domain, codomain, and range of a map is illustrated in Figure 8.2.

Examples 8.5 show that the codomain of a map need not equal its range. They also show that two different maps may have the same domain and the same graph, so the graph of a map does not always determine its codomain. However, the graph f of a map φ uniquely determines the domain of the map since $\operatorname{dmn} \varphi = \operatorname{dmn} f$; although the '$X$' in '$\langle X, f, Y \rangle$' is therefore redundant, it is nonetheless very useful to include it.

The following is a trivial observation but is worth stating just once:

8.6. LEMMA

Two maps

$$f: X \to Y, \qquad g: A \to B$$

are equal if and only if

$$X = A, \qquad Y = B,$$

and

$$f(x) = g(x) \qquad (x \in X).$$

The notation

$$f: X \to Y$$

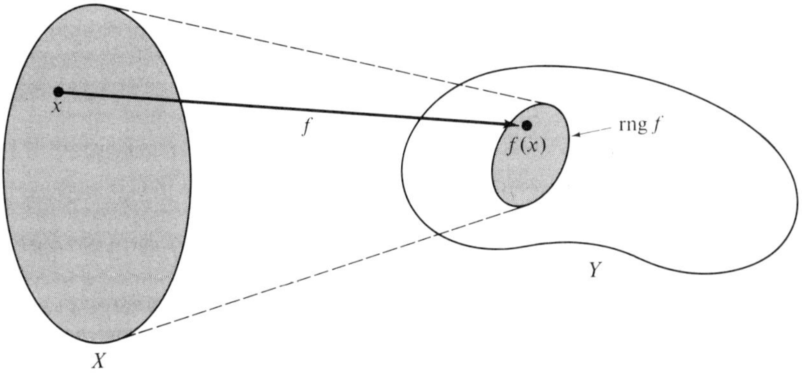

Figure 8.2

was defined to stand for the map $\langle X, f, Y\rangle$, a *class*. Sometimes it is used to stand for the *statement* that f is a map from X into Y. This notation tells only "where" f is but nothing about particular values of f other than their belonging to Y. Contrast it with the notation

$$f: x \mapsto y$$

which helps say "what" f is since it says $y = f(x)$.

METHODS OF CONSTRUCTING MAPS

Although the graph of a map is a class of ordered pairs, one does not usually define a map directly in terms of such a class. Particular maps are often defined by "formulas." In analysis one might discuss

$$\text{the map } f: \mathbb{R} \to \mathbb{R} \text{ given by } f(x) = x^4 + x^2$$

or in more elliptical yet typical language

$$\text{the function } x^4 + x^2 \text{ from } \mathbb{R} \text{ into } \mathbb{R}.$$

Here a map f has been specified by its domain ($\mathbb{R}$), its codomain ($\mathbb{R}$), and its value $f(x)$ at each member x of its domain, where $f(x)$ is given by a "formula" ($f(x) = x^4 + x^2$) involving an "expression in x" ($x^4 + x^2$). This procedure may be formalized.

8.7. METATHEOREM

Let $\mathbf{X}$ and $\mathbf{Y}$ be terms, let $\mathbf{v}$ be a term, and let $\mathbf{x}$ be a letter which is free in $\mathbf{X}$, $\mathbf{Y}$, and $\mathbf{v}$. Suppose

$$(\forall \mathbf{x} \in \mathbf{X})(\mathbf{v} \in \mathbf{Y})$$

is true. Then the following is true: There exists a unique map

$$\mathbf{f}: \mathbf{X} \to \mathbf{Y}$$

such that

$$\mathbf{f}(\mathbf{x}) = \mathbf{v} \qquad (\mathbf{x} \in \mathbf{X}).$$

Metaproof

To prove existence, take

$$\mathbf{f} = \{(\mathbf{x},\mathbf{y}) \mid \mathbf{x} \in \mathbf{X} \,\&\, \mathbf{y} = \mathbf{v}\}$$

where $\mathbf{y}$ is a letter not appearing in $\mathbf{v}$, $\mathbf{X}$, or $\mathbf{Y}$. To prove uniqueness, use 8.6. ▯

In the example above, $\mathbf{X}$ and $\mathbf{Y}$ are '$\mathbb{R}$' and $\mathbf{v}$ is the term '$x^4 + x^2$'.

The notation of 8.7 should not mislead one into believing the map has one and the same value at all elements of $\mathbf{X}$, for the letter $\mathbf{x}$ is free in $\mathbf{v}$.

In fact,

$$f(\mathbf{u}) = [\mathbf{u} \mid \mathbf{x}]\mathbf{v} \qquad (\mathbf{u} \in \mathbf{X}).$$

The map given by 8.7 is denoted by

$$\begin{aligned} f\colon \mathbf{X} &\to \mathbf{Y} \\ \mathbf{x} &\mapsto \mathbf{v} \qquad (\mathbf{x} \in \mathbf{X}). \end{aligned}$$

This usage of '$\mapsto$' is consistent with the usage explained earlier, for $\mathbf{u} \in \mathbf{X}$ implies

$$f\colon \mathbf{u} \mapsto [\mathbf{u} \mid \mathbf{x}]\mathbf{v}$$

in the sense of 8.2.

8.8. EXAMPLES

(1) In 8.7 take $\mathbf{X}$ and $\mathbf{Y}$ to be 'ω' and $\mathbf{v}$ to be 'x^+'. The resulting map

$$\begin{aligned} f\colon \omega &\to \omega \\ x &\mapsto x^+ \qquad (x \in \omega) \end{aligned}$$

satisfies $f(u) = u^+$ for all $u \in \omega$ and is called the *successor map of* ω.

(2) Let X be the class consisting of all continuous maps $f\colon [0,1] \to \mathbb{R}$; here $[0,1]$ is the usual closed unit interval in $\mathbb{R}$, and continuity is meant in the usual sense of elementary analysis. Then

$$\begin{aligned} I\colon X &\to \mathbb{R} \\ f &\mapsto \int_0^1 f(t)\,dt \qquad (f \in X) \end{aligned}$$

is the map from X to $\mathbb{R}$ for which

$$I(f) = \int_0^1 f(t)\,dt$$

for each continuous $f\colon [0,1] \to \mathbb{R}$.

(3) Let X and Y be classes. Let $c \in Y$. Take $\mathbf{v}$ to be 'c' in 8.7. Then

$$\begin{aligned} f\colon X &\to Y \\ x &\mapsto c \qquad (x \in X) \end{aligned}$$

is the map from X to Y such that

$$f(u) = c \qquad (u \in X)$$

and is called the *constant map from* X *to* Y *with value* c. A map from X to Y which is the constant map for some value $c \in Y$ is said to be *constant*.

(4) The inclusion map of a class X into a class Y containing X is just the map

$$\begin{aligned} j\colon X &\to Y \\ x &\mapsto x \qquad (x \in X). \end{aligned}$$

Here we take $\mathbf{v}$ in 8.7 to be 'x' itself.

(5) Let X and Y be classes. The maps

$$p: X \times Y \to X$$
$$z \mapsto \text{1st coord } z \qquad (z \in X \times Y),$$
$$q: X \times Y \to Y$$
$$z \mapsto \text{2nd coord } z \qquad (z \in X \times Y)$$

are called respectively the *first* and *second projection of* $X \times Y$. Thus $(x,y) \in X \times Y$ implies

$$p((x,y)) = x, \qquad q((x,y)) = y.$$

The projections p and q in the preceding example have a product of sets as their domains. If

$$f: X \times Y \to Z$$

is any map whose domain is the product of two classes X and Y, then f is said to be a *function of two variables*, and for any $(x,y) \in X \times Y$, one also denotes the value $f((x,y))$ of f at (x,y) simply by $f(x,y)$.

(6) Let R be a relation in a set X to a set Y. Then R gives rise to the map

$$R_*: \mathcal{P}(X) \to \mathcal{P}(Y)$$
$$A \mapsto R\langle A \rangle \qquad (A \in \mathcal{P}(X))$$

satisfying

$$R_*(A) = R\langle A \rangle$$

for each subset A of X. Now the new map

$$g: X \to \mathcal{P}(Y)$$
$$x \mapsto R_*(\{x\}) \qquad (x \in X)$$

satisfies

$$g(x) = R\langle \{x\} \rangle = \{y \mid xRy\}$$

for each $x \in X$. Hence this *map* g assigns to each $x \in \operatorname{dmn} R$ the set of all objects put into correspondence with x by the *relation* R.

(7) Let $f: X \to Y$ be any map. Then this equals the map

$$g: X \to Y$$
$$x \mapsto f(x) \qquad (x \in X)$$

obtained by taking **v** in 8.7 to be '$f(x)$', that is,

$$(\iota y)((x,y) \in f).$$

Thus any map can be defined by a formula.

The square-root function discussed in analysis may be defined as the map $f: X \to X$, where X is the set of all nonnegative real numbers, whose value y at any $x \in X$ is given by the rule

$$f(x) = y, \qquad \text{where } y \in X \text{ and } y^2 = x.$$

It is of course necessary to know that for each $x \in X$ there is exactly one $y \in X$ for which $y^2 = x$, so that the rule determines y uniquely for x. This procedure for defining a map by a "rule for determining $f(x)$" may be formalized as follows.

8.9. METATHEOREM

Let **X** and **Y** be terms and **P** be a formula in which the letters **x** and **y** are free. Suppose

$$(\forall \mathbf{x} \in \mathbf{X})(\exists!\, \mathbf{y} \in \mathbf{Y})(\mathbf{P})$$

is true. Then the following is true: There is a unique map $\mathbf{f}: \mathbf{X} \to \mathbf{Y}$ such that

$$\mathbf{y} = \mathbf{f}(\mathbf{x}) \Leftrightarrow \mathbf{P} \qquad (\mathbf{x} \in \mathbf{X}, \mathbf{y} \in \mathbf{Y}).$$

Metaproof

Apply 8.7 by taking **v** to be

$$(\iota \mathbf{y})(\mathbf{y} \in \mathbf{Y} \,\&\, \mathbf{P}). \quad \square$$

To construct the square-root map mentioned above, one takes **P** in 8.9 to be the formula '$y^2 = x$'.

In applying 8.9, one speaks about showing that $\mathbf{f}(\mathbf{x})$ is "well-defined" when proving

$$(\forall \mathbf{x} \in \mathbf{X})(\exists!\, \mathbf{y} \in \mathbf{Y})(\mathbf{P}).$$

8.10. EXAMPLES

(1) In 8.9 take **X** to be '$\omega \setminus \{0\}$', **Y** to be 'ω', and **P** to be '$y^+ = x$', to get the map $f: \omega \setminus \{0\} \to \omega$ such that

$$y = f(x) \Leftrightarrow y^+ = x \qquad (x \in \omega \setminus \{0\}, y \in \omega).$$

That this map is well-defined is the fact that

$$(\forall x \in \omega \setminus \{0\})(\exists!\, y \in \omega)(y^+ = x),$$

and the latter statement is true by 6.13.

(2) Let $\mathcal{A}$ be a class of sets and let E be a class such that $E \cap A$ is a singleton for each $A \in \mathcal{A}$. Take **X** to be '$\mathcal{A}$', **Y** to be '$\bigcup \mathcal{A}$', and **P** to be

$$y \in E \cap A$$

in 8.9. We obtain the map

$$f: \mathcal{A} \to \bigcup \mathcal{A}$$

such that

$$f(A) \in E \cap A \qquad (A \in \mathfrak{A}).$$

Thus f selects for each $A \in \mathfrak{A}$ the unique element y of the singleton $E \cap A$.

(3) Let $f: X \to Y$ be any map. Then 8.9 gives just $f: X \to Y$ itself when we take **P** to be the formula

$$(x,y) \in f.$$

Hence *any* map may be defined by a rule for determining $f(x)$, as well as by a formula.

The general concept of a function has taken long to evolve. For many centuries only those objects were called functions which could be constructed as in 8.7 by a formula. Not until the late nineteenth century did mathematicians generally accept as genuine functions those objects constructed as in 8.9 by a rule. The modern set-theoretic definition of a function in terms of ordered pairs shows that the superficially more general concept involving arbitrary rules leads to exactly the same functions as the concept involving only formulas.

COMPOSITION OF MAPS

In differential calculus one often analyzes a complicated function into simpler ones. To compute the derivative of

$$h: \mathbb{R} \to \mathbb{R}$$

$$x \mapsto \sin x^3 \qquad (x \in \mathbb{R})$$

one first decomposes h into the elementary functions

$$f: \mathbb{R} \to \mathbb{R} \qquad\qquad g: \mathbb{R} \to \mathbb{R}$$

$$x \mapsto x^3 \quad (x \in \mathbb{R}) \qquad\qquad y \mapsto \sin y \quad (y \in \mathbb{R})$$

so that h can be expressed as

$$h(x) = g(f(x)) \qquad (x \in \mathbb{R})$$

and then one applies the "chain rule" to obtain

$$h'(x) = g'(f(x)) \cdot f'(x) = \cos(x^3) \cdot (3x^2).$$

Putting things the other way around, h has been synthesized from f and g as their "composite" in the following sense.

8.11. DEFINITION

Let φ and ψ be maps for which

$$\operatorname{codmn} \varphi = \operatorname{dmn} \psi.$$

The *composite of* φ *with* ψ, denoted $\psi \circ \varphi$, is the map

$$\eta: \operatorname{dmn} \varphi \to \operatorname{codmn} \psi$$
$$x \mapsto \psi(\varphi(x)) \qquad (x \in \operatorname{dmn} \varphi).$$

In another notation, if φ and ψ are respectively

$$f: X \to Y, \qquad g: Y \to Z,$$

then $\psi \circ \varphi$ is the map

$$\psi \circ \varphi: X \to Z$$
$$x \mapsto g(f(x)) \qquad (x \in X).$$

In the notation of 8.11, if

$$\varphi: x \mapsto y, \qquad \psi: y \mapsto z,$$

then

$$\psi \circ \varphi: x \mapsto z.$$

Thus $(\psi \circ \varphi)(x)$ is found by first computing $\varphi(x)$ and then computing $\psi(y)$, where $y = \varphi(x)$.

Notice that the composite of $f: X \to Y$ with $g: Y' \to Z$ has been defined only in case $Y = Y'$. To explain this restriction, we consider the problem of building from

$$f: \mathbb{R} \to \mathbb{R} \qquad\qquad g: \mathbb{R}^+ \to \mathbb{R}$$
$$x \mapsto x^3 \quad (x \in \mathbb{R}) \qquad\qquad y \mapsto \log y \quad (y \in \mathbb{R}^+)$$

a new function h for which $h(x) = g(f(x))$; here $\mathbb{R}^+$ is the set of all positive real numbers. For $x \in \operatorname{dmn} f = \mathbb{R}$ one must have $f(x) = x^3 \in \operatorname{dmn} g = \mathbb{R}^+$ in order to compute $g(f(x)) = \log(f(x))$, but this will not be the case when $x < 0$. The way out of this difficulty is to compose $g: \mathbb{R}^+ \to \mathbb{R}$ not with $f: \mathbb{R} \to \mathbb{R}$ itself, but with the new function

$$f_0: \mathbb{R}^+ \to \mathbb{R}^+$$
$$x \mapsto x^3 \qquad (x \in \mathbb{R}^+),$$

the composite given by 8.11 being then

$$h: \mathbb{R}^+ \to \mathbb{R}$$
$$x \mapsto \log(x^3) \qquad (x \in \mathbb{R}^+).$$

Composition of maps is intimately related to composition of their graphs.

8.12. LEMMA

Let φ and ψ be maps with codmn φ = dmn ψ. Then

$$\text{graph } (\psi \circ \varphi) = (\text{graph } \psi) \circ (\text{graph } \varphi).$$

In other words, if φ, ψ are respectively

$$f: X \to Y, \qquad g: Y \to Z,$$

then

$$\text{graph } (\psi \circ \varphi) = g \circ f.$$

Proof

An element of either graph $(\psi \circ \varphi)$ or $g \circ f$ is an ordered pair $(x,z) \in X \times Z$. For $(x,z) \in X \times Z$,

$$\begin{aligned}(x,z) \in g \circ f &\Leftrightarrow (\exists\, y \in \text{dmn } g)((x,y) \in f \,\&\, (y,z) \in g)\\ &\Leftrightarrow (\exists\, y \in Y)(y = f(x) \,\&\, z = g(y))\\ &\Leftrightarrow z = g(f(x))\\ &\Leftrightarrow (x,z) \in \text{graph } (\psi \circ \varphi). \quad \square\end{aligned}$$

According to 8.12, the composite of

$$f: X \to Y, \qquad g: Y \to Z$$

is just the map

$$g \circ f: X \to Z.$$

With graphs written above arrows, this composite

$$X \xrightarrow{g \circ f} Z$$

is schematically obtained by merging the two diagrams

$$X \xrightarrow{f} Y, \qquad Y \xrightarrow{g} Z$$

into the single "linear diagram"

$$X \xrightarrow{f} Y \xrightarrow{g} Z$$

and by then erasing the Y and combining the two labeled arrows $\xrightarrow{f}$, $\xrightarrow{g}$ into the single labeled arrow $\xrightarrow{g \circ f}$.

The "associative law" for composition below follows both directly from definition 8.11 and indirectly from the analogous result 7.17 for relations together with 8.12.

8.13. THEOREM

Let φ, ψ, and η be maps with codmn φ = dmn ψ and codmn ψ = dmn η. Then

$$\eta \circ (\psi \circ \varphi) = (\eta \circ \psi) \circ \varphi.$$

Let X be a class. Recall that $1_X(x) = \Delta_X(x) = x$ for each $x \in X$. Then for any map

$$f: X \to Y$$

whose domain is X we have

$$f \circ 1_X = f$$

$$X \xrightarrow{1_X} X \xrightarrow{f} Y.$$

For any map

$$g: Z \to X$$

whose codomain is X, we have

$$1_X \circ g = g$$

$$Z \xrightarrow{g} X \xrightarrow{1_X} X.$$

The two maps

$$f: \mathbb{R} \to \mathbb{R} \qquad\qquad g: \mathbb{R}^+ \to \mathbb{R}$$

$$x \mapsto x^2 \quad (x \in \mathbb{R}) \qquad\qquad x \mapsto x^2 \quad (x \in \mathbb{R}^+)$$

are different, for the domain $\mathbb{R}^+ = \{x \mid x \in \mathbb{R} \;\&\; x > 0\}$ of the second is only a strict subset of the domain of the first. The second is obtained by "cutting down," or "restricting," the domain of the first.

8.14. DEFINITION

Let

$$f: X \to Y$$

be a map and let $A \subset X$. The *restriction of f to A*, denoted by $f|_A$ or $f \mid A$, is defined to be the map

$$f \circ j: A \to Y,$$

where

$$j: A \to X$$

is the inclusion map. (More generally, if also $B \subset Y$ and $f\langle A \rangle \subset B$, then the map $f \circ j: A \to B$ is called the *restriction of f to A and B*.)

In the notation of 8.14, $j(a) = a$ for each $a \in A$, so

$$f|_A(a) = f(a) \qquad (a \in A).$$

Here are two examples of restriction: Let c be a set and let $f: \mathfrak{U} \to \{c\}$ be the constant map with value c; then for any class A, $f|_A: A \to \{c\}$ is the constant map on A with value c. If X is any class and $A \subset X$, then $1_X|_A: A \to X$ is the inclusion map of A into X. (As a trivial example of the more general type of restriction, let A be any class; then the restriction of $1_{\mathfrak{U}}: \mathfrak{U} \to \mathfrak{U}$ to A and A is $1_A: A \to A$.)

DIRECT AND INVERSE IMAGES UNDER A MAP

Let $f: X \to Y$ be a map and let $A \subset X$. If also $A \in X$, one must carefully distinguish between $f(A)$—the value of f at the member A of its domain—and

$$f\langle A\rangle = \{y \mid (\exists x \in A)(y = f(x))\}$$

the direct image of A under the functional relation f. For example, if $f: \omega \to \omega$ is the successor map, then $2 \subset \omega$ and $2 \in \omega$,

$$f\langle 2\rangle = f\langle\{0,1\}\rangle = \{f(0), f(1)\} = \{1,2\},$$

$$f(2) = 3 = \{0,1,2\}.$$

Hence we scrupulously avoid the all-too-common practice of writing $f(A)$ to mean $f\langle A\rangle$. When A is a set, $f\langle A\rangle$ is of course the value at A of a map closely related to f, namely, the value $f_*(A)$ at A of the map

$$f_*: \mathcal{P}(X) \to \mathcal{P}(Y)$$

$$E \mapsto f\langle E\rangle \qquad (E \in \mathcal{P}(X)).$$

The direct image $f\langle A\rangle$ of a class $A \subset X$ under a map

$$f: X \to Y$$

is the range of a map, for

$$f\langle A\rangle = \operatorname{rng}(f|_A).$$

Moreover, the range of f is the direct image of its domain under f, that is,

$$\operatorname{rng} f = f\langle X\rangle.$$

Hence theorems concerning direct images have as corollaries corresponding theorems concerning ranges, and vice versa.

Since the graph of a map is a relation, results in Chapter 7 give in particular the following properties for direct images under a map $f: X \to Y$:

$$f\langle A \cup B\rangle = f\langle A\rangle \cup f\langle B\rangle,$$

$$f\langle A \cap B\rangle \subset f\langle A\rangle \cap f\langle B\rangle,$$

$$f\langle A \setminus B\rangle \supset f\langle A\rangle \setminus f\langle B\rangle,$$

for all $A, B \subset X$. The last two inclusions may be strict; consider the constant map $f: \{0,1\} \to \{0\}$ and the sets $A = \{0\}$, $B = \{1\}$. Things work more pleasantly for inverse images.

8.15. PROPOSITION

Let $f: X \to Y$ be a map, and let $A, B \subset Y$. Then

$$f^{-1}\langle A \cup B\rangle = f^{-1}\langle A\rangle \cup f^{-1}\langle B\rangle,$$

$$f^{-1}\langle A \cap B\rangle = f^{-1}\langle A\rangle \cap f^{-1}\langle B\rangle,$$

$$f^{-1}\langle A \setminus B\rangle = f^{-1}\langle A\rangle \setminus f^{-1}\langle B\rangle.$$

Proof

The first equation above as well as the inclusions

$$f^{-1}\langle A \cap B\rangle \subset f^{-1}\langle A\rangle \cap f^{-1}\langle B\rangle,$$

$$f^{-1}\langle A \setminus B\rangle \supset f^{-1}\langle A\rangle \setminus f^{-1}\langle B\rangle$$

follow from 7.26.

We show

$$f^{-1}\langle A\rangle \cap f^{-1}\langle B\rangle \subset f^{-1}\langle A \cap B\rangle.$$

Let $x \in f^{-1}\langle A\rangle \cap f^{-1}\langle B\rangle$. Then $f(x) \in A$ and $f(x) \in B$, so $f(x) \in A \cap B$. Hence $x \in f^{-1}\langle A \cap B\rangle$. □

In the proof of 8.15 we used implicitly the representation

$$f^{-1}\langle E\rangle = \{x \mid x \in X \,\&\, f(x) \in E\}$$

for the inverse image of a class $E \subset Y$ under the map $f: X \to Y$.

8.16. PROPOSITION

Let $f: X \to Y$ be a map. Then:

(1) If $A \subset X$, then

$$A \subset f^{-1}\langle f\langle A\rangle\rangle.$$

(2) If $B \subset Y$, then

$$f\langle f^{-1}\langle B\rangle\rangle \subset B,$$

and

$$B \subset \operatorname{rng} f \Rightarrow f\langle f^{-1}\langle B\rangle\rangle = B.$$

Proof

(1) Let $A \subset X$. If $x \in A$, then $f(x) \in f\langle A\rangle$ and hence $x \in f^{-1}\langle f\langle A\rangle\rangle$.

(2) Let $B \subset Y$. Let $y \in f\langle f^{-1}\langle B\rangle\rangle$. There exists $x \in f^{-1}\langle B\rangle$ such that $y = f(x)$. Now $x \in f^{-1}\langle B\rangle$ means $f(x) \in B$. Hence $y \in B$. This shows $f\langle f^{-1}\langle B\rangle\rangle \subset B$.

Assume now $B \subset \operatorname{rng} f$. We show $B \subset f\langle f^{-1}\langle B\rangle\rangle$. Let $y \in B$. There exists $x \in X$ such that $f(x) = y$. Then $x \in f^{-1}\langle B\rangle$ and so $y = f(x) \in f\langle f^{-1}\langle B\rangle\rangle$. □

The inclusion in (1) may be strict. For example, let X be a set having at least two distinct elements a and a', and let $f\colon X \to Y$ be a constant map having some value $c \in Y$. If $A = \{a\}$, then

$$A \subsetneq X = f^{-1}\langle\{c\}\rangle = f^{-1}\langle f\langle A\rangle\rangle.$$

Note that the inverse f^{-1} of this functional relation f is not itself functional, for

$$(c,a) \in f^{-1},\ (c,a') \in f^{-1},\ a \neq a'.$$

COMMUTATIVE DIAGRAMS

From now on, we will allow a map $f\colon X \to Y$ to be denoted by any "diagram" formed by writing X and Y and drawing from X to Y an arrow labeled with f.

Three maps

$$f\colon X \to Y, \qquad g\colon Y \to Z, \qquad h\colon X \to Z$$

may be displayed together in a single triangular diagram or "triangle":

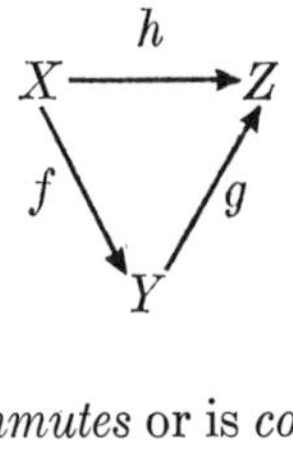

One says that this triangle *commutes* or is *commutative* if

$$h = g \circ f.$$

In this triangle one can trace out two "paths" from X to Z by moving along arrows in the directions they point. One path consists of $\xrightarrow{h}$ alone, the other of $\xrightarrow{f}$ followed by $\xrightarrow{g}$. To say the triangle commutes means that the result of applying to an element $x \in X$ in turn the various maps labeling the arrows on one path yields the same element of Z as is obtained by using the other path.

8.17. EXAMPLES

(1) Let $f: X \to Y$ be a map and let $A \subset X$. If $j: A \to X$ is the inclusion map, then there is a unique map $g: A \to Y$ making the following diagram commute:

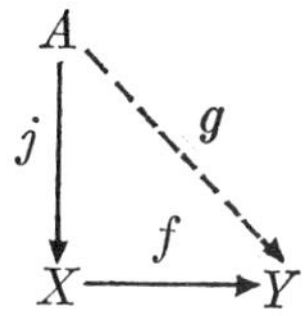

In fact, the commutativity of the triangle means $g = f \circ j$, that is,

$$g = f\,|_A.$$

The arrow labeled by g is dashed rather than solid to indicate the existence of g is under discussion.

(2) Let $g: A \to Y$ be a map, let $X \supset A$, and let $j: A \to X$ again be the inclusion map. Then a map $f: X \to Y$ making the diagram

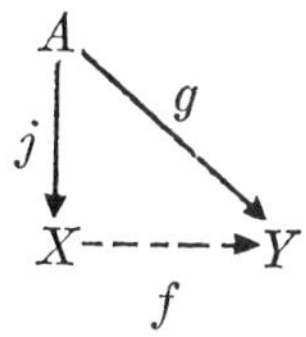

commute is one for which

$$f\,|_A = g.$$

Such an f is called an *extension of g to X*. An extension need not be unique. For example, the successor map

$$g: \omega \setminus \{0\} \to \omega$$
$$n \mapsto n^+ \qquad (n \in \omega \setminus \{0\})$$

has as extensions to ω both the map $f_1: \omega \to \omega$ given by

$$m = f_1(n) \Leftrightarrow (n \in \omega \setminus \{0\} \,\&\, m = n^+) \vee (n = 0 \,\&\, m = 1)$$

and the map $f_2: \omega \to \omega$ given by

$$m = f_2(n) \Leftrightarrow (n \in \omega \setminus \{0\} \,\&\, m = n^+) \vee (n = 0 \,\&\, m = 8).$$

Any map $g: A \to Y$ can be extended to any class $X \supset A$ (see Chapter 8, Exercise K). In practice one is usually interested only in an extension having this or that additional property. In algebra one might inquire whether a given homomorphism $g: A \to Y$ from a group A into a group Y has an extension to a group X of which A is a subgroup for which the extension is also a homomorphism. In analysis one might require a continuous map which extends a given continuous map from one metric space to another.

Four maps

$$f: X \to Y, \qquad g: Z \to W, \qquad h: X \to Z, \qquad k: Y \to W$$

may be represented by a single square diagram or "square":

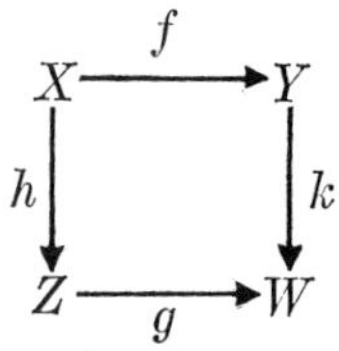

One says that this square *commutes* if

$$g \circ h = k \circ f.$$

To say the square commutes is to say that the result of applying to an element of X in turn the maps labeling the path $\xrightarrow{f}, \xrightarrow{k}$ from X to W yields the same element of W as is obtained by using the other path $\xrightarrow{h}, \xrightarrow{g}$.

8.18. EXAMPLE

Let $f: X \to Y$ be a given map. Consider the maps

$$h: X \to \mathcal{P}(X) \qquad\qquad k: Y \to \mathcal{P}(Y)$$

$$x \mapsto \{x\} \quad (x \in X) \qquad\qquad y \mapsto \{y\} \quad (y \in Y)$$

and

$$f_*: \mathcal{P}(X) \to \mathcal{P}(Y)$$

$$E \mapsto f\langle E\rangle \qquad (E \in \mathcal{P}(X)).$$

Then the square below commutes:

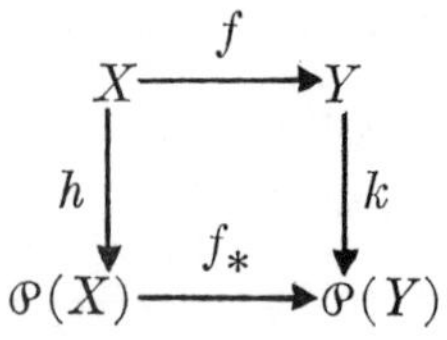

In fact, $x \in X$ implies

$$\begin{aligned}(f_* \circ h)(x) &= f_*(h(x)) = f_*(\{x\}) = f\langle\{x\}\rangle \\ &= \{f(x)\} = k(f(x)) = (k \circ f)(x).\end{aligned}$$

Linear, triangular, and square diagrams may be combined to form more complicated diagrams. For example:

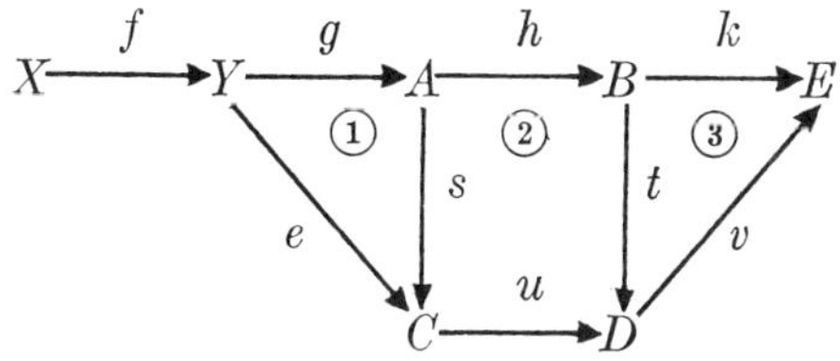

Such a composite diagram is said to *commute* if each of its component triangles and squares commutes. As usual, the effect of applying in turn maps on one path is the same as applying in turn those on another path (by analogy with the fact that the integral of an analytic complex function over a path joining two points is independent of the particular path used, a commutative diagram is also said to be *analytic*). For example, if the triangles ① and ③ and the square ② in the above diagram are commutative, then one shows that

$$v \circ u \circ e \circ f = k \circ h \circ g \circ f$$

by "diagram chasing":

$$\begin{aligned} v \circ u \circ e \circ f &= v \circ u \circ (s \circ g) \circ f \\ &= v \circ (u \circ s) \circ g \circ f \\ &= v \circ (t \circ h) \circ g \circ f \\ &= (v \circ t) \circ h \circ g \circ f \\ &= k \circ h \circ g \circ f. \end{aligned}$$

Diagrams provide an extremely useful—and often indispensable—tool in many branches of contemporary mathematics. They allow one to visualize geometrically relationships between various objects and maps.

A crucial property of the product of two classes is conveniently formulated through the use of a diagram.

8.19. THEOREM

Let X and Y be classes, and let

$$p: X \times Y \to X, \qquad q: X \times Y \to Y$$

be the projections. Suppose

$$f: Z \to X, \qquad g: Z \to Y$$

are two maps with the same domain Z. Then there exists a *unique* map h making the following diagram commute:

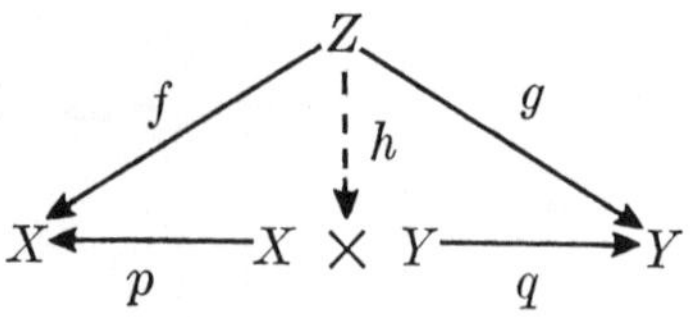

Proof

Suppose h makes the diagram commute, so that

$$p \circ h = f, \qquad q \circ h = g.$$

If $z \in Z$ and if $h(z) = (x,y)$, then

$$x = p((x,y)) = p(h(z)) = f(z),$$

$$y = q((x,y)) = q(h(z)) = g(z).$$

Hence

$$(*) \qquad h(z) = (f(z), g(z)) \qquad (z \in Z).$$

This proves the uniqueness of h.

To establish the existence of h, simply define h by (*). Then $z \in Z$ implies

$$(p \circ h)(z) = f(z), \qquad (q \circ h)(z) = g(z),$$

and hence $p \circ h = f$, $q \circ h = g$. □

Theorem 8.19 says that two maps $f: Z \to X$, $g: Z \to Y$ with the same domain can be used in a natural fashion to determine a single map $h: Z \to X \times Y$ which determines the given maps. The map h is called the *natural* (or *canonical*) *map of* Z *into* $X \times Y$ *induced by* f *and* g.

The next theorem says that two maps, not necessarily having the same domain, can be used to determine a single map on the product of their domains to the product of their codomains. Although 8.20 may be proved directly, we prefer to deduce it from 8.19, for the values of maps at various elements of their domains need not be explicitly mentioned.

8.20. THEOREM

Consider two maps

$$f: X' \to X, \qquad g: Y' \to Y.$$

Then there exists a *unique* map h making both of the following diagrams commute:

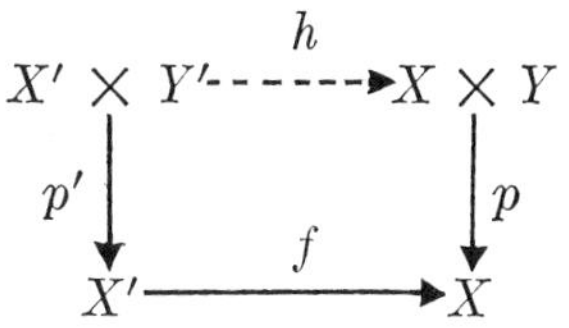

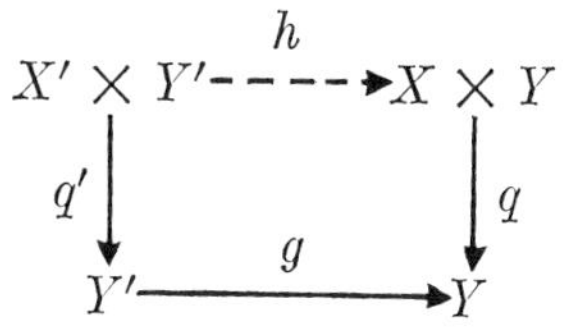

where p', p, q', q are the projections.

Proof

We apply 8.19 to the maps

$$X' \times Y' \xrightarrow{f \circ p'} X, \qquad X' \times Y' \xrightarrow{g \circ q'} Y.$$

There exists a unique map h making the diagram

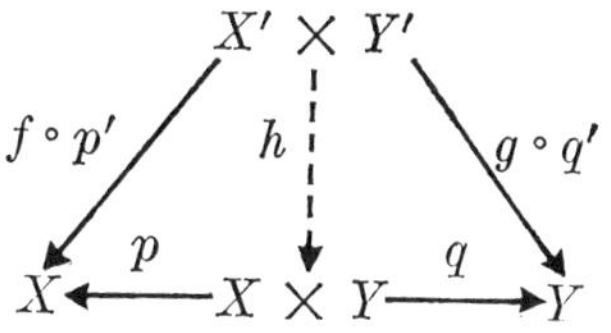

commute. Hence there is a unique h making the diagram

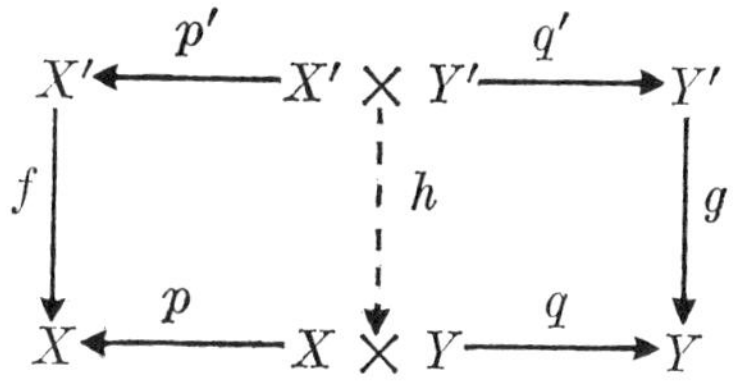

commute. $\square$

In the notation of 8.20, one calls h the *natural* (or *canonical*) *map of* $X' \times Y'$ *into* $X \times Y$ *induced by* f *and* g, or simply the *product of* f *and* g.

CLASSES OF MAPS

We next use diagrams to describe maps whose domains and codomains are themselves sets of maps.

8.21. DEFINITION

Let X and Y be sets. The class

$$\{\varphi \mid \varphi \text{ is a map from } X \text{ to } Y\}$$

is denoted by

$$\text{Map } (X,Y).$$

[A common notation for Map (X,Y) is Y^X, but this is a typographical nuisance and conflicts with the notation introduced later for powers of numbers.]

If X and Y are sets and $x \in X$, then the map

$$e_x \colon \text{Map } (X,Y) \to Y$$

$$f \mapsto f(x) \qquad (f \in \text{Map } (X,Y))$$

which assigns to each map $f \colon X \to Y$ its value $f(x)$ at the given $x \in X$ is often called the *evaluation map on* Map (X,Y) *at* x.

8.22. THEOREM

Let X, X', Y, Y' be sets, and let

$$f \colon X \to X', \qquad g \colon Y \to Y'$$

be maps. Then there exists a unique map $\varphi_g{}^f$ such that the following diagram commutes for each $x \in X$:

$$
\begin{array}{ccc}
\text{Map}(X', Y) & \xdashrightarrow{\;\varphi_g{}^f\;} & \text{Map}(X, Y') \\
\scriptstyle e_{f(x)} \big\downarrow & & \big\downarrow \scriptstyle e_x \\
Y & \xrightarrow{\;\;g\;\;} & Y'
\end{array}
$$

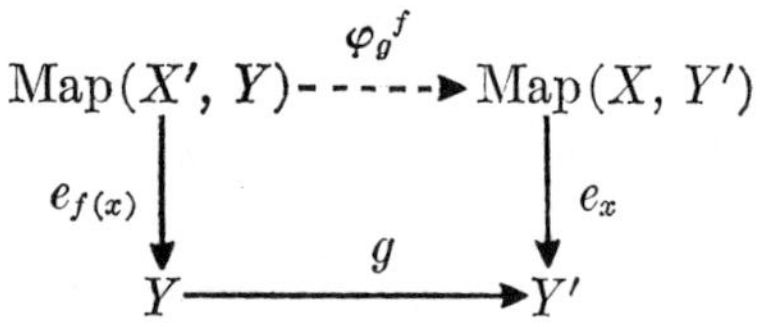

Proof

For simplicity denote $\varphi_g{}^f$ by φ. To say the above diagrams commute means that $x \in X$ and $h \in \text{Map } (X',Y)$ implies

$$\varphi(h)(x) = e_x(\varphi(h)) = (e_x \circ \varphi)(h)$$

$$= (g \circ e_{f(x)})(h) = g(e_{f(x)}(h))$$

$$= g(h(f(x))) = (g \circ h \circ f)(x)$$

and hence that

$$(*) \qquad\qquad \varphi(h) = g \circ h \circ f \qquad (h \in \text{Map } (X',Y)).$$

This proves uniqueness of φ. If now φ is defined by $(*)$, then reversal of the preceding computations shows that $g \circ e_{f(x)} = e_x \circ \varphi$ for each $x \in X$, and this proves existence. $\square$

If we change f and g we get a new $\varphi_g{}^f$, so we have a map

$$\text{Map } (X,X') \times \text{Map } (Y,Y') \to \text{Map } (\text{Map } (X',Y), \text{Map } (X,Y'))$$

$$(f,g) \mapsto \varphi_g{}^f.$$

When $Y' = Y$ and $g = 1_Y$, then for each map

$$f: X \to X'$$

we obtain the map $\varphi^f = \varphi_{1_X}{}^f$

$$\varphi^f: \text{Map}\ (X',Y) \to \text{Map}\ (X,Y)$$

satisfying

$$\varphi^f(h) = h \circ f \qquad (h \in \text{Map}\ (X',Y)).$$

MAPS VERSUS FUNCTIONAL RELATIONS

We have defined 'function' as a synonym for 'map'. Many authors use 'function' to mean what we have called a 'functional relation'. Of course, a map and the functional relation which is its graph are different things, for a map has a codomain but a functional relation does not. That there is a practical need to study maps and not just functional relations is illustrated by a situation arising in linear algebra.

For a vector space V, the "dual" V^* of V is the set of all linear transformations $h: V \to \mathbb{R}$ from V into the one-dimensional vector space $\mathbb{R}$; then

$$V^* \subset \text{Map}\ (V, \mathbb{R}).$$

The set V^* is made into a vector space as follows: if $h, k \in V^*$, then $h + k: V \to \mathbb{R}$ is the linear transformation defined "pointwise" by

$$(h + k)(v) = h(v) + k(v) \qquad (v \in V);$$

if $h \in V^*$ and $\alpha \in \mathbb{R}$, then $\alpha h: V \to \mathbb{R}$ is the linear transformation defined by

$$(\alpha h)(v) = \alpha \cdot h(v) \qquad (v \in V).$$

It is well known that the vector space V^* will be n-dimensional if V is.

For a linear transformation

$$f: V \to W$$

from one vector space to another, the "transpose" or "adjoint" of f is the map

$$f^*: W^* \to V^*$$
$$k \mapsto k \circ f \qquad (k \in W^*)$$

which thus satisfies

$$(f^*(k))(v) = k(f(v)) \qquad (k \in W^*, v \in V).$$

In the notation of the preceding section, f^* is the restriction to W^* and V^* of the map

$$\varphi^f: \text{Map}\ (W, \mathbb{R}) \to \text{Map}\ (V, \mathbb{R}).$$

Actually, f^* is itself a linear transformation, but this is irrelevant for our purposes.

Suppose now that W is n-dimensional and that the range of the linear transformation

$$f: V \to W$$

is an m-dimensional vector subspace W_0 of W with $m < n$. Let us just suppose for the moment we had never heard of maps, so that there were no such thing as a codomain. In short, let us confuse a map with its graph.

We have the functional relation

$$f: V \to W$$

in V to W, and since $f\langle V \rangle \subset W_0$, the *same* functional relation

$$f: V \to W_0$$

is a relation in V to W_0. From these we obtain their transposes

$$f^*: W^* \to V^*,$$

$$f^*: W_0{}^* \to V^*$$

which must be the same since they came from the same f. Yet the two functional relations denoted by f^* cannot be the same, for they have the respective domains W^* and $W_0{}^*$, and $W^* \neq W_0{}^*$ since

$$\dim W^* = \dim W = n \neq m = \dim W_0 = \dim W_0{}^*.$$

The resolution of this paradox lies simply in the observation that f^* is determined not just by the functional relation f, but also by a specific vector space in which f takes its values. In short, we need codomains and maps after all!

FAMILIES

In certain contexts it is useful to conceive of a map as labeling its various values by the elements of its domain at which it takes these values. For example, when the elements of the domain are given in some order, then the values can be regarded as being arranged in a corresponding order by their labels. In such contexts the special terminology and notation of families is applied to a map. To be sure, there is no formal distinction between 'family' and 'map', for by definition the two notions are synonymous; however, there is a difference in connotation, for the language of families suggests a special emphasis upon the values of the map.

8.23. DEFINITION

By a *family* is meant a map (we never use 'family' as a synonym for 'class' or 'set'). More specifically, a map $x: I \to X$ is called a *family in X* (or *of elements of X*) *indexed by I*, an element of I is said to be an *index* of the family, and I is called the *index class* of the family. The value $x(i)$ of the family x at the index i is called the *ith coordinate of x* and is denoted by means of a subscript as x_i, and then x itself is denoted by

$$(x_i \mid i \in I)$$

or

$$(x_i)_{i \in I}.$$

Customarily one thinks of the index i as running through the various elements of I in succession, so that x_i runs through the corresponding values of the family $(x_i \mid i \in I)$.

8.24. EXAMPLES

(1) If $\mathcal{A}$ is any class of sets, then the family $(A \mid A \in \mathcal{A})$ is the identity map of $\mathcal{A}$.

(2) If $f: X \to Y$ is a map and if $(A_i \mid i \in I)$ is a family in $\mathcal{P}(X)$, then the family $(f\langle A_i\rangle \mid i \in I)$ is the map

$$I \to \mathcal{P}(Y)$$

$$i \mapsto f\langle A_i\rangle \qquad (i \in I).$$

(3) Suppose the index class I of a family x in X is the set

$$\{i \mid i \in \omega \;\&\; 1 \leq i \leq n\}$$

for some natural number $n > 0$. Then the family is also denoted $(x_i \mid 1 \leq i \leq n)$ or $(x_i)_{i=1,2,\ldots,n}$. It may also be called a *list* of *length n* and denoted informally by

$$x_1, \ldots, x_n$$

or may be called an *(ordered) n-tuple* and denoted by

$$(x_1, \ldots, x_n);$$

here the ith coordinate x_i of x is to be written in the ith position for each index i. For example, both

$$(2,5,0)$$

and

$$2, 5, 0$$

denote the family indexed by $\{1,2,3\}$ taking values 2, 5, 0 at 1, 2, 3, respectively. The notational similarity of ordered pairs and 2-tuples is not accidental and will be explained later (see 9.19).

A standard notation for the class of all families in X indexed by $\{i \mid i \in \omega \,\&\, 1 \le i \le n\}$ is X^n. Thus

$$X^n = \text{Map}\,(\{i \mid i \in \omega \,\&\, 1 \le i \le n\}, X).$$

(4) A family $(x_n \mid n \in \omega)$ in a class X indexed by the set ω is called a *sequence* in X. The notations

$$(x_n \mid n = 0, 1, 2, \ldots), \qquad (x_n)_{n=0,1,2,\ldots}$$

are then used, and the family may be thought of as an "infinite list"

$$x_0, x_1, x_2, \ldots, x_n, \ldots .$$

For example, the sequence $(n^+ \mid n = 0, 1, 2, \ldots)$ in ω may be represented by the list

$$1, 2, 3, \ldots, n^+, \ldots .$$

The meaning of such notations as $(x_n)_{3 \le n \le 8}$ and $(x_n)_{n=2,3,\ldots}$ should now require no explanation.

Without necessarily meaning to disparage makers of intelligence tests, we point out that from the first few entries in an infinite list one cannot determine the entire sequence represented by the list. Question: What number "comes next" in the list 13, 15, 19, 22, 30, ...? Answer: 33—these numbers are the stops made by the Market St. subway-surface train in Philadelphia.

(5) The composite of a family $(i_j \mid j \in J)$ in I with a family $(x_i \mid i \in I)$ indexed by I may be denoted by "subsubscripts" as $(x_{i_j} \mid j \in J)$. For example, if $(i_j \mid j \in \omega)$ is a sequence in ω and if $(x_i \mid i \in \omega)$ is a sequence, then $(x_{i_j} \mid j \in \omega)$ is a "subsequence" of $(x_i \mid i \in \omega)$.

(6) The ordinary differential equation $dx/dt = t$ has, for each real number c, the particular solution x_c defined by $x_c(t) = t^2/2 + c$ satisfying the initial condition $x_c(0) = c$. These individual solutions combine to form the family $(x_c \mid c \in \mathbb{R})$.

When, as in the last example, the index class of a family is the set of all real numbers, then one speaks of a *one-parameter family*. More generally, the index i in the notation for any family is often called the *parameter*.

Colorful language is often used to specify properties of the index class or codomain of a family. For example, a "family of sets" is a family in $\mathcal{U}$. A "nonempty family" is a family whose index class is nonempty. A "family of subsets of E" is a family in $\mathcal{P}(E)$.

When the family language is applied to a map, new notation is used for its range.

8.25. DEFINITION

If $(x_i \mid i \in I)$ is a family, then $\{x_i \mid i \in I\}$ is defined to be rng $(x_i \mid i \in I)$, that is,

$$\{x_i \mid i \in I\} = \{y \mid (\exists i \in I)(y = x_i)\}.$$

This "indexed class" notation $\{x_i \mid i \in I\}$ must be carefully distinguished from the family notation $(x_i \mid i \in I)$, despite the regrettable practice of some authors of confusing the two. Compare the distinction between doubletons and ordered pairs.

Let **I**, **X**, and **v** be terms in which the letters **i** and **x** are free, and suppose $(\forall \mathbf{i} \in \mathbf{I})(\mathbf{v} \in \mathbf{X})$ is true. Then the map

$$\mathbf{f}: \mathbf{I} \to \mathbf{X}$$

$$\mathbf{i} \mapsto \mathbf{v} \qquad (\mathbf{i} \in \mathbf{I})$$

defined as in 8.7 by the formula $\mathbf{f}(\mathbf{i}) = \mathbf{v}\ (\mathbf{i} \in \mathbf{I})$ is just the family $(\mathbf{v} \mid \mathbf{i} \in \mathbf{I})$. Its range

$$\{\mathbf{v} \mid \mathbf{i} \in \mathbf{I}\} = \{\mathbf{x} \mid (\exists \mathbf{i} \in \mathbf{I})(\mathbf{x} = \mathbf{v})\}$$

is the class of all sets $\mathbf{x} \in \mathbf{X}$ "of the form **v** for some $\mathbf{i} \in \mathbf{I}$." For example, for a map $f: E \to X$ and a set $A \subset E$,

$$f\langle A \rangle = \{ f(a) \mid a \in A\}.$$

If $\mathcal{A}$ is a class of subsets of a set X, then

$$\{X \setminus A \mid A \in \mathcal{A}\}$$

is the range of the map

$$\mathcal{A} \to \mathcal{P}(X)$$

$$A \mapsto X \setminus A \qquad (A \in \mathcal{A}).$$

Unions and intersections of families are defined in terms of unions and intersections of classes.

8.26. DEFINITION

Let $(X_i \mid i \in I)$ be a family of sets. The *union of* $(X_i \mid i \in I)$, denoted $\bigcup_{i \in I} X_i$, is the class $\bigcup\{X_i \mid i \in I\}$, and the *intersection of* $(X_i \mid i \in I)$, denoted by $\bigcap_{i \in I} X_i$, is the class $\bigcap\{X_i \mid i \in I\}$. Thus,

$$\bigcup_{i \in I} X_i = \{x \mid (\exists i \in I)(x \in X_i)\},$$

$$\bigcap_{i \in I} X_i = \{x \mid (\forall i \in I)(x \in X_i)\}.$$

When for some $n \in \omega$, $n > 0$, the index class I of $(X_i \mid i \in I)$ is $\{i \mid i \in \omega \,\&\, 1 \leq i \leq n\}$, then the notation $\bigcup_{i=1}^{n} X_i$ is also used. When $I = \omega$, the notation $\bigcup_{i=0}^{\infty} X_i$ is also used. Similarly for intersections.

Discussion of computational rules for unions and intersections requires the notion of the product of a family and so is deferred until Chapter 12. We mention here just one rule.

8.27. THEOREM (DE MORGAN'S LAWS)

Let $(A_i \mid i \in I)$ be a family of subsets of a set X. Then

$$\bigcup_{i \in I} X \setminus A_i = X \setminus \bigcap_{i \in I} A_i$$

and if $I \neq \varnothing$,

$$\bigcap_{i \in I} X \setminus A_i = X \setminus \bigcup_{i \in I} A_i.$$

PIECEWISE CONSTRUCTION OF A MAP

The absolute value function $f\colon \mathbb{R} \to \mathbb{R}$ of analysis is defined by

$$f(x) = \begin{cases} x & \text{if } x \geq 0, \\ -x & \text{if } x \leq 0. \end{cases}$$

It is obtained by "piecing together" the two maps

$$f_1\colon A_1 \to \mathbb{R} \qquad\qquad f_2\colon A_2 \to \mathbb{R}$$

$$x \mapsto x \quad (x \in A_1) \qquad\qquad x \mapsto -x \quad (x \in A_2)$$

where

$$A_1 = \{x \mid x \in \mathbb{R} \,\&\, x \geq 0\}, \qquad A_2 = \{x \mid x \in \mathbb{R} \,\&\, x \leq 0\},$$

so that $f \mid A_1 = f_1$ and $f \mid A_2 = f_2$. For this to give a map with domain $\mathbb{R} = A_1 \cup A_2$, it is of course necessary that f_1 and f_2 agree on the overlap $A_1 \cap A_2 = \{0\}$ of their domains, so that for $x \in A_1 \cap A_2$ the same value $f(x)$ is obtained whether one uses $f(x) = f_1(x)$ or $f(x) = f_2(x)$. This process of "piecewise construction" of a map can be used starting with any number of pieces.

8.28. THEOREM

Let $(X_i \mid i \in I)$ be a family of sets, let Y be a class, and let $(f_i \mid i \in I)$ be a family of maps with

$$f_i\colon X_i \to Y$$

for each $i \in I$. Suppose

$$f_i \mid (X_i \cap X_j) = f_j \mid (X_i \cap X_j) \qquad (i, j \in I).$$

Then there exists a *unique* map

$$f\colon \bigcup_{i \in I} X_i \to Y$$

such that

$$f \mid X_i = f_i \qquad (i \in I).$$

Proof

Let $X = \bigcup_{i\in I} X_i$. The requirement $f \mid X_i = f_i$ $(i \in I)$ says that for $x \in X$ and $y \in Y$,

$$y = f(x) \Leftrightarrow (\exists i \in I)(x \in X_i \,\&\, y = f_i(x)).$$

The unique existence of f will follow from 8.9 once we prove

$$(\forall x \in X)(\exists! y \in Y)(\mathbf{P})$$

where $\mathbf{P}$ is the statement

$$(\exists i \in I)(x \in X \,\&\, y = f_i(x)).$$

Let $x \in X$. We show $(\exists! y \in Y)(\mathbf{P})$. Since $x \in X$, $x \in X_i$ for some $i \in I$, and then $f_i(x) \in Y$ with $f_i(x) = f_i(x)$. This proves existence of y. To prove uniqueness of y, suppose $y, y' \in Y$ and $i, j \in I$ with

$$x \in X_i \,\&\, y = f_i(x),$$

$$x \in X_j \,\&\, y' = f_j(x).$$

Then $x \in X_i \cap X_j$, so

$$\begin{aligned} y' = f_j(x) &= (f_j \mid X_i \cap X_j)(x) \\ &= (f_i \mid X_i \cap X_j)(x) = f_i(x) = y, \end{aligned}$$

and hence $y = y'$. $\square$

A case in which the hypothesis on the f_i is automatically satisfied is that $X_i \cap X_j = \varnothing$ for $i \neq j$.

8.29. DEFINITION

A family $(X_i \mid i \in I)$ of sets is said to be *(pairwise) disjoint* if $i, j \in I$ and $i \neq j$ implies $X_i \cap X_j = \varnothing$. (A class $\mathcal{A}$ of sets is said to be *disjoint* if the family $(A \mid A \in \mathcal{A})$ is disjoint, that is, if $A, B \in \mathcal{A}$ and $A \neq B$ implies $A \cap B = \varnothing$.)

8.30. EXAMPLE

Let X be a set and let $A \subset X$. The *characteristic function of A in X* is the map

$$\chi_A: X \to \{0,1\}$$

such that

$$\chi_A \mid A = c_1, \qquad \chi_A \mid X \setminus A = c_0,$$

where

$$c_1: A \to \{0,1\}, \qquad c_0: X \setminus A \to \{0,1\}$$

are constant maps with values 1, 0 respectively. Thus $x \in X$ implies

$$\chi_A(x) = \begin{cases} 1 & \text{if } x \in A, \\ 0 & \text{if } x \notin A. \end{cases}$$

Notice that

$$A = \chi_A^{-1}\langle\{1\}\rangle, \qquad X \setminus A = \chi_A^{-1}\langle\{0\}\rangle.$$

Let I be a set. Then the characteristic function $\delta: I \times I \to \{0,1\}$ of Δ_I in $I \times I$ is called the *Kronecker delta of I*. Writing $\delta_{i,j}$ for $\delta((i,j))$, we have

$$\delta_{i,j} = \begin{cases} 1 & \text{if } i = j, \\ 0 & \text{if } i \neq j. \end{cases}$$

EXERCISES

A. Let f be a functional relation. If $x \in \operatorname{dmn} f$, express $f\langle\{x\}\rangle$ as a singleton. If $y \in \operatorname{rng} f$, must $f^{-1}\langle\{y\}\rangle$ be a singleton?

B. Determine whether the relation

$$\{(x,y) \mid x \in \mathbb{R} \;\&\; y \in \mathbb{R} \;\&\; \mathbf{P}\}$$

is functional for each statement **P**:
(a) $x^2 + y^2 = 1 \;\&\; y \geq 0$.
(b) $x^2 + y^2 = 1 \;\&\; x > 0$.
(c) $x = \sin y$.
(d) $x = e^y$.
(e) $y^3 = x^2$.

C. Show that the composite $g \circ f$ of two functional relations f and g is always functional and compute $\operatorname{dmn}(g \circ f)$ and $\operatorname{rng}(g \circ f)$.

D. Let $\mathfrak{C}$ be the class of all continuous maps $f: \mathbb{R} \to \mathbb{R}$ and let $\mathfrak{D}$ be the class of all differentiable maps $g: \mathbb{R} \to \mathbb{R}$. Is the relation

$$\{(f,g) \mid f \in \mathfrak{C} \;\&\; g \in \mathfrak{D} \;\&\; g'(x) = f(x) \quad (x \in \mathbb{R})\}$$

functional?

E. Must the inverse of a functional relation be functional?

F. Let f be a relation in X to Y. Show that each of the following conditions is sufficient for f to be functional:
(a) $B \subset Y \Rightarrow f\langle f^{-1}\langle B\rangle\rangle \subset B$.
(b) $A, B \subset Y \Rightarrow f^{-1}\langle A \cap B\rangle = f^{-1}\langle A\rangle \cap f^{-1}\langle B\rangle$.
(c) $A, B \subset Y \;\&\; A \cap B = \varnothing \Rightarrow f^{-1}\langle A\rangle \cap f^{-1}\langle B\rangle = \varnothing$.

G. Construct the following maps explicitly via 8.7 or 8.9:
(a) The map from the class $\mathfrak{C}$ of all continuous maps $f: \mathbb{R} \to \mathbb{R}$ to the class Map $(\mathbb{R},\mathbb{R})$ which assigns to each such f the

map $g \in$ Map $(\mathbb{R},\mathbb{R})$ given by

$$g(x) = \int_0^x f(t)\,dt \qquad (x \in \mathbb{R}).$$

(b) The map taking values in ω which assigns to each ordered pair (m,n) of natural numbers the greater of the two numbers m and n.

H. Let $f: X \to Y$ be a map. Construct a map f^* such that $f^*(E) = f^{-1}\langle E\rangle$ for each subset E of rng f.

I. If $f: X \to Y$ is a map and $A \subset X$, show that

$$\text{graph}\,(f \mid A) = \text{graph}\,f \cap (A \times Y).$$

J. (a) If $f: X \to Y$ is a map and $A \subset B \subset X$, show that

$$(f \mid B) \mid A = f \mid A,$$

$$(f \mid B)\langle A\rangle = f\langle A \cap B\rangle.$$

(b) If $f: X \to Y$ and $g: Y \to Z$ are maps and $A \subset X$, show that

$$(g \circ f) \mid A = g \circ (f \mid A).$$

K. Let $f: A \to Y$ be a map. If $X \supset A$, show there always exists an extension g of f to X. When is g unique?

L. For each map φ, let

$$R(\varphi) = 1_{\text{dmn}\,\varphi}, \qquad L(\varphi) = 1_{\text{codmn}\,\varphi}.$$

If $f: X \to Y$, $g: Y \to Z$ are maps, what are $R(g \circ f)$ and $L(g \circ f)$?

M. Let X and Y be classes, and let

$$X + Y = (X \times \{1\}) \cup (Y \times \{2\}).$$

Consider the maps

$$j: X \to X + Y \qquad\qquad k: Y \to X + Y$$

$$x \mapsto (x,1) \quad (x \in X) \qquad\qquad y \mapsto (y,2) \quad (y \in Y).$$

Given maps $f: X \to Z$, $g: Y \to Z$, show there is a unique map h making the following diagram commute:

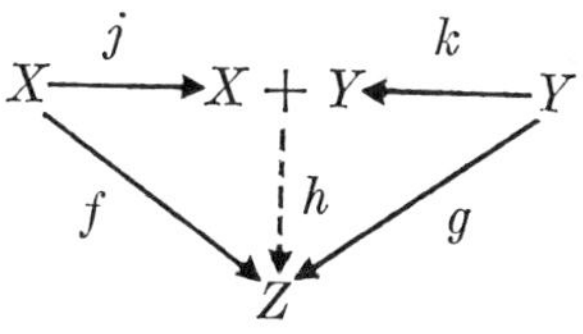

N. Let $f: X \to Z$, $g: Y \to Z$ be maps with the same nonempty codomain. Construct a nonempty class $E \subset X \times Y$ and a map h making the following diagram commute:

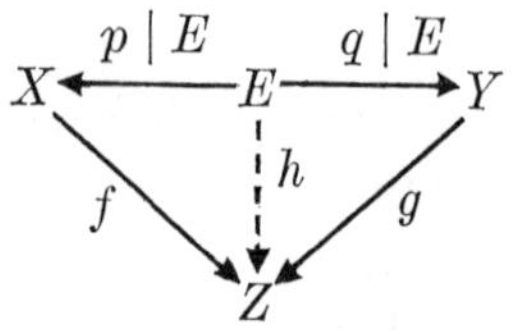

where $p: X \times Y \to X$, $q: X \times Y \to Y$ are the projections.

P. Let X be a class, and let $p: X \times X \to X$, $q: X \times X \to X$ be the projections. According to 8.19, there is a unique map δ making the following diagram commute:

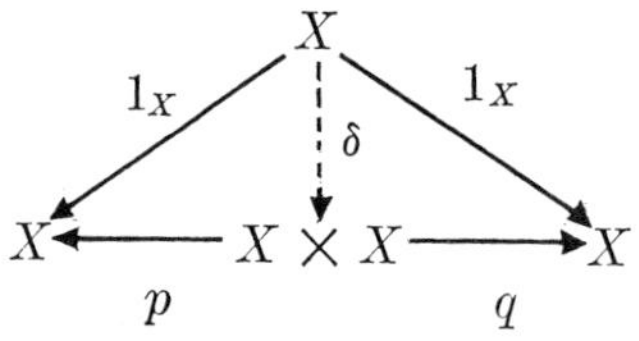

For $x \in X$ express $\delta(x)$ explicitly in terms of x. (The map δ is called the *diagonal map*.)

Q. Let X and Y be sets. Prove that Map (X,Y) is a set. [*Hint*: First prove that the class $\mathfrak{F}$ of all functional relations f with $\operatorname{dmn} f = X$, $\operatorname{rng} f \subset Y$ is a set. Then construct a map whose domain is $\mathfrak{F}$ and range is Map (X,Y), and apply the axiom of replacement.]

R. If Map $(X,Y) =$ Map (Y,X), prove that $X = Y$.

S. Let X, Y, Z be sets. Construct maps

$$\varphi: \text{Map}\,(X \times Y, Z) \to \text{Map}\,(Y, \text{Map}\,(X,Z)),$$

$$\psi: \text{Map}\,(Y, \text{Map}\,(X,Z)) \to \text{Map}\,(X \times Y, Z)$$

such that $\psi \circ \varphi$ and $\varphi \circ \psi$ are the identity maps on their respective domains.

T. For a map $g: Y \to Y'$ and a set X, define the map

$$g^X: \text{Map}\,(X,Y) \to \text{Map}\,(X,Y')$$

to be $\varphi_g{}^{1_X}$ in the notation of 8.22.

(a) If $g: Y \to Y'$ and $h \in$ Map (X,Y), express $g^X(h)$ explicitly in terms of h.

(b) Show that

$$(1_Y)^X = 1_{\text{Map}\,(X,Y)}.$$

(c) If

$$g: Y \to Y', \qquad g': Y' \to Y''$$

are maps, show that

$$(g' \circ g)^X = g'^X \circ g^X.$$

U. Let $\mathfrak{M}$ be the class of all maps whose domains and codomains are sets. Define a map

$$*: \mathfrak{M} \to \mathfrak{M}$$

$$\varphi \mapsto \varphi_*$$

as follows. If φ is a map $f: X \to Y$, then φ_* is the map

$$f_*: \mathcal{P}(X) \to \mathcal{P}(Y)$$

$$E \mapsto f\langle E\rangle \qquad (E \in \mathcal{P}(X)).$$

Prove:

(a) If X is a set, then

$$(1_X)_* = 1_{\mathcal{P}(X)}.$$

(b) If $\varphi, \psi \in \mathfrak{M}$ and if codmn φ = dmn ψ, then codmn φ_* = dmn ψ_* and

$$(\psi \circ \varphi)_* = \psi_* \circ \varphi_*.$$

V. Let $\mathfrak{M}$ be as in Exercise U. Define a map

$$*: \mathfrak{M} \to \mathfrak{M}$$

$$\varphi \mapsto \varphi^*$$

as follows. If φ is a map $f: X \to Y$, then φ^* is the map

$$f^*: \mathcal{P}(Y) \to \mathcal{P}(X)$$

$$E \mapsto f^{-1}\langle E\rangle \qquad (E \in \mathcal{P}(Y)).$$

State and prove for * results similar to Exercise U (a) and (b).

W. Let $(X_i \mid i \in I)$ be a nonempty family of sets. Why can one not legitimately write

$$\bigcup_{i \in I} X_i{}^c = (\bigcap_{i \in I} X_i)^c?$$

X. (a) Exhibit a sequence $(X_n \mid n \in \omega)$ of nonempty sets such that $X_{n^+} \subsetneqq X_n$ for each $n \in \omega$.

(b) Exhibit a sequence $(X_n \mid n \in \omega)$ of sets such that $X_n \in X_{n^+}$ for each $n \in \omega$.

(c) Prove there exists no sequence $(X_n \mid n \in \omega)$ of sets such that $X_{n^+} \in X_n$ for each $n \in \omega$.

Y. Let X be a nonempty set. Denote by cov X the class of all families $(A_i \mid i \in I)$ of subsets of X for which $\bigcup_{i \in I} A_i = X$. An element of cov X is said to *cover* X and is called a *covering of* X.

(a) Show that cov X is a nonempty set.
(b) For $\alpha = (A_i \mid i \in I)$, $\beta = (B_j \mid j \in J) \in \text{cov } X$, let $\alpha \leq \beta$ mean

$$(\forall j \in J)(\exists i \in I)(B_j \subset A_i).$$

Given $\alpha, \beta \in \text{cov } X$, prove there exists $\gamma \in \text{cov } X$ for which $\alpha \leq \gamma$ and $\beta \leq \gamma$.

Z. Let $f: X \to Y$ be a map and let $(A_i \mid i \in I)$ be a nonempty family of subsets of X. Prove:

$$f\langle \textstyle\bigcup_{i\in I} A_i \rangle = \textstyle\bigcup_{i\in I} f\langle A_i \rangle,$$

$$f\langle \textstyle\bigcap_{i\in I} A_i \rangle \subset \textstyle\bigcap_{i\in I} f\langle A_i \rangle.$$

Prove also analogous results for inverse images under f (compare 8.15).

AA. Let $(A_n \mid n \in \omega)$ be a sequence of sets. The sets

$$\liminf_{n\in\omega} A_n = \textstyle\bigcup_{i=0}^{\infty} \bigcap_{j=i}^{\infty} A_j,$$

$$\limsup_{n\in\omega} A_n = \textstyle\bigcap_{i=0}^{\infty} \bigcup_{j=i}^{\infty} A_j$$

are called the *limit inferior, limit superior of* $(A_n \mid n \in \omega)$.
(a) Show that $\liminf_{n\in\omega} A_n \subset \limsup_{n\in\omega} A_n$.
(b) Calculate both $\liminf_{n\in\omega} A_n$ and $\limsup_{n\in\omega} A_n$ in case $(A_n \mid n \in \omega)$ is "increasing," that is, $m, n \in \omega$ and $m \leq n$ implies $A_m \subset A_n$.
(c) Repeat (b) in case $(A_n \mid n \in \omega)$ is "decreasing," that is, $m, n \in \omega$ and $m \leq n$ implies $A_m \supset A_n$.

BB. Find maps

$$f: [0,\tfrac{1}{2}] \to [0,1], \qquad g: [\tfrac{1}{2},1] \to [0,1]$$

with the following property: Given any set X and any two maps

$$\alpha: [0,1] \to X, \qquad \beta: [0,1] \to X$$

with $\alpha(1) = \beta(0)$, there exists

$$\gamma: [0,1] \to X$$

with

$$\gamma \mid [0,\tfrac{1}{2}] = \alpha \circ f, \qquad \gamma \mid [\tfrac{1}{2},1] = \beta \circ g.$$

$$\gamma(0) = \alpha(0), \qquad \gamma(1) = \beta(1),$$

$$\text{rng } \gamma = \text{rng } \alpha \cup \text{rng } \beta.$$

CC. Establish the following facts concerning characteristic functions of subsets A, B of a set X:
(a) $\chi_{X\setminus A}(x) = 1 - \chi_A(x)$ for all $x \in X$.
(b) $\chi_{A\cap B}(x) = \chi_A(x) \cdot \chi_B(x)$ for all $x \in X$.
(c) $\chi_{A\cup B}(x) = \chi_A(x) + \chi_B(x) - \chi_{A\cap B}(x)$ for all $x \in X$.

DD. Show that the graph f of the map from $\bigcup_{i\in I} X_i$ into Y given by 8.28 is just $\bigcup_{i\in I} f_i$.

EE. Construct a sequence $(X_n \mid n \in \omega)$ of nonempty sets which is not pairwise disjoint and for which $\bigcap_{n\in\omega} X_n = \varnothing$.

FF. Let $(X_i \mid i \in I)$ be a family of sets.

(a) Show that the family $(X_i \times \{i\} \mid i \in I)$ is disjoint.

The class

$$\bigcup_{i\in I} X_i \times \{i\}$$

is called the *sum* or *disjoint union of* $(X_i \mid i \in I)$ and is denoted by $+_{i\in I} X_i$. For each $j \in I$ consider the map

$$q_j\colon X_j \to +_{i\in I} X_i$$
$$x \mapsto (x, j) \qquad (x \in X_j).$$

(b) Show there is a unique map

$$q\colon \bigcup_{i\in I} X_i \to +_{i\in I} X_i$$

such that

$$q \mid X_j = q_j \qquad (j \in I),$$
$$\operatorname{rng} q = +_{i\in I} X_i.$$

When will $x, y \in \bigcup_{i\in I} X_i$ with $x \neq y$ always imply $q(x) \neq q(y)$?

(c) Let Y be a class and $(f_i \mid i \in I)$ be a family of maps with

$$f_i\colon X_i \to Y \qquad (i \in I).$$

Show there is a unique map g making the following diagrams commute for all $j \in I$:

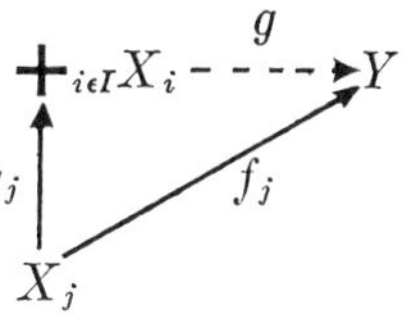

GG. A *pointed set* is an ordered pair (X,x), where X is a set and $x \in X$; one calls x the *base-point* of (X,x). Given pointed sets (X,x) and (Y,y), one says that a map $f\colon X \to Y$ is *base-point preserving* and writes $f\colon (X,x) \to (Y,y)$ to mean $f(x) = y$. For example, if (X,x) is a pointed set, then $1_X\colon (X,x) \to (X,x)$.

(a) If $f\colon (X,x) \to (Y,y)$ and $g\colon (Y,y) \to (Z,z)$ are base-point preserving maps of pointed sets, verify that $g \circ f$ is also.

(b) If (X,x) and (Y,y) are pointed sets, find a $z \in X \times Y$ such that the projections $p\colon X \times Y \to X$, $q\colon X \times Y \to Y$ are base-point preserving for $(X \times Y, z)$.

(c) State and prove an analog of 8.19 for pointed sets and base-point preserving maps.

9. Injections and Surjections

Two special kinds of maps will now be discussed: injections and surjections. Also considered are bijections, which are simultaneously injections and surjections. The attempt to give a characterization of surjections "dual" to a characterization of injections leads naturally to our final axiom, the axiom of choice.

INJECTIONS

An injection is a map which assigns distinct values to distinct elements of its domain:

9.1. DEFINITION

A map $f: X \to Y$ is said to be *injective* (or *one-to-one*) and is called an *injection* in case

$$x \neq x' \Rightarrow f(x) \neq f(x') \qquad (x, x' \in X).$$

A map $f: X \to Y$ is injective if and only if for each $y \in Y$, either $f^{-1}\langle\{y\}\rangle$ is a singleton (in case $y \in \operatorname{rng} f$) or else $f^{-1}\langle\{y\}\rangle = \varnothing$ (in case $y \notin \operatorname{rng} f$). Thus an injective map is one which assigns each element of its codomain to *at most one* element of its domain.

9.2. EXAMPLES

(1) If X is any class, then $1_X: X \to X$ is injective.

(2) If $f: X \to Y$ is an injection and $A \subset X$, then $f \mid A$ is injective. In particular, the inclusion map of A into X is injective.

(3) A constant map $c: X \to Y$ is injective if and only if X is empty or is a singleton.

(4) The successor map $f: \omega \to \omega$ is injective. In fact, Peano postulate 6.6(4) says that $m \neq n \Rightarrow f(m) \neq f(n)$.

(5) The map $g: \omega \setminus \{0\} \to \omega$ such that $(g(n))^+ = n$ for each $n \in \omega \setminus \{0\}$ is injective. In fact, $g(m) = g(n)$ implies $m = (g(m))^+ = (g(n))^+ = n$.

The property that a map $f: X \to Y$ be injective does not depend on the codomain Y; if Y is replaced by any class Y' for which $\operatorname{rng} f \subset Y'$, then $f: X \to Y$ is injective if and only if $f: X \to Y'$ is injective. In fact:

9.3. LEMMA

Let $f: X \to Y$ be a map whose graph is f. Then this map is injective if and only if the relation f^{-1} is functional.

Proof

Suppose first $f: X \to Y$ is injective. Let (y,x), $(y,x') \in f^{-1}$. We must show $x = x'$. We have $(x,y) \in f$ and $(x',y) \in f$, that is, $y = f(x)$ and $y = f(x')$. Then $f(x) = f(x')$, so $x = x'$.

Conversely, suppose f^{-1} is functional. Let x, $x' \in X$ with $f(x) = f(x')$. We show $x = x'$. We have $(x, f(x)) \in f$ and $(x, f(x')) \in f$, so $(f(x),x) \in f^{-1}$ and $(f(x'), x') \in f^{-1}$. Since $f(x) = f(x')$ and f^{-1} is functional, $x = x'$. ▯

With the aid of 9.3, we can now give an entirely map-theoretic characterization of "injective."

9.4. THEOREM

Let $f: X \to Y$ be a map with nonempty domain. Then the following statements are equivalent:

(1) f is injective.

(2) There exists a map $r: Y \to X$ such that

$$r \circ f = 1_X.$$

Proof

Assume (2). We show (1). Let x, $x' \in X$ with $f(x) = f(x')$. Then

$$\begin{aligned} x = 1_X(x) &= (r \circ f)(x) = r(f(x)) \\ &= r(f(x')) = (r \circ f)(x') = 1_X(x') = x', \end{aligned}$$

so $x = x'$.

Assume (1). We show (2). Let $Z = \text{rng} f$. The relation f^{-1} is functional by 9.3, and $f^{-1}\langle Z\rangle = X$. Hence f^{-1} is the graph of a map from Z to X,

$$f^{-1}: Z \to X.$$

If $x \in X$ and $x' = f^{-1}(f(x))$, then $(f(x),x') \in f^{-1}$, also $(f(x),x) \in f^{-1}$, and $x' = x$ since f^{-1} is functional. This shows

$$f^{-1}(f(x)) = x \qquad (x \in X).$$

Thus except for the fact that its domain is Z and not all of Y, f^{-1} is the map r we seek. Let

$$r: Y \to X$$

be any extension of $f^{-1}: Z \to X$ to Y; such an r exists by Chapter 8, Exercise K. If $x \in X$, then

$$(r \circ f)(x) = r(f(x)) = (r \mid Z)(f(x)) = f^{-1}(f(x)) = x.$$

Hence $r \circ f = 1_X$, as needed. □

A map $r: Y \to X$ satisfying condition (2) above is called a *retraction* or *left inverse of f*.

A retraction of a given injection need not be unique. For example, let $X = \{0,1\}$, $Y = \{0,1,2\}$, and let $f: X \to Y$ be the inclusion map. Define maps

$$g_0, g_1: Y \to X$$

by

$$g_0(0) = 0, \qquad g_0(1) = 1, \qquad g_0(2) = 0,$$

$$g_1(0) = 0, \qquad g_1(1) = 1, \qquad g_1(2) = 1.$$

Then $g_0 \circ f = 1_X = g_1 \circ f$, but $g_0 \neq g_1$.

If $f: X \to Y$ is an inclusion map, then a retraction of f is just an extension of 1_X to Y.

9.5. COROLLARY

Let $f: X \to Y$, $g: Y \to Z$ be maps.

(1) If f and g are injective, then $g \circ f$ is injective.
(2) If $g \circ f$ is injective, then f is injective.

Proof

(1) This is clear if X or Y is empty. Suppose $X \neq \varnothing \neq Y$. Let $r: Y \to X$, $t: Z \to Y$ be retractions of f, g. Then $r \circ t: Z \to X$ and

$$(r \circ t) \circ (g \circ f) = r \circ (t \circ g) \circ f = r \circ 1_Y \circ f = r \circ f = 1_X.$$

Hence $r \circ t$ is a retraction of $g \circ f$, and $g \circ f$ is injective by 9.4.

(2) The result is obvious if $X = \varnothing$. Suppose $X \neq \varnothing$. Let $r\colon Z \to X$ be a retraction of $g \circ f$. Then $r \circ g\colon Y \to X$ and $(r \circ g) \circ f = r \circ (g \circ f) = 1_X$, so $r \circ g$ is a retraction of f. By 9.4, f is injective. □

9.6. EXAMPLE

Let $X = \omega \setminus \{0,1\}$, $Y = \omega$, $f\colon X \to Y$ be the inclusion map, and $g\colon Y \to X$ be the map such that

$$g(0) = g(1) = 2, \qquad g(n) = n \qquad (n \in X).$$

Then $g \circ f = 1_X$ is injective (and hence f is injective by 9.5), but g is not injective.

SURJECTIONS

For a map $f\colon X \to Y$, condition (2) of 9.4 concerns a map $g\colon Y \to X$ "going backwards" which yields an identity map when composed with f in the order $g \circ f$. We consider next the "dual" condition according to which g yields an identity map when composed with f in the opposite order $f \circ g$.

9.7. PROPOSITION

Let $f\colon X \to Y$ be a map. Suppose there is a map $s\colon Y \to X$ such that

$$f \circ s = 1_Y.$$

Then $Y = \operatorname{rng} f$.

Proof

We need only show $Y \subset \operatorname{rng} f$. Let $y \in Y$. Then $s(y) \in X$,

$$y = 1_Y(y) = (f \circ s)(y) = f(s(y)),$$

and $y \in \operatorname{rng} f$. □

A map satisfying the hypothesis of 9.7 is called a *section* or *right inverse of f*.

Proposition 9.7 says that if a map has a section, then it is "surjective" in the sense of the following definition. The converse of 9.7 holds when, for example, the domain of the map is a set—see 9.21 below.

9.8. DEFINITION

A map $f\colon X \to Y$ is said to be *surjective* and is called a *surjection* if its codomain Y equals the range $\operatorname{rng} f$ of its graph, that is, if $Y = f\langle X \rangle$. One also says that f maps X *onto* Y to mean that $f\colon X \to Y$ is surjective. (The frequently used phrase 'an onto map', used to mean 'a surjective map', is linguistically irritating—hence the use of the French 'surjective'.)

A map $f: X \to Y$ is surjective if and only if $f^{-1}\langle\{y\}\rangle \neq \varnothing$ for each $y \in Y$. Thus, a surjective map is one which assigns each element of its codomain to at least one element of its domain.

9.9. EXAMPLES

(1) The identity map on any class is both injective and surjective.
(2) The constant map on ω to $\{0\}$ is surjective, but not injective.
(3) The inclusion map of $\omega \setminus \{0\}$ into ω is injective but not surjective.
(4) The constant map on ω to ω with value 0 is neither surjective nor injective.

These four examples show that the notions 'injective', 'surjective' are logically independent.

(5) Let X and Y be nonempty classes. Then the projections p: $X \times Y \to X$ and $q: X \times Y \to Y$ are surjective. In fact, there exists $y \in Y$, so $x \in X$ implies $x = p((x,y))$; thus, p is surjective. A similar argument applies to q.

Although a functional relation may be injective in the sense that it is the graph of an injective map, it is meaningless to speak of a surjective functional relation, for the property 'surjective' always refers to a given codomain. (Admittedly, we sometimes appear to be violating this dictum, as in: "Let $f: X \to Y$ be a map such that .... Then f is surjective." Such an apparent violation merely results from our use of the same symbol to denote both a map and its graph.)

9.10. PROPOSITION

Let $f: X \to Y$ and $g: Y \to Z$ be maps.

(1) If f and g are surjective, then $g \circ f$ is surjective.
(2) If $g \circ f$ is surjective, then g is surjective.

Proof

(1) Assume f and g are surjective. Then $f\langle X\rangle = Y$ and $g\langle Y\rangle = Z$, so

$$(g \circ f)\langle X\rangle = g\langle f\langle X\rangle\rangle = g\langle Y\rangle = Z.$$

(2) Assume $g \circ f$ is surjective. Then

$$g\langle Y\rangle \supset g\langle f\langle X\rangle\rangle = (g \circ f)\langle X\rangle = Z,$$

and since $g\langle Y\rangle \subset Z$, $g\langle Y\rangle = Z$. □

If f and g are as in Example 9.6, then $g \circ f$ is surjective, but f is not surjective.

Since an identity map is both injective and surjective, 9.5 and 9.10 yield as a corollary:

9.11. COROLLARY

If $f: X \rightarrow Y$ and $g: Y \rightarrow X$ are maps such that $g \circ f = 1_X$, then f is injective and g is surjective.

9.12. COROLLARY

Let $f: X \rightarrow Y$ be a map. If $r: Y \rightarrow X$ is a retraction of f, then r is surjective. If $s: Y \rightarrow X$ is a section of f, then s is injective.

BIJECTIONS

Of central importance to the theory of cardinality developed in Chapters 13–19 are maps which are simultaneously injective and surjective.

9.13. DEFINITION

A map $f: X \rightarrow Y$ is said to be *bijective* and is called a *bijection* if it is both injective and surjective, and then one writes

$$f: X \simeq Y.$$

A bijection on X to Y is also called a *one-to-one correspondence between X and Y*. A bijection whose domain and codomain are the same class X is called a *permutation of X*.

A map $f: X \rightarrow Y$ is bijective if and only if $f^{-1}\langle\{y\}\rangle$ is a singleton for each $y \in Y$. Thus, a bijection is a map which assigns each element of its codomain to precisely one element of its domain.

9.14. EXAMPLES

(1) If X is any class, then $1_X: X \simeq X$.

(2) Let $\sigma: \omega \rightarrow \omega \setminus \{0\}$ be the successor function on ω. Then σ is a bijection from ω to $\omega \setminus \{0\}$.

(3) Let X and Y be sets. Define $\mathfrak{F}$ to be the class of all *functional relations* with domain X and range a subset of Y. Consider the map

$$\varphi: \mathfrak{F} \rightarrow \text{Map}\,(X,Y)$$

$$f \mapsto \langle X, f, Y\rangle \qquad (f \in \mathfrak{F}).$$

Then φ is a one-to-one correspondence between $\mathfrak{F}$ and Map (X,Y).

The next three examples, from analysis, concern maps defined on subsets of the set $\mathbb{R}$ of real numbers.

(4) The exponential map $x \mapsto e^x$ from $\mathbb{R}$ to the set of all positive real numbers is bijective.

(5) The squaring map $x \mapsto x^2$ from the set of all nonnegative real numbers to itself is bijective.

(6) The map $x \mapsto \tan x$ from the open interval $\{x \mid x \in \mathbb{R} \;\&\; -\pi/2 < x < \pi/2\}$ to $\mathbb{R}$ is bijective.

Suppose $\varphi = \langle X, f, Y \rangle$ is a bijection. Since f is injective, f^{-1} is a functional relation whose range is $\operatorname{dmn} f = X$. Since φ is surjective, $\operatorname{dmn} f^{-1} = \operatorname{rng} f = Y$. This justifies the following definition.

9.15. DEFINITION

If $\varphi = \langle X, f, Y \rangle$ is a bijection, then the map $\langle Y, f^{-1}, X \rangle$ is called the *inverse* of φ and is denoted by φ^{-1}. In the more usual notation, the inverse of a bijection $f\colon X \to Y$ is $f^{-1}\colon Y \to X$.

The inverses of the maps in 9.14(4)–(6) are respectively the natural logarithm map, the square-root map, and the map $x \mapsto \arctan x$.

9.16. THEOREM

Let $f\colon X \to Y$ be a map. Then the following statements are equivalent:

(1) f is bijective.

(2) There exist maps r, $s\colon Y \to X$ with

$$r \circ f = 1_X, \qquad f \circ s = 1_Y.$$

(3) There exists a map $g\colon Y \to X$ such that

$$g \circ f = 1_X, \qquad f \circ g = 1_Y.$$

Moreover, if f is bijective, then f^{-1} is the unique left inverse and the unique right inverse of f.

Proof

By 9.4 and 9.7, (2) implies (1). Clearly (3) implies (2).

Assume (1). According to the remarks above, we have a map $f^{-1}\colon Y \to X$. We show here that $f^{-1} \circ f = 1_X$ and $f \circ f^{-1} = 1_Y$, thereby proving (3). If $x \in X$, then $y = f(x)$ implies $x = f^{-1}(y)$, and hence $f^{-1}(f(x)) = x$. If $y \in Y$, then $x = f^{-1}(y)$ implies $y = f(x)$, and hence $f(f^{-1}(y)) = y$.

Still assuming (1), let $r\colon Y \to X$ be a left inverse of f. Then $r \circ f = 1_X$, so

$$\begin{aligned} r = r \circ 1_Y &= r \circ (f \circ f^{-1}) = (r \circ f) \circ f^{-1} \\ &= 1_X \circ f^{-1} = f^{-1}. \end{aligned}$$

If $s\colon Y \to X$ is a right inverse of f, then

$$\begin{aligned} s = 1_X \circ s &= (f^{-1} \circ f) \circ s = f^{-1} \circ (f \circ s) \\ &= f^{-1} \circ 1_Y = f^{-1}. \quad \square \end{aligned}$$

The two corollaries below illustrate the use of condition (3) in proving a given map is bijective.

9.17. COROLLARY

If $f: X \to Y$ is a bijection, then $f^{-1}: Y \to X$ is a bijection, and

$$(f^{-1})^{-1} = f.$$

Proof

If f is bijective, then $g = f$ is a map on X to Y such that $g \circ f^{-1} = 1_Y$ and $f^{-1} \circ g = 1_X$, so 9.16 applies. □

9.18. COROLLARY

If $f: X \to Y$ and $g: Y \to Z$ are bijections, then $g \circ f: X \to Z$ is a bijection, and

$$(g \circ f)^{-1} = f^{-1} \circ g^{-1}.$$

Proof

Assume f and g are bijective. By 9.16, there are maps $h: Y \to X$, $k: Z \to Y$ with

$$h \circ f = 1_X, \qquad f \circ h = 1_Y, \qquad k \circ g = 1_Y, \qquad g \circ k = 1_Z.$$

Let $m = h \circ k: Z \to X$. Then

$$\begin{aligned} m \circ (g \circ f) &= (h \circ k) \circ (g \circ f) = h \circ (k \circ g) \circ f \\ &= h \circ 1_Y \circ f = h \circ f = 1_X, \end{aligned}$$

and similarly, $(g \circ f) \circ m = 1_Z$. By 9.16, $g \circ f$ is bijective. □

Of course, 9.18 also follows directly from 9.5 and 9.10.

As a further illustrative example of 9.16, we show as promised earlier that each family $(x_i \mid i = 1,2)$ in a given class X can be regarded as an ordered pair of elements of X, and vice versa.

9.19. EXAMPLE

Let X be a class. Recall that

$$X^2 = \text{Map}(\{1,2\},X).$$

Consider the map

$$\varphi: X^2 \to X \times X$$

$$f \mapsto (f(1), f(2)) \qquad (f \in X^2)$$

and the map

$$\psi: X \times X \to X^2$$

$$(x,y) \mapsto c_{x,y} \qquad ((x,y) \in X \times X)$$

where for $(x,y) \in X \times X$,

$$c_{x,y}: \{1,2\} \to X$$

is the map defined by

$$c_{x,y}(1) = x, \qquad c_{x,y}(2) = y.$$

Then $\psi \circ \varphi$ and $\varphi \circ \psi$ are the identity maps on X^2 and $X \times X$. By 9.16, φ is a one-to-one correspondence between X^2 and $X \times X$.

NEED FOR THE AXIOM OF CHOICE

Consider the converse of 9.7, namely: If $f: X \to Y$ is a surjection, then there exists a map $s: Y \to X$ such that $f \circ s = 1_Y$. We have already proved this converse in case f is injective as well as surjective, for then we may take $s = f^{-1}$. In the general case, of course, another approach is needed.

Assume that $f: X \to Y$ is surjective. If $s: Y \to X$ is a map, then the condition $f \circ s = 1_Y$ is equivalent to

$$s(y) \in f^{-1}\langle\{y\}\rangle \qquad (y \in Y).$$

Then in order to construct a right inverse of f, we need only "choose" for each $y \in Y$ some element of $f^{-1}\langle\{y\}\rangle$ and define $s(y)$ to be that element. Now for $y \in Y$, there certainly exists some element of $f^{-1}\langle\{y\}\rangle$, for the assumption that f is surjective guarantees $f^{-1}\langle\{y\}\rangle \neq \varnothing$. The trouble is that for a given $y \in Y$, $f^{-1}\langle\{y\}\rangle$ need not be singleton, so there is no preferred way to choose one specific element out of the many belonging to $f^{-1}\langle\{y\}\rangle$.

Let us examine more closely the intuitively appealing notion of choosing an element from each $f^{-1}\langle\{y\}\rangle$, with the aim of formulating it in a mathematically meaningful way. In order to consider simultaneously all the classes $f^{-1}\langle\{y\}\rangle$ for various elements y of Y, it is convenient to form the class $\mathcal{A} = \{f^{-1}\langle\{y\}\rangle \mid y \in Y\}$. Of course, we need here the assumption that $f^{-1}\langle\{y\}\rangle$ is a set for each $y \in Y$, and we now so assume. The act of choosing an element of each member of $\mathcal{A}$ may be regarded as the assignment to each element $A \in \mathcal{A}$ of some member of A, that is, as a function c with domain $\mathcal{A}$ such that $c(A) \in A$ for each $A \in \mathcal{A}$. This suggests a definition.

9.20. DEFINITION

Let $\mathcal{A}$ be a class of sets. A *choice function for* $\mathcal{A}$ is by definition a map

$$c: \mathcal{A} \to \bigcup \mathcal{A}$$

such that

$$c(A) \in A \qquad (A \in \mathcal{A}).$$

Clearly a necessary condition for a class $\mathcal{A}$ of sets to have a choice function is that no member of $\mathcal{A}$ be empty. In order to prove the converse of 9.7 in the way suggested above, we want this condition to be sufficient as well. Hence we now state our final axiom of set theory.

AXIOM 8 (AXIOM OF CHOICE)

If $\mathcal{A}$ is a class of nonempty sets, then there exists a choice function for $\mathcal{A}$.

At last our theorem can be proved.

9.21. THEOREM

Let $f: X \to Y$ be a map. Suppose $f^{-1}\langle\{y\}\rangle$ is a set for each $y \in Y$. Then the following statements are equivalent:

(1) f is surjective.
(2) There exists a map $s: Y \to X$ such that

$$f \circ s = 1_Y.$$

Proof

According to 9.7, (2) implies (1), without any hypothesis concerning the classes $f^{-1}\langle\{y\}\rangle$.

Assume (1). We show (2). Let

$$\mathcal{A} = \{f^{-1}\langle\{y\}\rangle \mid y \in Y\}.$$

Since f is surjective, each member of $\mathcal{A}$ is nonempty. By the axiom of choice, there exists a choice function c for $\mathcal{A}$. Define $s: Y \to X$ to be the map such that

$$s(y) = c(f^{-1}\langle\{y\}\rangle) \qquad (y \in Y).$$

Then $y \in Y$ implies $s(y) \in f^{-1}\langle\{y\}\rangle$ and so $f(s(y)) = y$. Hence $f \circ s = 1_Y$. □

Note that the hypothesis of 9.21 is satisfied automatically in case X is itself a set.

9.22. THEOREM

Let X be a nonempty class and Y be a set. Then the following statements are equivalent:

(1) There exists an injection $f: X \to Y$.
(2) There exists a surjection $g: Y \to X$.

Moreover, if either (1) or (2) holds, then X is also a set.

Proof

Assume (2). By the axiom of replacement, X is a set. By 9.21, there exists a section $f: X \to Y$ of g. By 9.12, f is injective.

Assume (1). By 9.4, there exists a retraction $g: Y \to X$ of f. By 9.12, g is surjective. □

THE AXIOM OF CHOICE

The axiom of choice, due essentially to E. Zermelo, was introduced above for the specific purpose of guaranteeing that certain surjections have sections. It has many other significant consequences. In this section we shall deduce a few of the simpler ones not involving the notion of order.

It is appropriate to comment first on the status of the axiom of choice in contemporary mathematics. Despite the axiom's unquestioned utility, mathematicians have often been reluctant to use it on grounds that it is not intuitively acceptable or that it is too profound in contrast to the other axioms. In fact, some texts apologetically attach a special label to each theorem whose proof employs the axiom of choice.

In 1938, K. Gödel proved that the axiom of choice is consistent with the other axioms of set theory, in the sense that if a contradiction can be deduced from the axiom of choice together with the other axioms, then a contradiction can be deduced from the other axioms alone. In 1963, P. J. Cohen showed that the negation of the axiom of choice is also consistent with the other axioms, so that it is independent of the other axioms. (For an expository account of Gödel's results, see [27]. For a more detailed treatment of both Gödel's and Cohen's results, see [8], [11], [15].)

These deep metamathematical results of Gödel and Cohen tell us that the decision to include the axiom of choice as one of our primitive assumptions can be made only on grounds of taste and convenience. Because the body of mathematics would be drastically impoverished were the axiom of choice not at our disposal, we adopt it here without regrets or apologies.

Notice that the axiom of choice mentions no procedure for explicitly constructing any specific choice function and so is purely existential in character. For example, there is no general procedure for distinguishing between the elements of a doubleton, and so one cannot write down a rule for choosing simultaneously a member of each doubleton; nevertheless, the axiom of choice says the class

$$\{\{x,y\} \mid x \in \mathcal{U} \;\&\; y \in \mathcal{U}\}$$

has a choice function. Of course, one may be able to exhibit a choice function for a specific given class. For example, the assignment to each ordered pair (x,y) of sets of its element $\{x\}$ defines a choice function for the class $\mathcal{U} \times \mathcal{U}$.

The contrast between the two examples may be illuminated by a metaphor. Since the two socks of a matched "pair" are indistinguishable, there is no way to prescribe a choice of one sock from each pair. However, one can easily choose from each pair of shoes the one fitting the left foot.

Even when the axiom of choice tells us a choice function for a class exists, it does not guarantee its uniqueness. For example, let $\mathcal{A} = \omega \setminus \{\varnothing\}$. Define a function $c\colon \mathcal{A} \to \bigcup\mathcal{A}$ by the rule: $(c(n))^+ = n$ for each $n \in \mathcal{A}$. Then c is a choice function for $\mathcal{A}$. Many other choice functions for $\mathcal{A}$ can be constructed from c. If $k \in \mathcal{A}$, then the function c_k such that

$$c_k \mid (\mathcal{A} \setminus \{k\}) = c \mid (\mathcal{A} \setminus \{k\}), \qquad c_k(k) = 0$$

is a choice function for $\mathcal{A}$. If $k \neq 1$, then $c_k \neq c$. If $k \neq j$, then $c_k \neq c_j$.

A particular instance of Axiom 8 is the theorem:

(*) There exists a choice function for $\mathcal{U} \setminus \{\varnothing\}$.

Now if one assumes (*) but not Axiom 8, then the statement of Axiom 8 can be deduced using the other seven axioms. In fact, if c is a choice function for $\mathcal{U} \setminus \{\varnothing\}$, and if $\mathcal{A}$ is a class of nonempty sets, then $c \mid \mathcal{A}$ is a choice function for $\mathcal{A}$. Hence in the system obtained by deleting Axiom 8 from our list of axioms, statement (*) becomes equivalent to the statement of Axiom 8.

It so happens that many other theorems which are proved through use of the axiom of choice, including significant ones outside set theory proper, are actually equivalent to it. [*Caution*: A number of theorems said to be equivalent to the axiom of choice imply not Axiom 8 as formulated here, but rather the weaker statement that each *set* of nonempty sets has a choice function.] In view of the results of Gödel and Cohen, the only interest in establishing such equivalences—aside from the amusement in constructing proofs—is in knowing what theorems one must do without were one to abandon the axiom of choice. Readers interested in these considerations are referred to the little book by Rubin and Rubin [33].

A notion of choice functions for families is often useful.

9.23. DEFINITION

Let $(X_i \mid i \in I)$ be a family of sets. A function

$$c\colon I \to \bigcup_{i \in I} X_i$$

such that

$$c(i) \in X_i \qquad (i \in I)$$

is called a *choice function for* $(X_i \mid i \in I)$. Thus a choice function for $(X_i \mid i \in I)$ is just a family $(x_i \mid i \in I)$ in $\bigcup_{i \in I} X_i$ such that $x_i \in X_i$ for each $i \in I$.

If $\mathcal{A}$ is a class of sets, then a choice function for $\mathcal{A}$ in the sense of 9.20 is nothing but a choice function for the family $(A \mid A \in \mathcal{A})$.

9.24. THEOREM

Let $(X_i \mid i \in I)$ be a family of nonempty sets. Then there exists a choice function for $(X_i \mid i \in I)$.

Proof

Let $\mathcal{A} = \{X_i \mid i \in I\}$. By Axiom 8 there exists a choice function c for the class $\mathcal{A}$. Then

$$i \mapsto c(X_i) \qquad (i \in I)$$

is the desired choice function for $(X_i \mid i \in I)$. □

If $f: X \to Y$ is a map with $f^{-1}\langle\{y\}\rangle$ a set for each $y \in Y$, then a choice function for the family $(f^{-1}\langle\{y\}\rangle \mid y \in Y)$ is a section of f (and vice versa), so 9.24 may be used to prove 9.21 directly.

9.25. THEOREM

Let R be a relation such that $R\langle\{x\}\rangle$ is a set for each $x \in \operatorname{dmn} R$. Then there exists a function

$$f: \operatorname{dmn} R \to \operatorname{rng} R$$

such that

$$f(x) \in R\langle\{x\}\rangle \qquad (x \in \operatorname{dmn} R).$$

Proof

Take for f any choice function for the family $(R\langle\{x\}\rangle \mid x \in \operatorname{dmn} R)$; such exists by 9.24. □

Thus, given a rule R assigning to each element of a class X a whole set of elements, one can choose simultaneously for each $x \in X$ one particular element assigned to x by R.

We formulate now one more formal analog of the act of choice.

9.26. DEFINITION

Let $\mathcal{A}$ be a class of sets. A class C is called a *choice class for* $\mathcal{A}$ if for each $A \in \mathcal{A}$ there is some set x such that $C \cap A = \{x\}$.

Thus, a choice class for a class $\mathcal{A}$ has for each $A \in \mathcal{A}$ exactly one member chosen from A.

9.27. EXAMPLES

(1) If $x \in \bigcap \mathcal{A}$, then $\{x\}$ is a choice class for $\mathcal{A}$. In particular, $\{0\}$ is a choice class for $\omega \setminus \{0\}$.

(2) If $\mathcal{A}$ is a class with $\varnothing \in \mathcal{A}$, then there is no choice class for $\mathcal{A}$.

(3) Let $A = \mathcal{P}(\omega) \setminus \{\varnothing\}$. Then there does not exist a choice class for $\mathcal{A}$, even though no member of $\mathcal{A}$ is empty. In fact, just suppose C is a choice class for $\mathcal{A}$. If $n \in \omega$, then $\{n\} \in \mathcal{A}$, $C \cap \{n\} = \{x\}$ for some set x, and so $n \in C$. Hence, $\omega \subset C$. But $\omega \in \mathcal{A}$ and $\omega \cap C = \omega \neq \{x\}$ for any set x. Notice that $\mathcal{A}$ is not disjoint.

9.28. THEOREM

Let $\mathcal{A}$ be a disjoint class of nonempty sets. Then there exists a choice class for $\mathcal{A}$.

Proof

By Axiom 8, there exists a choice function c for $\mathcal{A}$. Let

$$C = \operatorname{rng} c.$$

We claim that C is a choice class for $\mathcal{A}$.

Let $A \in \mathcal{A}$. We show that

$$C \cap A = \{c(A)\}.$$

Clearly $c(A) \in C \cap A$. Suppose $x \in C \cap A$. There exists $B \in \mathcal{A}$ such that $x = c(B)$. If $x \neq c(A)$, then $A \neq B$ since $c(A) \neq c(B)$, but $x \in A \cap B$, a contradiction. Hence $x = c(A)$. ▯

A class may have a choice class even without being disjoint. For example, $\{0,2\}$ is a choice class for $\{\{0,1\}, \{1,2\}\}$.

In the sequel, a use of any of Theorems 9.24, 9.25, 9.28, as well as of Axiom 8, is indicated simply by reference to the axiom of choice.

CATEGORIES

This chapter concludes our presentation of generalities concerning maps. No discussion of maps would be complete, however, without some mention of categories.

Roughly speaking, a category consists of a particular class of sets (vector spaces, topological spaces, differentiable manifolds, etc.) having some type of "structure," together with specified maps (linear transformations, continuous functions, differentiable functions, and so on) between these sets "preserving" their structure. More precisely, a (concrete) *category* $\mathcal{C}$

consists of a class Obj $\mathcal{C}$, whose elements are called *objects* of $\mathcal{C}$, and for each pair (X,Y) of objects of $\mathcal{C}$, a set

$$\text{map}_{\mathcal{C}}(X,Y) \subset \text{Map}(X,Y),$$

whose elements are called *morphisms from X to Y in $\mathcal{C}$*; it is required that

$$(*) \qquad 1_X \in \text{map}_{\mathcal{C}}(X,X) \qquad (X \in \text{Obj } \mathcal{C}),$$

$$(**) \qquad f \in \text{map}_{\mathcal{C}}(X,Y) \ \&\ g \in \text{map}_{\mathcal{C}}(Y,Z) \Rightarrow$$

$$g \circ f \in \text{map}_{\mathcal{C}}(X,Z) \qquad (X, Y, Z \in \text{Obj } \mathcal{C}).$$

Take Obj $\mathcal{C} = \mathfrak{U}$, and for $(X,Y) \in \mathfrak{U} \times \mathfrak{U}$, take $\text{map}_{\mathcal{C}}(X,Y) = \text{Map}(X,Y)$; conditions (*) and (**) are trivially satisfied, so we obtain a category. Next, take Obj $\mathcal{C}$ to be the class of all vector spaces (over the scalar field $\mathbb{R}$, say), and for $X, Y \in \text{Obj } \mathcal{C}$, take $\text{map}_{\mathcal{C}}(X,Y)$ to be the set of all linear transformations from X to Y; clearly (*) holds, and (**) is just a way of saying that the composite of two linear maps is again linear. Similarly, one has the category of topological spaces and continuous maps, and so on.

Category theory provides a language for stating complicated relationships between maps quite concisely. However, any meaningful treatment of categories requires more algebra than is appropriate here, so we refer the reader to MacLane and Birkhoff [24, Chapters 1, 2 (Section 10), 15].

Category theory is but one manifestation of the tendency today to consider maps at least as important as sets. Another is the recent attempt at an axiom system for set theory in which "map" is the primitive concept and "set" is a defined concept—see Hatcher [15] or Lawvere [22].

EXERCISES

A. Construct all injections from $\{0,1\}$ to $\{0,1,2\}$ and all injections from $\{0,1,2\}$ to $\{0,1\}$.

B. (a) Let X be the set of all sequences in $\{0,1\}$ and let $\sigma: X \to X$ be the map such that $(x_n \mid n \in \omega) \in X$ implies

$$\sigma((x_n \mid n \in \omega)) = (x_{n^+} \mid n \in \omega).$$

Is σ injective? Is it surjective?

(b) Let X now be the set of all families in $\{0,1\}$ indexed by the set $\mathbb{Z}$ of all integers, and let $\sigma: X \to X$ be the map such that $(x_n \mid n \in \mathbb{Z}) \in X$ implies

$$\sigma((x_n \mid n \in \mathbb{Z})) = (x_{n+1} \mid n \in \mathbb{Z}).$$

Is σ injective? Is it surjective?

C. Let $\mathcal{C}$ be the set of all continuous maps from $\mathbb{R}$ to $\mathbb{R}$. Let $I: \mathcal{C} \to \mathcal{C}$

be the map such that

$$I(f)(x) = \int_0^x f(t)\,dt \qquad (f \in \mathcal{C},\ x \in \mathbb{R}).$$

Is I injective? Is it surjective?

D. Let $f\colon X \to Y$ be a map from a set X to a set Y. Consider the map

$$f^*\colon \mathcal{P}(Y) \to \mathcal{P}(X)$$
$$B \mapsto f^{-1}\langle B\rangle \qquad (B \in \mathcal{P}(Y)).$$

Prove that f^* is surjective (respectively, injective) if and only if f is injective (respectively, surjective).

E. Is there an analog of Exercise D for the map

$$f_*\colon \mathcal{P}(X) \to \mathcal{P}(Y)$$
$$A \mapsto f\langle A\rangle \qquad (A \in \mathcal{P}(X))?$$

F. Let $f\colon X \to Y$ be a map. Prove that each of the following conditions is necessary for f to be injective:
(a) $E \subset X \Rightarrow f^{-1}\langle f\langle E\rangle\rangle = E$.
(b) $A, B \subset X \Rightarrow f\langle A \cap B\rangle = f\langle A\rangle \cap f\langle B\rangle$.
(c) $A, B \subset X \Rightarrow f\langle A \setminus B\rangle = f\langle A\rangle \setminus f\langle B\rangle$.
(d) $(A_i \mid i \in I)$ a nonempty family of subsets of X $\Rightarrow f\langle \bigcap_{i\in I} A_i\rangle = \bigcap_{i\in I} f\langle A_i\rangle$.

G. Prove that each of the conditions (a)–(d) in Exercise F is also sufficient for f to be injective.

H. Exhibit all permutations of $\{0,1\}$ and of $\{0,1,2\}$.

I. Construct a one-to-one correspondence between ω and $\omega \setminus E$ in case:
(a) $E = \{n\}$ for some $n \in \omega$.
(b) $E = \{m,n\}$ for distinct $m, n \in \omega$.

J. Let X and Y be classes, with X nonempty. When is the projection $p\colon X \times Y \to Y$ bijective?

K. Let $f\colon X \to Y$. Show that the following condition is both necessary and sufficient for f to be injective: If $g_1, g_2\colon Z \to X$ and $f \circ g_1 = f \circ g_2$, then $g_1 = g_2$ (that is, f is *left cancellable*).

L. Formulate and prove the analog of Exercise K for surjections.

M. Let $g\colon Z \to Y$ be injective, and let $f\colon X \to Y$. Prove that the following two statements are equivalent:
(i) There exists a map h making the following diagram commute:

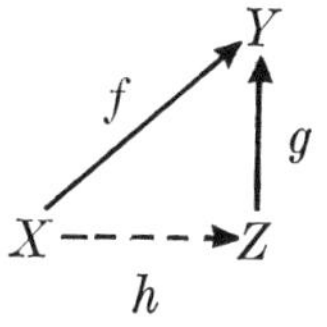

(ii) $f\langle X\rangle \subset g\langle Z\rangle$.

Prove, moreover, that an h satisfying (i) is necessarily unique.

N. Let $g: Y \to Z$ be surjective and let $f: Y \to X$. Prove that the following statements are equivalent:

(i) There exists a map h such that the following diagram commutes:

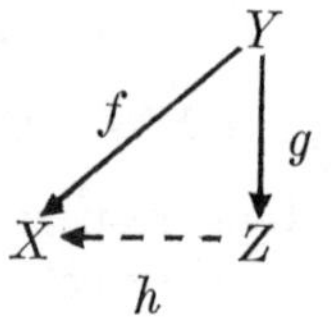

(ii) If $y_1, y_2 \in Y$ and $g(y_1) = g(y_2)$, then $f(y_1) = f(y_2)$.

Prove, moreover, that an h satisfying (i) is unique.

P. (a) Show that a map having a unique section must be bijective.

(b) Must a map having a unique retraction be bijective?

Q. Let $f: X \to X'$, $g: Y \to Y'$ be given maps. Consider the map

$$\varphi: \text{Map } (X',Y) \to \text{Map } (X,Y')$$

defined in 8.22. Prove that φ is injective (respectively, surjective, bijective) if f is surjective (respectively, injective, bijective) and g is injective (respectively, surjective, bijective).

R. Deduce 9.21 from 9.25.

S. Does there exist a choice class for $\mathcal{U} \setminus \{\varnothing\}$?

T. Deduce 9.21 from 9.28.

U. Without using Axiom 8, deduce the statement of Axiom 8 from the statement of 9.28. [*Hint*: If $\mathcal{A}$ is a class of nonempty sets, consider the class $\{\{A\} \times A \mid A \in \mathcal{A}\}$.]

V. Without using Axiom 8, deduce 9.28 from 9.21.

W. Let X be a set. Suppose $\pi: \mathbb{R} \times X \to X$ is a map such that $x \in X$ and $t, s \in \mathbb{R}$ implies

(i) $\pi(0,x) = x$,

(ii) $\pi(t, \pi(s,x)) = \pi(t + s, x)$.

For each $t \in \mathbb{R}$ consider the map

$$\pi_t: X \to X$$
$$x \mapsto \pi(x,t) \qquad (x \in X).$$

Show that $\pi_{t+s} = \pi_t \circ \pi_s$ for all $t, s \in \mathbb{R}$, and show that π_t is a permutation of X for each $x \in \mathbb{R}$. (The map $t \mapsto \pi_t$ of $\mathbb{R} \to$ Map (X,X) is called a *one-parameter group of permutations of* X; the map π itself is called a *flow on* X.)

X. Let X be a set. Establish a one-to-one correspondence between the set of all permutations of X and the set of all maps

π: $\mathbb{Z} \times X \to X$ satisfying conditions (i) and (ii) of Exercise W for all $x \in X$ and all t, $s \in \mathbb{Z}$; here $\mathbb{Z}$ is the set of all integers.

Y. Let I denote the closed unit interval $[0,1]$ in $\mathbb{R}$. If (X,x) is a pointed set (see Chapter 8, Exercise GG), a map $\sigma: I \to X$ for which $\sigma(0) = \sigma(1) = x$ is called a "loop at x in X," and the set of all loops at x in X is denoted $\Omega(X,x)$.

(a) Associate to each base-point preserving map f: $(X,x) \to (Y,y)$ of pointed sets a map

$$\Omega(f): \Omega(X,x) \to \Omega(Y,y)$$

in such a way that the following hold: For any pointed set (X,x), $\Omega(1_X)$ is the identity map of $\Omega(X,x)$. For any base-point preserving maps

$$f: (X,x) \to (Y,y), \qquad g: (Y,y) \to (Z,z)$$

we have

$$\Omega(g \circ f) = \Omega(g) \circ \Omega(f).$$

(b) If the base-point preserving map f: $(X,x) \to (Y,y)$ is injective (surjective), what can be said of $\Omega(f)$?

(c) Given pointed sets (X,x) and (Y,y), find an injection of $\Omega(X,x) \times \Omega(Y,y)$ into $\Omega(X \times Y, (x,y))$.

Z. Let (X,x) be a pointed set, and let τ be the "trivial" loop at x in X, that is, the constant map on I to X with value x. Establish a one-to-one correspondence between $\Omega(\Omega(X,x),\tau)$ and a suitable subset of Map $(I \times I,X)$.

10. Recursion

In this chapter a method for generating functions on ω is discussed. As an application, the arithmetic operations on the natural numbers are defined.

ORDINARY RECURSION AND ADDITION

Consider the problem of defining addition of natural numbers, that is, of assigning to each pair (m,n) of natural numbers a "sum" $m + n$. Fix a particular $m \in \omega$. We require that the assignment to each $n \in \omega$ of its sum $m + n$ with m satisfy the following conditions. First $m + 0 = m$. Next, $m + 1$ should be the first natural number greater than m, that is, $m + 1 = m^+$. Then $m + 2$ should be the second natural number greater than m, and hence the first natural number greater than $m + 1$, that is, $m + 2 = (m + 1)^+$. And so on. Thus we require

$$(*) \qquad \begin{cases} m + 0 = m, \\ m + n^+ = (m + n)^+ \end{cases} \qquad (n \in \omega).$$

It is tempting to believe that our problem is solved, for we can define $m + 0 = m$, and if $n \in \omega$ with $n \neq 0$, then $n = k^+$ for a unique $k \in \omega$ and we can define $m + n = (m + k)^+$. However, a difficulty arises here similar to the one encountered when we attempted to define natural numbers. The "definition" does not determine $m + n$ for every $n \in \omega$, but merely allows us to define $m + n$ only in case we have already defined $m + k$ for all $k < n$.

In order to overcome this difficulty, we reformulate our goal. To assign to each $n \in \omega$ its sum $m + n$ with m is to define a map $f: \omega \to \omega$ with $f(n) = m + n$ for each $n \in \omega$. Stated in terms of f, requirement (*) reads

$$(**) \qquad \begin{cases} f(0) = m, & \\ f(n^+) = (f(n))^+ & (n \in \omega). \end{cases}$$

Our problem is thus reduced to establishing the existence of a unique map $f: \omega \to \omega$ satisfying (**). If $G: \omega \to \omega$ is the map such that $G(n) = n^+$ for each $n \in \omega$, then we seek a map $f: \omega \to \omega$ such that

$$(***) \qquad \begin{cases} f(0) = m, & \\ f(n^+) = G(f(n)) & (n \in \omega). \end{cases}$$

The theorem below establishes the existence of a unique f satisfying (***).

10.1. THEOREM (ORDINARY RECURSION)

Let

$$G: X \to X$$

be a given map on a class X to itself, and let

$$z \in X.$$

Then there exists a unique map

$$f: \omega \to X$$

such that

$$(*) \qquad \begin{cases} f(0) = z, & \\ f(n^+) = G(f(n)) & (n \in \omega). \end{cases}$$

Proof

Uniqueness

Suppose maps

$$f_1: \omega \to X, \qquad f_2: \omega \to X$$

both satisfy (*). To show $f_1 = f_2$, we use induction on n to prove $f_1(n) = f_2(n)$ for all $n \in \omega$. First, $f_1(0) = z = f_2(0)$. If $n \in \omega$ and if $f_1(n) = f_2(n)$ is assumed, then

$$f_1(n^+) = G(f_1(n)) = G(f_2(n)) = f_2(n^+).$$

Existence

To construct an f satisfying (*), we construct for each $n \in \omega$ the restriction

$$f \mid n^+ : n^+ \to X$$

of f to $n^+ = \{0,1,\ldots,n\}$. For each $n \in \omega$, define M_n to be the class of all maps $h: n^+ \to X$ such that

$$h(0) = z,$$

$$h(k^+) = G(h(k)) \qquad (k \in \omega, k < n).$$

Then a map $f: \omega \to X$ satisfies (*) if and only if $f \mid n^+ \in M_n$ for each $n \in \omega$.
Note that

(1) $$h \in M_{n^+} \Rightarrow h \mid n^+ \in M_n \qquad (n \in \omega).$$

Induction on n with the aid of (1) shows

(2) $$h \in M_n \Rightarrow h \mid m^+ \in M_m \qquad (n, m \in \omega; m < n).$$

We use induction on n to show that for each $n \in \omega$, M_n has a *unique* member

$$h_n: n^+ \to X.$$

There is exactly one member of M_0, namely, the constant map $h_0: \{0\} \to X$ with value z. Now let $n \in \omega$ and assume M_n has a unique member h_n. Then the map

$$h_{n^+}: \{0,\ldots,n,n^+\} \to X$$

having $(n^+)^+$ as its domain and defined by

$$h_{n^+} \mid \{0,\ldots,n\} = h_n, \qquad h_{n^+}(n^+) = G(h_n(n))$$

certainly is a member of M_{n^+}. Moreover, h_{n^+} is the only member of M_{n^+}. In fact, if $h \in M_{n^+}$, then $h \mid n^+ \in M_n$ by (1), $h \mid n^+ = h_n$ by the inductive hypothesis, and

$$h(n^+) = G(h(n)) = G(h_n(n)) = h_{n^+}(n^+),$$

so $h = h_{n^+}$.

If $m, n \in \omega$ with $m \neq n$, say $m < n$, then $h_n \in M_n$, $h_n \mid m^+ \in M_m$ by (2), and

$$h_n \mid \operatorname{dmn} h_m = h_m$$

since h_m is the only member of M_m. Now

$$\bigcup_{n \in \omega} \operatorname{dmn} h_n = \bigcup_{n \in \omega} n^+ = \omega.$$

By 8.28, there exists a necessarily unique map

$$f\colon \omega \to X$$

such that

$$f \mid n^+ = h_n \qquad (n \in \omega).$$

Then $f \mid n^+ \in M_n$ for each $n \in \omega$, and so f satisfies (*). ∎

For the remainder of this section,

$$\sigma\colon \omega \to \omega$$

will denote the successor map of ω.

The condition

$$f(n^+) = G(f(n)) \qquad (n \in \omega)$$

in 10.1 simply asserts the commutativity of the diagram

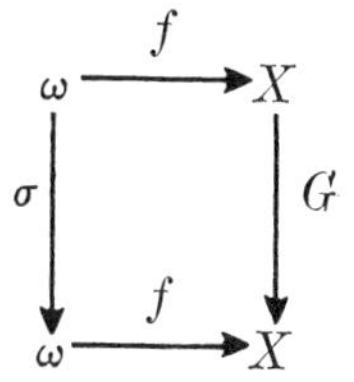

A function whose unique existence is guaranteed by 10.1, or by the recursion theorems stated below, is said "to be defined" *by recursion* or *recursively* (or *by iteration* or even *by induction*).

When desirable, the initial condition $f(0) = z$ in 10.1 may be replaced by $f(m) = z$ for some $m \in \omega$, and then the recursive condition required is

$$f(n^+) = G(f(n)) \qquad (n \in \omega,\ n \geq m).$$

The definition of addition of natural numbers may now be formalized. If $m \in \omega$, then by taking for X in 10.1 the set ω, for G the map σ, and for z the element m of X, we obtain the existence of a unique map

$$s_m\colon \omega \to \omega$$

such that

$$s_m(0) = m$$

$$s_m(n^+) = (s_m(n))^+ \qquad (n \in \omega).$$

10.2. DEFINITION

If $m, n \in \omega$, then the *sum of m and n*, denoted by $m + n$, is defined to be the natural number $s_m(n)$.

The map

$$s: \omega \times \omega \to \omega$$

$$(m,n) \mapsto m + n$$

is called *addition* (*of natural numbers*).

The map s is an "operation":

10.3. DEFINITION

If X is a class, then a map on $X \times X$ to X is called a (*binary*) *operation on* X.

In algebraic contexts involving this operation of addition (or the operation of multiplication defined below), the set ω of natural numbers is customarily denoted by $\mathbf{N}$.

According to the defining properties of the maps s_m,

$$m + 0 = m$$

$$m + n^+ = (m + n)^+$$

for all m, $n \in \omega$. In particular, $n + 1 = n + 0^+ = (n + 0)^+ = n^+$, so

$$n + 1 = n^+$$

for every $n \in \omega$. Then the defining properties of the maps s_m may be written

$$m + 0 = m,$$

$$m + (n + 1) = (m + n) + 1$$

for all m, $n \in \omega$.

The fundamental properties of addition are established next.

10.4. PROPOSITION

Let m, n, $k \in \omega$. Then:

(1) (identity element) $m + 0 = m$.
(2) (associative law) $(m + n) + k = m + (n + k)$.
(3) (commutative law) $m + n = n + m$.
(4) If $m < n$, then $m + k < n + k$; if $m \leq n$ then $m + k \leq n + k$.
(5) If $m + k = n + k$, then $m = n$.
(6) If $m \leq n$, then there exists a unique $d \in \omega$ such that $m + d = n$.

Proof

(2) Fix $m, n \in \omega$, and use induction on k. For $k = 0$,

$$(m + n) + 0 = m + n = m + (n + 0).$$

Now let $k \in \omega$ and assume $(m + n) + k = m + (n + k)$. Then

$$\begin{aligned}(m + n) + (k + 1) &= ((m + n) + k) + 1\\ &= (m + (n + k)) + 1\\ &= m + ((n + k) + 1)\\ &= m + (n + (k + 1)).\end{aligned}$$

(3) Easy inductions on n show first

(3.1) $$0 + n = n \qquad (n \in \omega),$$

(3.2) $$(m + 1) + n = (m + n) + 1 \qquad (m, n \in \omega).$$

To prove (3), fix n and use induction on m.

(4) Suppose $m < n$. We use induction on k to show $m + k < n + k$. The case $k = 0$ follows from (1). Let $k \in \omega$ and assume $m + k < n + k$. Then

$$(m + k) + 1 = (m + k)^+ \leq n + k < (n + k) + 1,$$

so

$$m + (k + 1) = (m + k) + 1 < (n + k) + 1 = n + (k + 1).$$

(5) follows from (4).

(6) If $m + d_1 = n$ and $m + d_2 = n$, then $d_1 = d_2$ by (5). This establishes uniqueness.

Fix $n \in \omega$. We use induction on m to show

$$m \leq n \Rightarrow (\exists d \in \omega)(m + d = n).$$

If $m = 0$ (and automatically $m \leq n$), $d = n$ will do. Let $m \in \omega$ and assume $m \leq n$ implies $m + d = n$ for some $d \in \omega$. Suppose $m + 1 \leq n$. Then $m \leq n$, and by the inductive hypothesis, $m + d = n$ for some $d \in \omega$. Now $d \neq 0$, for otherwise $n = m + 0 < m + 1$. There exists $h \in \omega$ with $h + 1 = d$. Then

$$\begin{aligned}(m + 1) + h &= (m + h) + 1 = m + (h + 1)\\ &= m + d = n. \quad \square\end{aligned}$$

The associative law (2) allows one to write $m + n + k$ unambiguously for both $(m + n) + k$ and $m + (n + k)$.

Property (6) permits a restricted "subtraction" of natural numbers.

10.5. DEFINITION

If $m, n \in \omega$ with $m \leq n$, then the unique $d \in \omega$ such that $m + d = n$ is called the *difference between n and m* and is denoted by $n - m$; thus

$$m + (n - m) = n.$$

Notice that $n - m$ is defined here only when $m \leq n$.

Let $n \in \omega$ with $n > 0$. Then $n - 1$ is the greatest natural number less than n. In fact, $n - 1 \geq n$ implies $n = (n - 1) + 1 \geq n + 1$, so $n - 1 < n$. If $k \in \omega$ with $k < n$, then $k > n - 1$ implies $n = (n - 1) + 1 < k + 1$ and $n \leq k$, so $k \leq n - 1$.

ITERATION OF MAPS

Before proceeding to multiplication, we give an important non-arithmetic application of recursion.

Let $f: X \to X$ be a map with the same set X as domain and codomain. For each $n \in \omega$, we want to define the nth power f^n of f in such a way that

$$f^n = \underbrace{f \circ f \circ \ldots \circ f}_{n \text{ times}}$$

is obtained by composing f with itself n times. Then we require

$$f^1 = f, \quad f^2 = f \circ f = f \circ f^1, \quad f^3 = f \circ f \circ f = f \circ f^2, \quad \ldots,$$

in other words,

$$(*) \qquad f^{n+1} = f \circ f^n$$

for each $n \geq 1$. If we agree to take $f^0 = 1_X$, then we require (*) to hold for every $n \in \omega$.

To construct all the maps f^n, $n \in \omega$, let

$$G: \text{Map }(X,X) \to \text{Map }(X,X)$$

be the map defined by

$$G(h) = f \circ h \qquad (h \in \text{Map }(X,X)).$$

By 10.1, there exists a unique map

$$I_f: \omega \to \text{Map }(X,X)$$

such that

$$I_f(0) = 1_X,$$

$$I_f(n + 1) = G(I_f(n)) = f \circ I_f(n) \qquad (n \in \omega).$$

10.6. DEFINITION

If $f: X \to X$ is a map with the same set X as domain and codomain, then for each $n \in \omega$ the *nth iterate* (or *power*) *of* f, denoted by f^n, is defined to be the map $I_f(n)$ of X into X.

The defining properties of I_f say

$$f^0 = 1_X,$$
$$f^{n+1} = f \circ f^n \qquad (n \in \omega).$$

In particular,

$$f^1 = f.$$

If X is a set, then $(1_X)^n = 1_X$ for each $n \in \omega$.

10.7. EXAMPLE

The maps $s_m: \omega \to \omega$ used to define addition are just the iterates of σ, namely,

$$s_m = \sigma^m \qquad (m \in \omega).$$

This is easily proved by induction on m.

This example shows that

$$m + n = \sigma^m(n) \qquad (m, n \in \omega).$$

Hence addition could have been defined solely in terms of iterates of σ.

10.8. THEOREM

Let $f: X \to X$ be a map, where X is a set. Then

(1) $$f^{m+n} = f^m \circ f^n \qquad (m, n \in \omega).$$

(2) $$f^m \circ f^n = f^n \circ f^m \qquad (m, n \in \omega).$$

If $g: X \to X$ is another map and if g "commutes with" f, that is,

$$g \circ f = f \circ g,$$

then

(3) $$g^m \circ f^n = f^n \circ g^m \qquad (m, n \in \omega).$$

(4) $$(g \circ f)^m = g^m \circ f^m \qquad (m \in \omega).$$

Proof

Suppose $g: X \to X$ satisfies $g \circ f = f \circ g$. By induction on m we prove first

(5) $$g^m \circ f = f \circ g^m \qquad (m \in \omega).$$

Clearly $g^0 \circ f = f = f \circ g^0$. If $m \in \omega$ and if one assumes $g^m \circ f = f \circ g^m$, then

$$\begin{aligned} g^{m+1} \circ f &= (g \circ g^m) \circ f = g \circ (g^m \circ f) = g \circ (f \circ g^m) \\ &= (g \circ f) \circ g^m = (f \circ g) \circ g^m = f \circ (g \circ g^m) \\ &= f \circ g^{m+1}. \end{aligned}$$

From (5), statements (3) and (4) now follow by induction on n and m, respectively.

For fixed $m \in \omega$, we use induction on n to prove (1). Clearly $f^{m+0} = f^m \circ f^0$. Let $n \in \omega$ and assume $f^{m+n} = f^m \circ f^n$. Taking $g = f$ in (5), we obtain $f \circ f^m = f^m \circ f$. Then

$$\begin{aligned} f^{m+(n+1)} &= f^{(m+n)+1} = f \circ f^{m+n} = f \circ (f^m \circ f^n) \\ &= (f \circ f^m) \circ f^n = (f^m \circ f) \circ f^n = f^m \circ (f \circ f^n) \\ &= f^m \circ f^{n+1}. \end{aligned}$$

Statement (2) follows either from (1) or from (3). □

Statement (2) says, in particular,

$$f \circ f^n = f^n \circ f \qquad (n \in \omega).$$

Hence the iterates of f could equally well have been defined by taking for G in 10.1 the map

$$h \mapsto h \circ f.$$

MULTIPLICATION AND EXPONENTIATION

Consider now the problem of defining multiplication on ω. To each pair (m,n) of natural numbers we wish to assign a product $m \cdot n$ in such a way that $0 \cdot n = 0$, $1 \cdot n = n$, $2 \cdot n = n + n$, and in general

$$m \cdot n = \underbrace{n + n + \ldots + n}_{m \text{ times}}$$

is obtained by adding n to itself m times. Now

$$\begin{aligned} 0 &= s_n{}^0(0), \\ n &= s_n{}^1(0), \\ n + n &= s_n{}^2(0), \\ n + n + n &= s_n{}^3(0), \end{aligned}$$

and so on. This suggests the appropriate definition.

10.9. DEFINITION

If $m, n \in \omega$, then the *product of m and n*, denoted by $m \cdot n$ or simply mn, is defined to be the natural number $s_n{}^m(0)$.

The operation

$$(m,n) \mapsto m \cdot n$$

on ω is called *multiplication*.

Since $s_n = \sigma^n$, the nth iterate of the successor map σ, we have

$$mn = (\sigma^n)^m(0).$$

Thus multiplication could have been defined solely in terms of iterates of σ.

10.10. LEMMA

Let $m, n \in \omega$. Then:

(1) $m \cdot 0 = 0$.
(2) $m(n + 1) = mn + m$.

Proof

(1) $m \cdot 0 = (\sigma^0)^m(0) = (1_\omega)^m(0) = 1_\omega(0) = 0$.
(2) We have

$$\begin{aligned} m(n+1) &= (\sigma^{n+1})^m(0) = (\sigma \circ \sigma^n)^m(0) = (\sigma^m \circ (\sigma^n)^m)(0) \\ &= \sigma^m((\sigma^n)^m(0)) = \sigma^m(mn) = m + mn \\ &= mn + m. \quad \square \end{aligned}$$

With the aid of this lemma, easy inductions establish the standard properties of multiplication:

10.11. PROPOSITION

Let $m, n, k \in \omega$. Then:

(1) (identity element) $1 \cdot n = n$.
(2) (associative law) $(mn)k = m(nk)$.
(3) (commutative law) $mn = nm$.
(4) (distributive law) $m(n + k) = mn + mk$.
(5) If $mn = 0$, then $m = 0$ or $n = 0$.
(6) If $m < n$ and $k \neq 0$, then $mk < nk$.
(7) (cancellation law) If $mk = nk$ and $k \neq 0$, then $m = n$.

The cancellation law (7) says that if $m, n \in \omega$ with $n \neq 0$, then there is at most one $k \in \omega$ for which $m = nk$.

10.12. DEFINITION

Let $m, n \in \omega$ with $n \neq 0$. If $m = nk$ for some (necessarily unique) $k \in \omega$, then one writes $n \mid m$ and says that n *divides* m. If $n \mid m$, then the unique $k \in \omega$ such that $m = nk$ is called the *quotient of* m *by* n and is denoted $\frac{m}{n}$.

The reader should verify the following laws for quotients, which hold for all $m, n, p, q, k \in \omega$ with $n \neq 0$, $q \neq 0$, $k \neq 0$:

(1) If $n \mid m$, then $kn \mid km$ and

$$\frac{km}{kn} = \frac{m}{n}.$$

(2) If $n \mid m$ and $q \mid p$, then $nq \mid (mq + pn)$ and

$$\frac{m}{n} + \frac{p}{q} = \frac{mq + pn}{nq}.$$

(3) If $n \mid m$ and $q \mid p$, then $nq \mid mp$ and

$$\frac{m}{n} \cdot \frac{p}{q} = \frac{mp}{nq}.$$

(4) If $n \mid m$ and $n \mid p$, then $n \mid (m + p)$ and

$$\frac{m + p}{n} = \frac{m}{n} + \frac{p}{n}.$$

Lemma 10.10 may also be used to prove properties of iterates supplementing those stated in 10.8.

10.13. THEOREM

Let $f: X \to X$ be a map, where X is a set. Then

(1) $f^{mn} = (f^m)^n$,
(2) $(f^m)^n = (f^n)^m$,

for all $m, n \in \omega$.

Proof

Fix $m \in \omega$. With the aid of 10.10 and 10.8, we use induction on n to establish both (1) and (2).

(1) First

$$f^{m \cdot 0} = f^0 = 1_X = (f^m)^0.$$

Now let $n \in \omega$ and assume $f^{mn} = (f^m)^n$. Then

$$f^{m(n+1)} = f^{mn+m} = f^{mn} \circ f^m = (f^m)^n \circ f^m = (f^m)^{n+1}.$$

(2) Clearly $(f^m)^0 = (f^0)^m$. Let $n \in \omega$ and assume $(f^m)^n = (f^n)^m$. Then

$$(f^m)^{n+1} = (f^m)^n \circ f^m = (f^n)^m \circ f^m = (f^n \circ f)^m$$
$$= (f^{n+1})^m. \quad \square$$

The final arithmetic operation we define is exponentiation. If $m, n \in \omega$, we want to define the nth power of m as

$$m^n = \underbrace{m \cdot m \cdot \ldots \cdot m}_{n \text{ times}},$$

the product of m with itself n times. Let p_m be multiplication by m, that is,

$$p_m(k) = mk \qquad (m, k \in \omega).$$

Then we require

$$m^1 = m = p_m(1),$$
$$m^2 = m \cdot m = p_m^2(1),$$
$$m^3 = m \cdot m \cdot m = p_m^3(1),$$

and so on.

10.14. DEFINITION

If $m, n \in \omega$, then the *nth power of m*, denoted by m^n, is defined to be $p_m^n(1)$.

The binary operation

$$(m,n) \mapsto m^n$$

on ω is called *exponentiation.*

10.15. PROPOSITION (LAWS OF EXPONENTS)

Let $m, n, k \in \omega$. Then:

(1) $m^0 = 1$; in particular, $0^0 = 1$.
(2) If $n \neq 0$, then $0^n = 0$.
(3) $m^1 = m$.
(4) $1^n = 1$.
(5) $m^{n+k} = m^n \cdot m^k$.
(6) $m^{nk} = (m^n)^k = (m^k)^n$.
(7) $(mn)^k = m^k \cdot n^k$.

SIMULTANEOUS AND PRIMITIVE RECURSION

We turn now to a form of recursion more complicated than 10.1. Consider the "Fibonacci sequence"

$$a = (0,1,1,2,3,5,8,13,\ldots)$$

given by

$$(*) \qquad a_0 = 0, \qquad a_1 = 1, \qquad a_{n+1} = a_n + a_{n-1} \qquad (n \geq 1).$$

How do we know such a sequence actually exists? Ordinary recursion cannot directly justify its existence, for the value a_{n+1} for $n \geq 1$ depends not only upon a_n but upon a_{n-1} as well. Now a solution of (*), if it exists, also satisfies

$$(**) \qquad \begin{cases} a_0 = 0, & a_{n+1} = a_n + b_n \quad (n \geq 0) \\ b_0 = 1, & b_{n+1} = a_n \quad (n \geq 0). \end{cases}$$

We have complicated the problem of justifying the existence of a by reducing it to the problem of justifying the simultaneous existence of two sequences a and b! However, this complication suggests a general result which solves the original problem. Indeed, condition (**) may be written

$$\begin{cases} a_0 = 0, \qquad b_0 = 1, \\ a_{n+1} = H(a_n,b_n) \qquad (n \in \omega), \\ b_{n+1} = K(a_n,b_n) \qquad (n \in \omega), \end{cases}$$

where $H: \omega \times \omega \to \omega$ is the operation of addition and $K: \omega \times \omega \to \omega$ is the first projection.

10.16. THEOREM (SIMULTANEOUS RECURSION)

Let X and Y be classes, let $z \in X$ and $w \in Y$, and let

$$H: X \times Y \to X, \qquad K: X \times Y \to Y$$

be maps. Then there exist unique maps

$$f: \omega \to X, \qquad g: \omega \to Y$$

such that

$$f(0) = z, \qquad g(0) = w$$

$$f(n+1) = H(f(n),g(n)) \qquad (n \in \omega),$$

$$g(n+1) = K(f(n),g(n)) \qquad (n \in \omega).$$

Proof

Define

$$G: X \times Y \to X \times Y$$

to be the natural map induced by H and K, whence

$$G(x,y) = (H(x,y), K(x,y)) \qquad (x \in X, y \in Y).$$

Let

$$p: X \times Y \to X, \qquad q: X \times Y \to Y$$

be the projections. Now any two maps $f: \omega \to X$, $g: \omega \to Y$ determine a map $F: \omega \to X \times Y$ given by

$$(*) \qquad\qquad F(n) = (f(n), g(n)) \qquad\qquad (n \in \omega);$$

conversely, for any map $F: \omega \to X \times Y$, the maps $f = p \circ F$, $g = q \circ F$ satisfy (*). Hence the desired conclusion is equivalent to the existence of a map $F: \omega \to X \times Y$ satisfying

$$\begin{cases} F(0) = (z,w), \\ F(n+1) = G(F(n)) \qquad (n \in \omega). \end{cases}$$

The unique existence of such an F follows from 10.1. $\square$

10.17. COROLLARY (PRIMITIVE RECURSION)

Let X be a class, let $z \in X$, and let

$$G: X \times \omega \to X$$

be a given map. Then there exists a unique map

$$f: \omega \to X$$

such that

$$f(0) = z,$$

$$f(n+1) = G(f(n), n) \qquad\qquad (n \in \omega).$$

Proof

In 10.16 take $Y = \omega$, $w = 0$, $H = G$, and K the map defined by

$$K(x,n) = n + 1 \qquad\qquad (x \in X, n \in \omega).$$

There exist unique maps $f: \omega \to X$, $g: \omega \to \omega$ satisfying

$$f(0) = z, \qquad g(0) = 0,$$

$$f(n+1) = G(f(n), g(n)) \qquad\qquad (n \in \omega),$$

$$g(n+1) = g(n) + 1 \qquad\qquad (n \in \omega).$$

Now 1_ω satisfies

$$1_\omega(0) = 0, \qquad 1_\omega(n+1) = 1_\omega(n) + 1 \qquad\qquad (n \in \omega).$$

By the uniqueness assertion of 10.1, $g = 1_\omega$. Hence

$$f(n+1) = G(f(n), n) \qquad\qquad (n \in \omega). \quad \square$$

As an example of primitive recursion, consider the problem of constructing a map $F: \omega \to \omega$ such that

$$F(n) = 1 \cdot 2 \cdot \ldots \cdot (n-1) \cdot n,$$

the product of the first n nonzero natural numbers, for each $n > 0$. By requiring also $F(0) = 1$, F is to satisfy

$$F(0) = 1,$$

$$F(n+1) = (n+1) \cdot F(n) \qquad (n \in \omega).$$

That a unique F satisfying these conditions exists follows from the corollary: Take $X = \omega$, $z = 1$, and $G: \omega \times \omega \to \omega$ the map defined by

$$G(m,n) = (n+1) \cdot m \qquad (m, n \in \omega).$$

10.18. DEFINITION

For each $n \in \omega$, the natural number $F(n)$ is called *factorial* n and is denoted by $n!$.

Thus,

$$0! = 1,$$

$$(n+1)! = (n+1) \cdot (n!) \qquad (n \in \omega).$$

Here is an elementary property of the factorial function.

10.19. LEMMA

If $m, n \in \omega$ with $m \leq n$, then

$$m!(n-m)! \mid n!.$$

Proof

We use induction on n. For $n = 0$, $m \in \omega$ and $m \leq n$ implies $m = 0$, so $m!(n-m)! = (0!)(0!) = 1 = 0! = n!$.

Let $n \in \omega$. Assume

$$k \in \omega \;\&\; k \leq n \Rightarrow k!(n-k)! \mid n!.$$

Let $m \in \omega$ with $m \leq n+1$. We must show that $m!(n+1-m)! \mid (n+1)!$. If $m = 0$ or $m = n+1$, this is obvious. Suppose $0 < m < n+1$. Since $m \leq n$, the inductive hypothesis yields

$$m!(n-m)! \mid n!,$$

so

$$(*) \qquad m!(n+1-m)! \mid (n!)(n+1-m).$$

Since $m - 1 \leq n$, the inductive hypothesis also yields

$$(m-1)!(n-(m-1))! \mid n!,$$

so

$$(**) \qquad m!(n+1-m)! \mid (n!)m.$$

Since

$$(n!)(n+1-m) + (n!)m = n!(n+1) = (n+1)!,$$

(*) and (**) together show

$$m!(n+1-m)! \mid (n+1)!. \quad \square$$

This lemma justifies the following definition.

10.20. DEFINITION

For $m, n \in \omega$ with $m \leq n$, the *binomial coefficient n above m*, denoted by $\binom{n}{m}$, is defined to be the natural number

$$\frac{n!}{m!(n-m)!}.$$

This definition will be used later in counting the number of elements in certain sets. Note that

$$\binom{n}{0} = 1, \quad \binom{n}{n} = 1,$$

and if $n > 0$,

$$\binom{n}{1} = n.$$

GENERALIZED SUMS AND PRODUCTS

As a final application of recursion, we define generalized sums and products of more than two numbers, thereby giving formal status to such expressions as $n + n + \ldots + n$ and $1 \cdot 2 \cdot \ldots \cdot n$ used informally above.

For each $n \geq 1$, we want to assign to each n-tuple $(x_1, \ldots, x_n)$ of natural numbers a generalized sum $x_1 + \ldots + x_n$ and a generalized product $x_1 \cdot \ldots \cdot x_n$. In particular, the generalized sum and product of a 2-tuple (x_1, x_2) are to be the ordinary sum $x_1 + x_2$ and product $x_1 \cdot x_2$ as defined earlier. Once generalized sums and products of arbitrary n-tuples have been defined, then the generalized sum and product of a given $(n + 1)$-tuple

$(x_1,\ldots,x_n,x_{n+1})$ are to be expressed in terms of the generalized sum and product of the n-tuple $(x_1,\ldots,x_n)$ by

$$x_1 + \ldots + x_{n+1} = (x_1 + \ldots + x_n) + x_{n+1},$$

$$x_1 \cdot \ldots \cdot x_{n+1} = (x_1 \cdot \ldots \cdot x_n) \cdot x_{n+1}.$$

The procedure suggested here can be carried out just as easily if we start with any binary operation.

10.21. DEFINITION

Let $n \geq 1$ be a natural number. If X is a class, then a map

$$f: X^n \to X$$

is called an *n-ary operation on X.*

Any 2-ary operation on a class X may be regarded as a binary operation on X, and vice versa. More precisely, there is a one-to-one correspondence $f \mapsto \bar{f}$ between the class of all 2-ary operations on X and the class of all binary operations on X, given as follows. Consider the bijection

$$k: X^2 \simeq X \times X.$$

If f is a 2-ary operation on X, then the corresponding binary operation $\bar{f}: X \times X \to X$ is the unique operation making the following triangle commute:

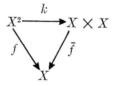

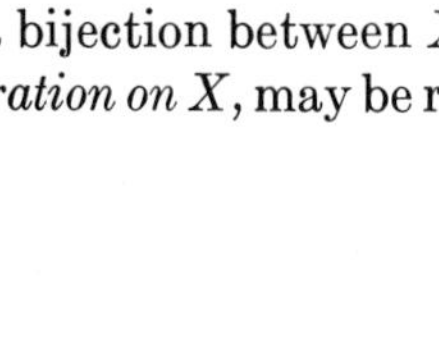

Similarly, since we have a bijection between X^1 and X, a 1-ary operation on X, also called a *unary operation on X*, may be regarded as a map $f: X \to X$.

10.22. THEOREM

Let

$$*: X \times X \to X$$

$$(x,y) \mapsto x * y \qquad (x, y \in X)$$

be a given binary operation on a set X. Then there exists a unique family $(A_n \mid n = 1,2,\ldots)$ with

$$A_n: X^n \to X$$

an n-ary operation on X for each $n \geq 1$, such that:

$$A_1(x_1) = x_1 \qquad ((x_1) \in X^1),$$

$$A_2(x_1,x_2) = x_1 * x_2 \qquad ((x_1,x_2) \in X^2),$$

$$A_{n+1}(x_1,\ldots,x_n,x_{n+1}) = A_n(x_1,\ldots,x_n) * x_{n+1}$$
$$((x_1,\ldots,x_n,x_{n+1}) \in X^{n+1},\ n \geq 2).$$

Proof

For $n \geq 1$ define

$$D_n = \{i \mid i \in \omega \,\&\, 1 \leq i \leq n\},$$

so that D_n is the domain of any element of X^n. The requirements for the A_n's may be written

$$(') \qquad A_1(x) = x_1 \qquad (x \in X^1),$$

$$('') \qquad A_{n+1}(x) = A_n(x \mid D_n) * x_{n+1} \qquad (x \in X^{n+1},\ n \geq 1).$$

In fact, for $n \geq 1$,

$$x = (x_1,\ldots,x_n,x_{n+1}) \in X^{n+1} \Rightarrow x \mid D_n = (x_1,\ldots,x_n) \in X^n.$$

Then the stated requirement for A_{n+1} takes the form ($''$) in case $n \geq 2$; since ($'$) yields

$$x_1 * x_2 = A_1(x \mid D_1) * x_2 \qquad (x \in X^2),$$

the stated requirement for A_2 takes the form ($''$) for $n = 1$ as well.

Since an n-ary operation on X is just an element of Map (X^n,X), the desired A_n's form a family in the set

$$Y = \bigcup_{n=1}^{\infty} \text{Map } (X^n,X).$$

Construct a map

$$g: Y \to Y$$

assigning to each n-ary operation on X an $(n + 1)$-ary operation on X as follows. If $H \in$ Map (X^n,X) for a certain $n \geq 1$, then

$$H: X^n \to X,$$

and we define

$$g(H): X^{n+1} \to X$$

to be the map such that

$$g(H)(x) = H(x \mid D_n) * x_{n+1} \qquad (x \in X^{n+1}).$$

In terms of g the requirement ($''$) reads

$$(''') \qquad A_{n+1} = g(A_n) \qquad (n \geq 1).$$

By ordinary recursion (starting with $n = 1$), there exists a unique map

$$A: \omega \setminus \{0\} \to Y$$

satisfying (′) and (‴). □

Suppose $*$ is the operation of addition on ω. If $n \in \omega$ and $(x_1,\ldots,x_n) \in X^n$, then $A_n(x_1,\ldots,x_n)$ is denoted by $\sum_{i=1}^n x_i$, $\sum_{j=1}^n x_j$, and so on. Thus

$$\textstyle\sum_{i=1}^1 x_i = x_i \qquad (x \in X^1),$$

$$\textstyle\sum_{i=1}^{n+1} x_i = (\sum_{i=1}^n x_i) + x_{n+1} \qquad ((x_1,\ldots,x_{n+1}) \in X^{n+1},\ n \geq 1).$$

When $*$ is the operation of multiplication on ω, we write $\prod_{i=1}^n x_i$ for $A_n(x_1,\ldots,x_n)$.

Suppose, finally, that $*$ is the operation

$$(f,g) \mapsto f \circ g$$

of composition on Map (X,X) for a set X. If $f \in$ Map (X,X) and if $x_i = f$ for all $i \geq 1$, then for $n \geq 1$

$$A_n(x_1,\ldots,x_n) = f^n,$$

the nth iterate of f.

EXERCISES

A. Let $m \in \omega$. The map $p_m: \omega \to \omega$ was defined by

$$p_m(n) = mn \qquad (n \in \omega).$$

Obtain p_m directly by ordinary recursion.

B. If X is a set in 10.1, show that f is given by

$$f(n) = G^n(z) \qquad (n \in \omega).$$

C. A natural number n is said to be *even* in case 2 divides n, and *odd* otherwise. Prove that $n \in \omega$ is odd if and only if $n = 2k + 1$ for some $k \in \omega$.

D. In 10.17, take $X = \omega$, $z = 0$ and $G: \omega \times \omega \to \omega$ the second projection. What connection is there between f and σ?

E. If $m, n \in \omega$ with $m \geq 1$, $n \geq 1$, show

$$m \cdot n = \textstyle\sum_{i=1}^m n, \qquad m^n = \prod_{i=1}^n m, \qquad n! = \prod_{i=1}^n i.$$

F. If $n \geq 1$ and $(x_1,\ldots,x_n) \in \omega^n$, show that

$$\textstyle\sum_{i=1}^n x_i = \sum_{i=1}^k x_i + \sum_{i=1}^{n-k} x_{i+k} \qquad (1 \leq k \leq n-1).$$

G. Prove the "general commutative law" for addition: If $n \in \omega$ and $x = (x_1,\ldots,x_n) \in \omega^n$, then

$$\textstyle\sum_{i=1}^{n} x_{\pi(i)} = \sum_{i=1}^{n} x_i$$

for each permutation π of $\{1,\ldots,n\}$.

H. Prove the *principle of double recursion*: Let

$$H: X \to X, \qquad K: X \to X$$

be given maps, and let $z \in X$. Then there exists a unique map

$$f: \omega \times \omega \to X$$

such that

$$\begin{aligned} f(0,0) &= z, && \\ f(0, n+1) &= H(f(0,n)) && (n \in \omega), \\ f(m+1, n) &= K(f(m,n)) && (m, n \in \omega). \end{aligned}$$

I. Use double recursion to obtain directly the operations of addition and multiplication on ω.

J. Use the expression for $m \cdot n$ in terms of σ, 10.8, 10.13, and 10.10 to give a new proof of 10.11. (Note that the proof given for 10.13 did not use 10.11.)

K. Let $\bar{\omega}$ be a set, $\bar{\sigma}: \bar{\omega} \to \bar{\omega}$ a map, and $\bar{0} \in \bar{\omega}$. Suppose for each map $G: X \to X$ and each $z \in X$ there exists a unique map $\bar{f}: \bar{\omega} \to X$ such that $\bar{f}(\bar{0}) = z$ and $\bar{f}(\bar{\sigma}(x)) = G(\bar{f}(x))$ for all $x \in \bar{\omega}$. Establish the existence of a unique bijection

$$\varphi: \omega \simeq \bar{\omega}$$

such that

$$\varphi(0) = \bar{0}$$

and such that the diagram

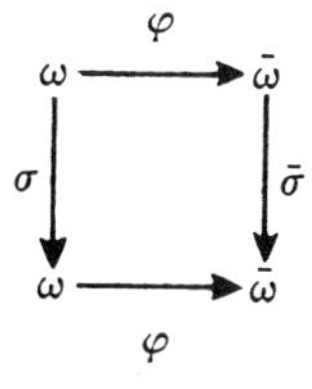

commutes. (This result says that ω is essentially unique as a set on which maps can be defined by ordinary recursion.)

L. Here is another version of recursion: Let z be a given element of a class X. Let

$$Y = \textstyle\bigcup_{n=1}^{\infty} X^n,$$

and let

$$G: Y \to X$$

be a given map, so that G assigns an element of X to each n-tuple of elements of X, $n = 1, 2, \ldots$. Then there exists a unique map

$$f: \omega \to X$$

such that

$$f(0) = z,$$

$$f(n+1) = G((f(0), f(1), \ldots, f(n))) \qquad (n \in \omega).$$

(a) Prove this. [*Hint*: Define a map $H: Y \to Y$ as follows: If $x = (x_1, \ldots, x_n) \in X^n$ with $n \geq 1$, let

$$H(x) = (x_1, \ldots, x_n, G(x)).$$

Obtain by ordinary recursion a map $g: \omega \to Y$ such that

$$g(0) = (z) \in X^1,$$

$$g(n+1) = H(g(n)) \qquad (n \in \omega).$$

Then take

$$f(n) = g(n)(n) \qquad (n \in \omega).]$$

(b) Apply this to obtain anew the Fibonacci sequence.

Another application is given in Example 13.2(11).

M. Prove: If x is a set, then $x \subset y$ for some full set y. (This statement was used as an assumption in Chapter 6, Exercise Q.) [*Hint*: For a full set y to contain x one needs $\bigcup x \subset y$ and hence $\bigcup(\bigcup x) \subset y$ and hence....]

11. Equivalence Relations

In this chapter we discuss from several points of view the general features of schemes for classifying objects into like kinds.

Recall that a relation R is said to be *in* a class X if dmn $R \cup$ rng $R \subset X$. Given a relation R, recall also the infix notation xRy standing for $(x,y) \in R$.

EQUIVALENCE RELATIONS

One way to classify the objects in a given collection is by a rule declaring certain of the objects to be alike and others unlike. The connotations of "like" suggest the following properties for such a rule: each of the objects is like itself; if x is like y, then y is like x; if x is like y and y is like z, then x is like z. Now the given collection may be represented by a class X and the classifying rule by a relation R in X, xRy being interpreted to mean that x is like y. Then the relation R should have the three properties named in the next definition.

11.1. DEFINITION

Let R be a relation in a class X. The relation is said to be *symmetric* if

$$xRy \Rightarrow yRx,$$

transitive if

$$xRy \;\&\; yRz \Rightarrow xRz,$$

and *reflexive* if

$$x \in \text{dmn } R \Rightarrow xRx.$$

11.2. DEFINITION

If a relation R in a class X is reflexive, symmetric, and transitive, then R is called an *equivalence relation*, and if in addition $X = \text{dmn } R$, then R is said to be an equivalence relation *on* X.

11.3. EXAMPLES

(1) If X is a class, then Δ_X is an equivalence relation on X, and for $x, y \in X$,

$$x \Delta_X y \Leftrightarrow x = y.$$

Under this equivalence relation, no object is classified as like another distinct from itself.

(2) If X is a class, then $R = X \times X$ is an equivalence relation on X, and xRy for all $x, y \in X$. This equivalence relation is at the opposite extreme from Δ_X, for any two objects in X are classed as alike under R.

(3) If R is the relation

$$\{(X,Y) \mid X \in \mathfrak{U} \;\&\; Y \in \mathfrak{U} \;\&\; (\exists f)(f\colon X \simeq Y)\},$$

then R is an equivalence relation on $\mathfrak{U}$. This relation R will be studied in great detail in Chapter 13 and the sequel.

(4) The relation

$$\{(X,Y) \mid X \in \mathfrak{U} \setminus \{\varnothing\} \;\&\; Y \in \mathfrak{U} \setminus \{\varnothing\} \;\&\; X \cap Y \neq \varnothing\}$$

is a symmetric, reflexive relation with domain $\mathfrak{U} \setminus \{\varnothing\}$, but it is not transitive.

(5) The relation

$$\{(m,n) \mid m \in \omega \;\&\; n \in \omega \;\&\; m \leq n\}$$

with domain ω is transitive and reflexive, but not symmetric.

(6) Let X be the set of all triangles in the plane. The relation R in X such that xRy if and only if x is congruent to y is an equivalence relation on X. Another equivalence relation on X is the relation S such that xSy if and only if x is similar to y.

(7) Let k be a fixed positive integer. Let R_k be the relation in the set $\mathbb{Z}$ of all integers defined as follows: mR_kn if and only if k divides $m - n$, in other words, m and n leave the same remainder when divided by k. Then R_k is an equivalence relation on $\mathbb{Z}$. In number theory, one says that m is *congruent to n modulo k* if mR_kn, and then one writes

$$m \equiv n \pmod{k}.$$

By analogy with the number-theoretic Example (7), for an arbitrary equivalence relation R on a class X, one says that x is *equivalent* (or *con-*

gruent) to y modulo R and writes

$$x \equiv y \pmod{R}$$

to mean xRy. In this notation the reflexive, symmetric, and transitive properties of R read:

$$x \in X \Rightarrow x \equiv x \pmod{R},$$

$$x \equiv y \pmod{R} \Rightarrow y \equiv x \pmod{R},$$

$$x \equiv y \pmod{R} \;\&\; y \equiv z \pmod{R} \Rightarrow x \equiv z \pmod{R}.$$

Given an equivalence relation on a class X, we may group together with each $x \in X$ all those elements of X which are like x.

11.4. DEFINITION

Let R be an equivalence relation on a class X. For each $x \in X$, the subclass

$$R\langle\{x\}\rangle = \{y \mid y \in X \;\&\; x \equiv y \pmod{R}\}$$

of X is called the *equivalence class of x under R*. The class

$$\{R\langle\{x\}\rangle \mid x \in X\}$$

of all those equivalence classes under R which are sets is called the *quotient of X under R* or simply *X modulo R* and is denoted by X/R. If $A \in X/R$ and if $y \in A$, then y is called a *representative of A*.

11.5. EXAMPLES

(1) If X is a class, then

$$\{y \mid y \in X \;\&\; x\Delta_X y\} = \{x\}$$

for each $x \in X$, so

$$X/\Delta_X = \{\{x\} \mid x \in X\}.$$

(2) If X is a nonempty set, then the equivalence class of any element of X under $X \times X$ is all of X, so

$$X/(X \times X) = \{X\}.$$

(3) Let R be the equivalence relation on the set $\mathbb{Z}$ of all integers such that

$$mRn \Leftrightarrow m \equiv n \pmod{2}$$

[see 11.3(7)]. Temporarily fix $m \in \mathbb{Z}$. If m is even, that is, if 2 divides m, then 2 divides $m - n$ if and only if n is also even; if m is odd, that is, if 2 does not divide m, then 2 divides $m - n$ if and only if n is also odd. Let

E denote the set of all even integers and O the set of all odd integers. Then

$$R\langle\{m\}\rangle = \begin{cases} E & \text{if } m \in E, \\ O & \text{if } m \in O. \end{cases}$$

Hence

$$X/R = \{E,O\}.$$

PARTITIONS

In each of the Examples 11.5, the given equivalence relation divides its domain into disjoint pieces, the equivalence classes, with all of the elements of a given equivalence class being equivalent to one another but not equivalent to the elements in any other equivalence class. This division of the domain occurs for all equivalence relations, as we are going to show.

11.6. DEFINITION

Let X be a class. A family $(X_i \mid i \in I)$ of sets is said to *cover* X if $X \subset \bigcup_{i \in I} X_i$, that is, each $x \in X$ belongs to X_i for some $i \in I$. The family $(X_i \mid i \in I)$ is said to *partition* X if $(X_i \mid i \in I)$ is a pairwise disjoint family of nonempty subsets of X which covers X.

A class $\mathcal{A}$ of sets is said to *cover* X if the family $(A \mid A \in \mathcal{A})$ covers X, that is, if $X \subset \bigcup \mathcal{A}$. The class $\mathcal{A}$ is said to *partition* X and is called a *partition* of X if $(A \mid A \in \mathcal{A})$ partitions X, that is, if $\mathcal{A}$ is a pairwise disjoint class of nonempty subsets of X whose union is X.

A class $\mathcal{A}$ of nonempty subsets of a class X is a partition of X if and only if

$$(\forall x \in X)(\exists! A \in \mathcal{A})(x \in A).$$

Now let R be an equivalence relation on a nonempty set X. We show that *the class X/R of all equivalence classes under R is a partition of X*. Let $x \in X$. Since $x \equiv x \pmod{R}$, $x \in R\langle\{x\}\rangle \in X/R$. Suppose $A \in X/R$ with $x \in A$ also; we must show $A = R\langle\{x\}\rangle$. For some $y \in X$, $A = R\langle\{y\}\rangle$. Then $x \in R\langle\{y\}\rangle$, and $x \equiv y \pmod{R}$. If $z \in R\langle\{x\}\rangle$, then $x \equiv z \pmod{R}$, $z \equiv y \pmod{R}$ by the symmetry and transitivity of R, and so $z \in R\langle\{y\}\rangle$. Hence $R\langle\{x\}\rangle \subset R\langle\{y\}\rangle$. Similarly, $R\langle\{y\}\rangle \subset R\langle\{x\}\rangle$, and $A = R\langle\{x\}\rangle$ as required.

The same technique as used in this argument establishes:

11.7. LEMMA

Let R be an equivalence relation on a class X. If $x, y \in X$, then

$$x \equiv y \pmod{R}$$

if and only if

$$R\langle\{x\}\rangle = R\langle\{y\}\rangle.$$

In other words, not only does the class

$$X/R = \{R\langle\{x\}\rangle \mid x \in X\}$$

partition X in case X is a nonempty set, but even the family

$$(R\langle\{x\}\rangle \mid x \in X)$$

partitions X.

We began this chapter by interpreting an equivalence relation on a class X as a rule for classifying elements of X as like or unlike one another. We have just seen that an equivalence relation R on X (at least when X is a nonempty set) gives rise to a partition X/R of X. To give *any* partition of X is to give a scheme for classifying the elements of X, for such a partition divides all of X into mutually exclusive subclasses. Now a partition $\mathcal{P}$ of X determines anew a rule for classifying elements of X as like or unlike one another: All the elements of a given member of $\mathcal{P}$ may be regarded as being like one another but unlike any element of any other member of $\mathcal{P}$.

11.8. DEFINITION

Let $\mathcal{P}$ be a partition of a class X. Then the relation

$$\bigcup\{A \times A \mid A \in \mathcal{P}\}$$

in X is said to be *induced by* $\mathcal{P}$ and is denoted by $R_{\mathcal{P}}$. We have

$$xR_{\mathcal{P}}y \Leftrightarrow (\exists A \in \mathcal{P})(x \in A \;\&\; y \in A).$$

In view of the intuitive interpretation of equivalence relations and partitions as classifying schemes, it is hardly surprising that the relation $R_{\mathcal{P}}$ induced by a partition $\mathcal{P}$ turns out to be an equivalence relation.

11.9. THEOREM

Let X be a nonempty set.

(1) If R is an equivalence relation on X, then X/R is a partition of X and

$$R_{X/R} = R.$$

(2) If $\mathcal{P}$ is a partition of X, then $R_{\mathcal{P}}$ is an equivalence relation on X and

$$X/R_{\mathcal{P}} = \mathcal{P}.$$

Proof

(1) Let R be an equivalence relation on X. We have already shown that X/R is a partition of X. Let $S = R_{X/R}$, so that

$$S = \bigcup_{x \in X} R\langle\{x\}\rangle \times R\langle\{x\}\rangle.$$

We show $S = R$. If xRy, then $R\langle\{x\}\rangle = R\langle\{y\}\rangle$ by 11.7, $(x,y) \in R\langle\{x\}\rangle \times R\langle\{x\}\rangle$, and xSy. Conversely, if xSy, then $x, y \in R\langle\{z\}\rangle$ for some $z \in X$, $x \equiv z \pmod R$ and $y \equiv z \pmod R$, $x \equiv y \pmod R$, and xRy.

(2) Let $\mathcal{P}$ be a partition of X. Recall that

$$xR_{\mathcal{P}}y \Leftrightarrow (\exists\, A \in \mathcal{P})(x \in A \;\&\; y \in A).$$

We show that $R_{\mathcal{P}}$ is an equivalence relation on X. Since $\mathcal{P}$ covers X, $X = \operatorname{dmn} R_{\mathcal{P}}$. Clearly $R_{\mathcal{P}}$ is symmetric and reflexive. To see it is transitive, suppose $xR_{\mathcal{P}}y$ and $yR_{\mathcal{P}}z$. There exist $A, B \in \mathcal{P}$ with $x, y \in A$ and $y, z \in B$. Then $y \in A \cap B$, $A = B$ since $\mathcal{P}$ is disjoint, x and z both belong to A, and so $xR_{\mathcal{P}}z$.

We now show that $X/R_{\mathcal{P}} = \mathcal{P}$. For each $x \in X$ there exists a unique $A \in \mathcal{P}$ with $x \in A$, and we denote this A by $A(x)$.

Let $x \in X$. We show

$$R_{\mathcal{P}}\langle\{x\}\rangle = A(x).$$

If $y \in R_{\mathcal{P}}\langle\{x\}\rangle$, then $x, y \in B$ for some $B \in \mathcal{P}$, $B = A(x)$ by definition of $A(x)$, and $y \in A(x)$. If $y \in A(x)$, then $x, y \in A(x)$, $xR_{\mathcal{P}}y$, and $y \in R_{\mathcal{P}}\langle\{x\}\rangle$.

Because $A(x) \in \mathcal{P}$ for each $x \in X$, it follows from the preceding paragraph that $X/R_{\mathcal{P}} \subset \mathcal{P}$. To prove the opposite inclusion, let $A \in \mathcal{P}$. Choose any $x \in A$. Since $x \in A(x)$,

$$A = A(x) = R_{\mathcal{P}}\langle\{x\}\rangle,$$

and $A \in X/R_{\mathcal{P}}$. □

Theorem 11.9 shows that the two approaches to classification—via equivalence relations and via partitions—are essentially the same. More precisely:

11.10. COROLLARY

Let X be a nonempty set. Define $\mathcal{R}$ to be the class of all equivalence relations on X and Π to be the class of all partitions of X. Then the map

$$f: \mathcal{R} \to \Pi$$

$$R \mapsto X/R \qquad (R \in \mathcal{R})$$

is a one-to-one correspondence whose inverse

$$f^{-1}: \Pi \simeq \mathcal{R}$$

is given by

$$\mathcal{P} \mapsto R_{\mathcal{P}} \qquad (\mathcal{P} \in \Pi).$$

QUOTIENT MAPS

The third approach to classification is map-theoretic.

11.11. DEFINITION

Let R be an equivalence relation on a class X such that $R\langle\{x\}\rangle$ is a set for each $x \in X$, so that $R\langle\{x\}\rangle \in X/R$ for each $x \in X$. The surjection

$$p: X \to X/R$$

given by

$$p(x) = R\langle\{x\}\rangle \qquad (x \in X)$$

is called the *projection* or *quotient map* of X onto X/R *induced by* R.

With the notation of 11.11, in terms of p, 11.7 states

$$x \equiv y \pmod R \Leftrightarrow p(x) = p(y).$$

Thus R induces a surjection which in turn completely determines R.

11.12. THEOREM

Let R be an equivalence relation on a set X, and let

$$f: X \to Y$$

be a map. Suppose f is constant on each equivalence class under R, that is,

$$x \equiv y \pmod R \Rightarrow f(x) = f(y).$$

Then there exists a *unique* map

$$\bar{f}: X/R \to Y$$

making the diagram below commutative, where p is the projection.

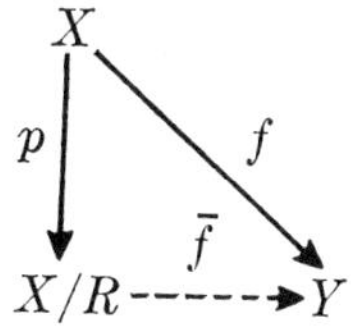

Moreover, $\bar{f}$ is surjective if f is.

Proof

Since p is surjective, there exists

$$s: X/R \to X$$

with $p \circ s = 1_{X/R}$.

If $g: X/R \to Y$ with $g \circ p = f$, then

$$g = g \circ 1_{X/R} = g \circ (p \circ s) = (g \circ p) \circ s = f \circ s.$$

This proves the uniqueness of $\bar{f}$.

To prove existence, we define

$$\bar{f} = f \circ s \colon X/R \to Y.$$

Let $x \in X$. We show $(\bar{f} \circ p)(x) = f(x)$. Now

$$(\bar{f} \circ p)(x) = (f \circ s)(p(x)) = f((s \circ p)(x)),$$

so it suffices to show

$$(*) \qquad f((s \circ p)(x)) = f(x).$$

But

$$p \circ (s \circ p)(x) = (p \circ s)(p(x)) = 1_{X/R}(p(x)) = p(x),$$

that is,

$$(s \circ p)(x) \equiv x \pmod{R},$$

and (*) follows from the fact that f is constant on each equivalence class.
Finally, if $f = \bar{f} \circ p$ is surjective, then $\bar{f}$ is surjective by 9.10(2). ☐

In the notation of 11.12, $\bar{f}$ is said to be *obtained from f by passing to quotients modulo R*. The relationship between f, R, and $\bar{f}$ is represented in Figure 11.1.

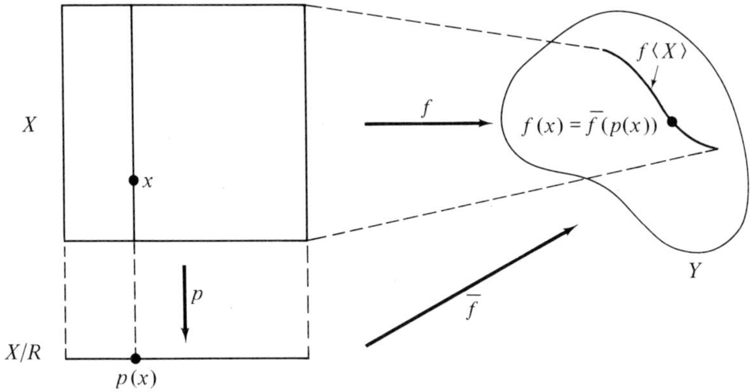

Figure 11.1

11.13. EXAMPLE

Consider the unit circle

$$S = \{(x, y) \mid x \in \mathbb{R} \;\&\; y \in \mathbb{R} \;\&\; x^2 + y^2 = 1\}$$

in the Euclidean plane, the map

$$f \colon \mathbb{R} \to S$$

given by

$$f(t) = (\cos 2\pi t, \sin 2\pi t) \qquad (t \in \mathbb{R}),$$

and the equivalence relation R on $\mathbb{R}$ such that

$$t \equiv s \pmod{R} \Leftrightarrow t - s \in \mathbb{Z}.$$

Because the sine and cosine functions are periodic of period 2π, f is constant on each equivalence class under R. Hence we have the commutative diagram below

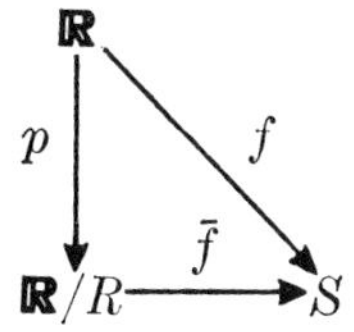

where p is the projection and $\bar{f}$ is obtained from f by passing to quotients modulo R. Elementary properties of sine and cosine imply that f is surjective, so $\bar{f}$ is surjective. We shall show that $\bar{f}$ is also injective, so that $\bar{f}$ is a one-to-one correspondence between $\mathbb{R}/R$ and S.

Let $A, B \in \mathbb{R}/R$ with $\bar{f}(A) = \bar{f}(B)$. We show $A = B$. Choose any $t \in A$, $s \in B$, so that

$$A = p(t), \qquad B = p(s).$$

Since $\bar{f} \circ p = f$, it follows that $f(t) = f(s)$. Considering the second coordinates of $f(t)$ and $f(s)$, we have in particular

$$\sin 2\pi t = \sin 2\pi s.$$

Then $2\pi t = 2\pi(s + k)$ for some integer k, that is, $t \equiv s \pmod R$. Hence

$$A = p(t) = p(s) = B.$$

This argument actually shows that

$$t \equiv s \pmod R \Leftrightarrow f(t) = f(s),$$

so that R is completely determined by f.

By analogy with 11.13, any map determines an equivalence relation on its domain:

11.14. DEFINITION

Let

$$f\colon X \to Y$$

be a map. Then the relation

$$\{(x,y) \mid x \in X \;\&\; y \in X \;\&\; f(x) = f(y)\},$$

which is obviously an equivalence relation on X, is denoted by R_f, is said to be *induced by* f, and is called the *equivalence kernel of* f.

Suppose X is a set. In case R is an equivalence relation on X, $Y = X/R$, and $f\colon X \to Y$ is the projection, then clearly $R_f = R$. We show next that for any surjection f, f looks like the projection of X onto X/R_f.

11.15. COROLLARY

Let X be a set and

$$f: X \to Y$$

be a map. Then f is constant on each equivalence class under R_f, and so we have the commutative diagram below

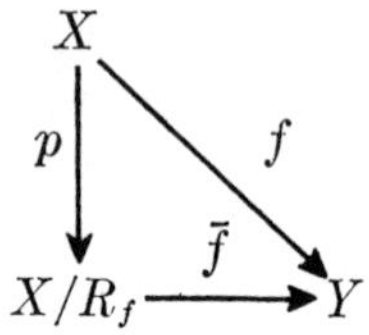

where p is the projection and $\bar{f}$ is obtained from f by passing to quotients modulo R_f. Moreover, $\bar{f}$ is injective. If f is surjective, then

$$\bar{f}: X/R_f \simeq Y.$$

Proof

By definition of R_f,

$$x \equiv y \pmod{R_f} \Leftrightarrow f(x) = f(y).$$

In particular, f is constant on each equivalence class under R_f.

We show that $\bar{f}$ is injective. Let $A, B \in X/R_f$, and suppose

$$\bar{f}(A) = \bar{f}(B).$$

Choose $x \in A$, $y \in B$. Then

$$A = p(x), \qquad B = p(y),$$

$$f(y) = \bar{f}(p(y)) = \bar{f}(B) = \bar{f}(A) = \bar{f}(p(x)) = f(x),$$

$x \equiv y \pmod{R_f}$, and

$$A = p(x) = p(y) = B.$$

If f is surjective, then $\bar{f}$ is surjective by 11.12, so $\bar{f}$ is in fact bijective. $\square$

One interesting consequence of 11.15 is that any map whose domain is a set can be written as the composite of a surjection, a bijection, and an injection!

11.16. COROLLARY

Let

$$f: X \to Y$$

be a map whose domain X is a set. Define

$$p: X \to X/R_f$$

to be the projection of X onto the quotient of X under the equivalence relation R_f induced by f. Let

$$Z = \operatorname{rng} f,$$

let

$$g: X \to Z$$

be the map having the same graph as f, and define

$$\bar{g}: X/R_g \to Z$$

to be the map obtained from g by passing to quotients modulo R_g. Finally, define

$$j: Z \to Y$$

to be the inclusion map. Then

$$R_f = R_g,$$

and

$$(*) \qquad f = j \circ \bar{g} \circ p$$

with p surjective, $\bar{g}$ bijective, and j injective. Diagrammatically:

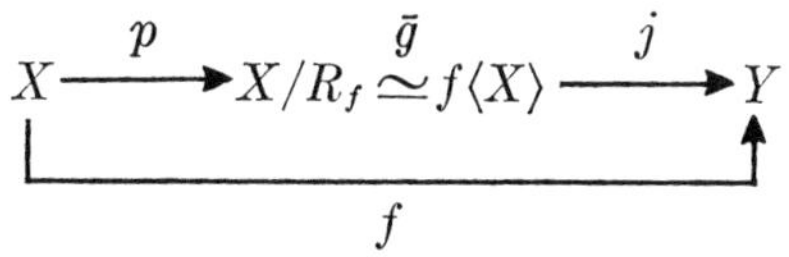

The representation (*) of f is called the *canonical factorization of f*.

To conclude this section, we show that under suitable conditions a map from one set to another determines a map between quotients of the two sets.

11.17. COROLLARY

Let R and S be equivalence relations on sets X and Y respectively. Let

$$f: X \to Y$$

be a map. Suppose f is "compatible with R and S" in the sense that

$$x \equiv y \pmod{R} \Rightarrow f(x) \equiv f(y) \pmod{S}.$$

Then there exists a unique map f' making the following diagram commute, where p and q are the projections:

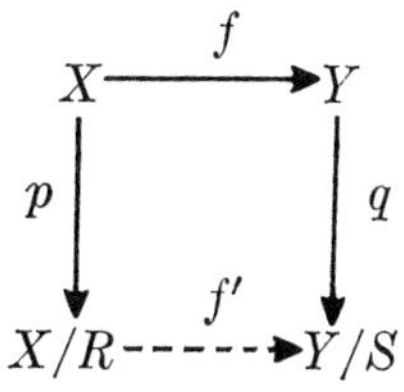

Proof

We have

$$x \equiv y \pmod{R} \Rightarrow q(f(x)) = q(f(y)).$$

Hence 11.12 applies to the map

$$q \circ f: X \to Y/S. \quad \square$$

For an application of 11.17, consider again Example 11.13. Let $s \in \mathbb{R}$ with $0 < s < 1$. We construct a map

$$h: S \to S$$

such that for all $t \in \mathbb{R}$,

$$(*) \quad h(\cos 2\pi t, \sin 2\pi t) = (\cos 2\pi(t + s), \sin 2\pi(t + s)).$$

In other words, h is the rotation of S through an angle of $2\pi s$ radians. Define

$$g: \mathbb{R} \to \mathbb{R}$$

by

$$g(t) = t + s \qquad (t \in \mathbb{R}).$$

Clearly

$$t_1 \equiv t_2 \pmod{R} \Rightarrow g(t_1) \equiv g(t_2) \pmod{R},$$

and 11.17 gives a commutative diagram

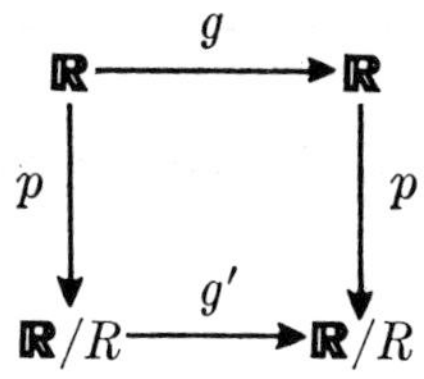

Using the bijection

$$\bar{f}: \mathbb{R}/R \simeq S,$$

we define

$$h = \bar{f} \circ g' \circ (\bar{f})^{-1}.$$

This gives the "three-dimensional" commutative diagram below.

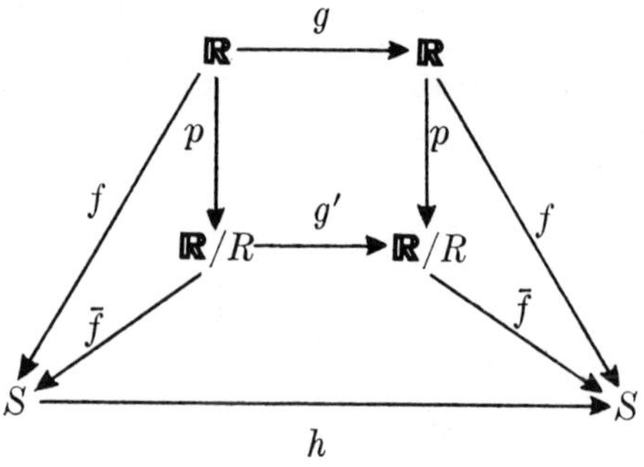

From this diagram we read off

$$h \circ f = f \circ g,$$

and this is precisely what (*) says.

In practice one would hardly give so much detail to define h. The typical way of "defining" h is: "Each point z of S has the form

$$z = (\cos 2\pi t, \sin 2\pi t)$$

for some $t \in \mathbb{R}$. If $t_1 \in \mathbb{R}$ with

$$z = (\cos 2\pi t_1, \sin 2\pi t_1)$$

also, then

$$(\cos 2\pi(t_1 + s), \sin 2\pi(t_1 + s)) = (\cos 2\pi(t + s), \sin 2\pi(t + s)),$$

so the map h defined by (*) is well defined."

EXERCISES

A. (a) If R is a symmetric relation, show that dmn R = rng R.
(b) Prove that any symmetric, transitive relation is necessarily reflexive. (Hence "reflexive" as part of the definition of "equivalence relation" is redundant!)

B. For a relation R, show:
R is symmetric $\Leftrightarrow R^{-1} = R$,
R is transitive $\Leftrightarrow R \circ R \subset R$,
R is reflexive $\Leftrightarrow \Delta_{\text{dmn } R} \subset R$,
R is an equivalence relation $\Leftrightarrow R \circ R^{-1} = R$.

C. In the notation of 11.3(7), construct a one-to-one correspondence between $\mathbb{Z}/R_k$ and $k = \{0,1,\ldots,k-1\}$.

D. Let A be a nonempty subset of a set X. Define a relation R in X by

$$xRy \Leftrightarrow (x = y) \text{ or } (x \in A \ \&\ y \in A).$$

(a) Verify that R is an equivalence relation on X, and determine X/R.
(b) If $p: X \to X/R$ is the projection, show that $p\langle A\rangle = \{A\}$ and that the map

$$f: X \setminus A \to (X/R) \setminus \{A\}$$
$$x \mapsto p(x) \qquad (x \in X \setminus A)$$

is bijective.

The set X/R is said to be "obtained from X by collapsing A to a point."

E. Let

$$I = \{x \mid x \in \mathbb{R} \,\&\, 0 \leq x \leq 1\},$$

$$D = I \times I,$$

$$S = \{(x,y) \mid x \in \mathbb{R} \,\&\, y \in \mathbb{R} \,\&\, x^2 + y^2 = 1\}.$$

In each case below, an equivalence relation $\sim$ on D and a set Y are defined; you are to construct a bijection from $D/\sim$ to Y.

(a) Here $\sim$ is the unique reflexive, symmetric relation on D such that

$$(0,y) \sim (1,y) \qquad (y \in I).$$

In other words, for (x,y), $(t,s) \in D$,

$$(x,y) \equiv (t,s) \pmod{\sim}$$

if and only if

$$((x,y) = (t,s)) \vee (x = 0 \,\&\, t = 1 \,\&\, y = s)$$
$$\vee\ (x = 1 \,\&\, t = 0 \,\&\, y = s).$$

Also, $Y = S \times I$. Note that $S \times I$ can be put into one-to-one correspondence with the cylindrical surface

$$\{(x,y,z) \mid x,y,z \in \mathbb{R} \,\&\, x^2 + y^2 = 1 \,\&\, 0 \leq z \leq 1\}$$

in Euclidean 3-space.

(b) Here $\sim$ is the unique reflexive, symmetric relation on D such that

$$(0,y) \sim (1,y) \qquad (y \in I),$$

$$(x,0) \sim (x,1) \qquad (x \in I).$$

Also, $Y = S \times S$. The set Y can be put into one-to-one correspondence with the "torus," the doughnut-shaped surface

$$\{(x,y,z) \mid x,y,z \in \mathbb{R} \,\&\, ((x^2 + y^2)^{1/2} - 2)^2 + z^2 = 1\}$$

in 3-space obtained by rotating the circle in the xz-plane with equation $(x - 2)^2 + z^2 = 1$ about the z-axis (draw a picture).

F. Let X be a set. In the class of all surjections having domain X define a relation $\sim$ as follows: For surjections $f: X \to Y$, $g: X \to Z$, $f \sim g$ means $h \circ f = g$ for some bijection $h: Y \to Z$. Verify that $\sim$ is an equivalence relation. Construct a one-to-one correspondence between the set of all equivalence relations on X

and a set consisting of exactly one representative of each equivalence class under $\sim$.

G. In the notation of 11.12, show that p is "unique up to a bijection" with respect to the mapping property given by 11.12. More precisely, suppose

$$q: X \to Z$$

is a surjection with the property that for each map $f: X \to Y$ which is constant on each equivalence class under R, there is a unique f^* making the diagram

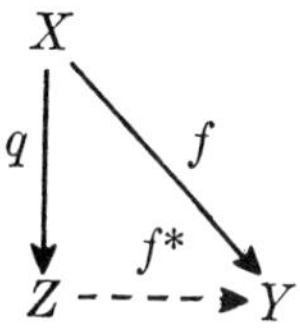

commute. Show that then there is a unique bijection

$$h: X/R \simeq Z$$

such that the following diagram commutes:

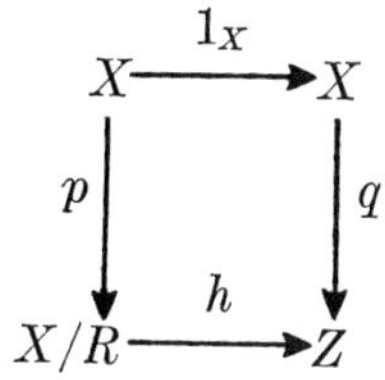

H. Let G be a set of permutations of a set X.
(a) What conditions on G are sufficient for the relation R_G in X given by

$$xR_Gy \Leftrightarrow (\exists f \in G)(f(x) = y)$$

to be an equivalence relation on X?
(b) If $X = \mathbb{R}$, find a set G of permutations of X satisfying these conditions such that X/R_G can be put into one-to-one correspondence with the unit circle

$$S = \{(x,y) \mid x,y \in \mathbb{R} \;\&\; x^2 + y^2 = 1\}.$$

Is this G unique?

I. Let R be an equivalence relation on a set X, let

$$p: X \to X/R$$

be the projection, and let

$$q: X \times X \to (X/R) \times (X/R)$$

be the map given by

$$q(x,y) = (p(x),p(y)).$$

(a) Verify that the relation R^* in $X \times X$ given by

$$(x,y)R^*(u,v) \Leftrightarrow xRu \;\&\; yRv$$

is an equivalence relation on $X \times X$, and show there is a unique bijection

$$h: (X/R) \times (X/R) \simeq (X \times X)/R^*$$

making the following diagram commute, where r is the projection:

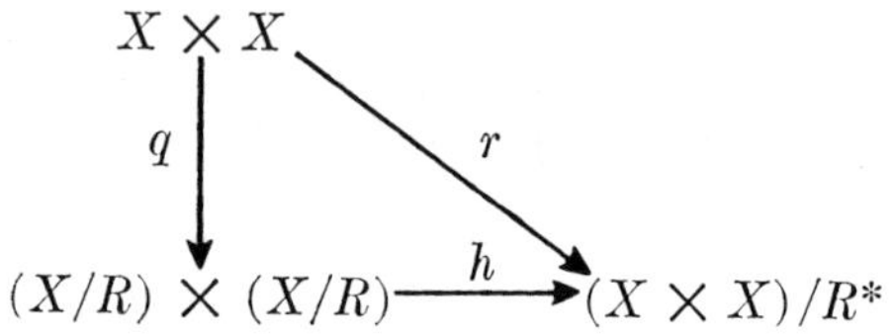

(b) Let $f: X \times X \rightarrow X$ be a map "compatible with R in each variable" in the sense that

$$x \equiv u \ (\text{mod } R) \;\&\; y \equiv v \ (\text{mod } R) \Rightarrow f(x,y) = f(u,v).$$

Show there is a unique map f^* making the following diagram commute:

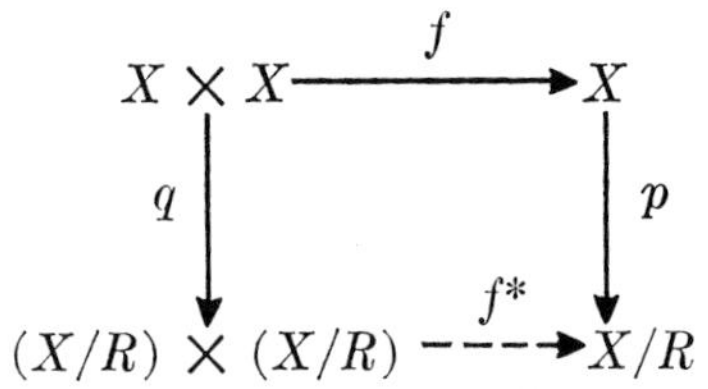

(c) Let S be a relation in X which is "compatible with R" in the sense that

$$xSy \;\&\; u \equiv x \ (\text{mod } R) \;\&\; v \equiv y \ (\text{mod } R) \Rightarrow uSz.$$

Construct a relation S^* in X/R such that

$$xSy \Leftrightarrow p(x)\,S^*p(y),$$

and show that S^* is unique; S^* is said to be *induced by* S.

The results of this exercise will be applied in the Appendix to construct the set $\mathbb{Q}$ of all rational numbers starting with the set $\mathbb{Z}$ of all integers. Another application, in the next exercise, can be used to provide a construction of $\mathbb{Z}$ different from the one given in the Appendix.

J. Let $\sim$ be the relation in $\omega \times \omega$ such that

$$(m,n) \sim (i,j) \Leftrightarrow m + j = n + i.$$

It is easily verified that $\sim$ is an equivalence relation on $\omega \times \omega$. Let

$$W = (\omega \times \omega)/\sim,$$

and for each $(m,n) \in \omega \times \omega$ let $(m,n)^*$ denote the equivalence class of (m,n) under $\sim$. Prove:

(a) There is a unique bijection

$$f \colon W \simeq \mathbb{Z}$$

such that

$$f((m,n)^*) = m - n \qquad (m, n \in \omega).$$

(b) There is a unique relation $\leq^*$ in W such that for (m,n), $(i,j) \in \omega \times \omega$,

$$(m,n)^* \leq^* (i,j)^* \Leftrightarrow f((m,n)^*) \leq f((i,j)^*).$$

(c) The map

$$g \colon \omega \to W$$

given by

$$g(n) = (n,0)^*$$

is injective and satisfies

$$m \leq n \Leftrightarrow g(m) \leq^* g(n) \qquad (m, n \in \omega).$$

(d) There are unique binary operations $+^*$, $\cdot^*$ in W such that for (m,n), $(i,j) \in \omega \times \omega$,

$$(m,n)^* +^* (i,j)^* = f((m,n)^*) + f((i,j)^*),$$

$$(m,n)^* \cdot^* (i,j)^* = f((m,n)^*) \cdot f((i,j)^*).$$

K. We use the notation of Example 11.13, except that we now denote the quotient set $\mathbb{R}/R$ by $\mathbb{R}/\mathbb{Z}$.

(a) Show that each element of $\mathbb{R}/\mathbb{Z}$ has the form

$$t + \mathbb{Z} = \{t + n \mid n \in \mathbb{Z}\}$$

for some $t \in \mathbb{R}$.

(b) Construct a binary operation $\oplus$ on $\mathbb{R}/\mathbb{Z}$ such that

$$(t + \mathbb{Z}) \oplus (s + \mathbb{Z}) = (t + s) + \mathbb{Z} \qquad (t, s \in \mathbb{R}).$$

(c) Let $\cdot$ be the binary operation on S defined by

$$(x,y) \cdot (x',y') = (xx' - yy', xy' + yx') \quad ((x,y), (x',y') \in S).$$

[Under the usual definition of complex numbers as ordered pairs

$(x,y) \in \mathbb{R} \times \mathbb{R}$, this operation is precisely the usual operation of multiplication on the set of complex numbers.] Verify that the bijection $\bar{f}: \mathbb{R}/\mathbb{Z} \to S$ satisfies

$$\bar{f}((t + \mathbb{Z}) \oplus (s + \mathbb{Z})) = \bar{f}(t + \mathbb{Z}) \cdot \bar{f}(s + \mathbb{Z}) \qquad (t, s \in \mathbb{R}).$$

L. Let $\mathfrak{D}$ be the set of all maps $f: \mathbb{R} \to \mathbb{R}$ such that $f \mid I$ is differentiable for some open interval I in $\mathbb{R}$ containing 0. Let $\sim$ be the relation in $\mathfrak{D}$ such that $f \sim g$ if and only if $f \mid I = g \mid I$ for some open interval I containing 0.
(a) Show that $\sim$ is an equivalence relation on $\mathfrak{D}$.
(b) If $\mathcal{G} = \mathfrak{D}/\sim$, show that there is a unique map $D: \mathcal{G} \to \mathbb{R}$ such that $G \in \mathcal{G}$ implies $D(G) = f'(0)$ for each $f \in \mathcal{G}$; here $f'(0)$ denotes the derivative of f at 0.

M. Let $\mathfrak{D}$ be the set of all differentiable maps from $\mathbb{R}$ to $\mathbb{R}$ whose derivatives are continuous. For $f \in \mathfrak{D}$, let $f': \mathbb{R} \to \mathbb{R}$ denote the derivative of f. Let $\sim$ be the equivalence relation on $\mathfrak{D}$ such that $f \sim g$ if and only if $f' = g'$. Establish a one-to-one correspondence between $\mathfrak{D}/\sim$ and a well-known set.

N. Let R be a relation on a set X. Define a relation E in X as follows: xEy if and only if there exists $n \in \omega \setminus \{0\}$ and an ordered n-tuple $(x_1, \ldots, x_n)$ in X such that

$$x_1 = x, \qquad x_n = y,$$

$$x_i R x_{i+1} \text{ or } x_{i+1} R x_i \text{ or } x_i = x_{i+1} \qquad (1 \leq i \leq n - 1).$$

(a) Show that E is the least equivalence relation on X containing R. One calls E the *transitive closure of* R.
(b) Let $p: X \to X/E$ be the projection. Show that $xRy \Rightarrow p(x) = p(y)$.
(c) Let $f: X \to Y$ be a map such that xRy implies $f(x) = f(y)$. Show there is a unique map $f^*: X/E \to Y$ such that $f^* \circ p = f$.

P. Let I be a nonempty set on which there is given a reflexive, transitive relation $\leq$ satisfying

$$(\forall i, j \in I)(\exists k \in I)(i \leq k \,\&\, j \leq k);$$

then one says that $\leq$ *directs* I and calls $(I, \leq)$ a *directed set.* For example, the usual ordering $\leq$ of ω directs ω.

Let $X = (X_i \mid i \in I)$ be a family of sets and $f = (f_i{}^j \mid i, j \in I \,\&\, i \leq j)$ a family of maps with

$$f_i{}^j: X_i \to X_j \qquad (i \leq j)$$

such that

$$f_i{}^i = \text{identity map of } X_i \qquad (i \in I),$$

$$f_j{}^k \circ f_i{}^j = f_i{}^k \qquad (i \leq j \leq k);$$

then one calls (X,f) a *direct* (or *inductive*) *system on* $(I,\leq)$.

(a) Let $(X_i \mid i \in \omega)$ be a sequence of sets and $(f_i \mid i \in \omega)$ a sequence of maps with $f_i\colon X_i \to X_{i+1}$ for each $i \in I$. Construct a direct system (X,f) on $(\omega,\leq)$ such that $f_i{}^{i+1} = f_i \quad (i \in \omega)$.

(b) Let $(I,\leq)$ be a directed set. Let $X = (X_i \mid i \in I)$ be a family of sets such that $X_i \subset X_j$ whenever $i \leq j$, and for $i \leq j$ let $f_i{}^j\colon X_i \to X_j$ be the inclusion map. Verify that (X,f) is a direct system.

(c) Let $(I,\leq)$ be a directed set, let $(A_i \mid i \in I)$ be a family of subsets of a set A such that $A_i \supset A_j$ whenever $i \leq j$, and let B be a set. Construct a direct system (X,f) for which

$$X_i = \operatorname{Map}(A_i,B) \qquad (i \in I).$$

Q. (Continuation of P.) Consider the disjoint union $+_{i\in I} X_i$ of $(X_i \mid i \in I)$—see Chapter 8, Exercise FF. Define a relation $\sim$ in $+_{i\in I} X_i$ by

$$(x,i) \sim (y,j) \Leftrightarrow (\exists k \in I)(i \leq k \;\&\; j \leq k \;\&\; f_i{}^k(x) = f_j{}^k(y)).$$

(a) Verify that $\sim$ is an equivalence relation on $+_{i\in I} X_i$.

Let

$$X_\infty = +_{i\in I} X_i/\sim.$$

One calls X_∞ the *direct limit of* (X,f) and denotes it by dir lim (X,f).

(b) If (X,f) is the direct system of Exercise P(b), construct a bijection between dir lim (X,f) and $\bigcup_{i\in I} X_i$.

Let

$$p\colon +_{i\in I} X_i \to X_\infty$$

be the quotient map and for each $i \in I$ let

$$q_i\colon X_i \to +_{i\in I} X_i$$

be the map defined in Chapter 8, Exercise FF. For each $i \in I$, let

$$f_i{}^\infty = p \circ q_i\colon X_i \to X_\infty.$$

(c) Show that

$$f_j{}^\infty \circ f_i{}^j = f_i{}^\infty \qquad (i,j \in I, i \leq j).$$

[Note that if we let $I^* = I \cup \{\infty\}$, where ∞ is some set with $\infty \notin I$, and if we set $i \leq \infty$ for all $i \in I^*$, then $(I^*,\leq)$ is also a directed set. Then (X^*,f^*) is also a direct system, where $X^* = (X_i \mid i \in I^*)$, $f^* = (f_i{}^j \mid i,j \in I^* \;\&\; i \leq j)$, $f_\infty{}^\infty =$ identity map of X_∞.]

(d) If $f_i{}^j$ is injective for all $i, j \in I$ with $i \leq j$, then X_∞ can be put into one-to-one correspondence with $\bigcup_{i \in I} X_i$. [*Hint*: Show that f_i^{∞} is injective for each i.]

R. (Continuation of Q.) Let X be a set and $(h_i \mid i \in I)$ be a family of maps with $h_i \colon X_i \to Y$ for each $i \in I$ satisfying

$$h_j \circ f_i^j = h_i \qquad (i, j \in I, i \leq j).$$

Prove:

(a) There is a unique map $h \colon X_\infty \to Y$ such that

$$h_i = h \circ f_i^{\infty} \qquad (i \in I).$$

(b) The map h is injective if and only if for each $i \in I$,

$$x, y \in X_i \,\&\, h_i(x) = h_i(y) \Rightarrow (\exists j \in I)(i \leq j \,\&\, f_i^j(x) = f_i^j(y)).$$

(c) The map h is surjective if and only if

$$Y = \bigcup_{i \in I} h_i\langle X_i \rangle.$$

S. (Continuation of R.) Let (Y,g) be a second direct system on $(I,\leq)$. Let $\alpha = (\alpha_i \mid i \in I)$ be a family of maps with

$$\alpha_i \colon X_i \to Y_i \qquad (i \in I)$$

such that

$$g_i^j \circ \alpha_i = \alpha_j \circ f_i^j \qquad (i \leq j);$$

one calls α a *direct system of maps from* (X, f) *to* (Y,g). Prove:

(a) There is a unique map

$$\alpha_\infty \colon X_\infty \to Y_\infty$$

such that

$$\alpha_\infty \circ f_i^{\infty} = g_i^{\infty} \circ \alpha_i$$

for all $i \in I$; one calls α_∞ the *direct limit of* α.

(b) The map α_∞ is injective (respectively, surjective) if each α_i is.

(c) If (Z,h) is a third direct system on $(I,\leq)$ and if $\beta = (\beta_i \mid i \in I)$ is a direct system of maps from (Y,g) to (Z,h), then $\gamma = (\beta_i \circ \alpha_i \mid i \in I)$ is a direct system of maps from (X, f) to (Z,h) and $\gamma_\infty = \beta_\infty \circ \alpha_\infty$.

12. Product of a Family

Unions and intersections of families of sets were defined in Chapter 8, where one of their crucial properties, expressed by the De Morgan laws, was established. In the present chapter the commutative, associative, and distributive laws for union and intersection given in Chapter 5 are generalized to the case of arbitrary families. The very formulation of a general distributive law requires the principal notion of this chapter—the product of an arbitrary family of sets. Such products generalize simultaneously the product of two sets and the set of all maps from one set to another.

PROPERTIES OF UNION AND INTERSECTION

12.1. PROPOSITION

Let $(X_i \mid i \in I)$ be a family of sets and let

$$f: J \to I$$

be a surjection. Then

$$\bigcup_{j \in J} X_{f(j)} = \bigcup_{i \in I} X_i$$

and

$$\bigcap_{j \in J} X_{f(j)} = \bigcap_{i \in I} X_i.$$

Proof

We have

$$\{i \mid i \in I\} = \{f(j) \mid j \in J\}.$$

Hence

$$x \in \bigcup_{j \in J} X_{f(j)} \Leftrightarrow (\exists j \in J)(x \in X_{f(j)})$$
$$\Leftrightarrow (\exists i \in I)(x \in X_i)$$
$$\Leftrightarrow x \in \bigcup_{i \in I} X_i.$$

This proves the first equality. The proof of the second is similar. □

By taking J to be equal to I and f to be a permutation of I, we obtain:

12.2. COROLLARY (COMMUTATIVE LAWS)

Let $(X_i \mid i \in I)$ be a family of sets. If σ is a permutation of I, then

$$\bigcup_{i \in I} X_{\sigma(i)} = \bigcup_{i \in I} X_i$$

and

$$\bigcap_{i \in I} X_{\sigma(i)} = \bigcap_{i \in I} X_i.$$

Thus the sets used to form a union or intersection may be rearranged in any order without changing the union or intersection.

12.3. PROPOSITION (ASSOCIATIVE LAWS)

Let $(X_i \mid i \in I)$ be a family of sets, and let $(I_j \mid j \in J)$ be a family of subsets of the index class I such that

$$I = \bigcup_{j \in J} I_j.$$

Then

$$\bigcup_{i \in I} X_i = \bigcup_{j \in J} \left(\bigcup_{i \in I_j} X_i\right)$$

and

$$\bigcap_{i \in I} X_i = \bigcap_{j \in J} \left(\bigcap_{i \in I_j} X_i\right).$$

Proof

We have

$$i \in I \Leftrightarrow (\exists j \in J)(i \in I_j).$$

Hence

$$x \in \bigcup_{i \in I} X_i \Leftrightarrow (\exists i \in I)(x \in X_i)$$
$$\Leftrightarrow (\exists j \in J)(\exists i \in I_j)(x \in X_i)$$
$$\Leftrightarrow (\exists j \in J)(x \in \bigcup_{i \in I_j} X_i)$$
$$\Leftrightarrow x \in \bigcup_{j \in J} \left(\bigcup_{i \in I_j} X_i\right).$$

This proves the associative law for union. The law for intersection is proved similarly. □

These associative laws are applied most often in the case that $(I_j \mid j \in J)$ actually partitions I. In that case they say that the sets used to form a union or intersection may be grouped in any manner.

As already mentioned, the distributivity of union over intersection and intersection over union will employ the notion of product we are about to define.

PRODUCT OF A FAMILY

Let X be a set. According to 9.19, there is a one-to-one correspondence between the product $X \times X$ of X with itself and the set of all functions on $\{1,2\}$ to X, given by

$$(1) \qquad (x,y) \mapsto c_{x,y}$$

where

$$(2) \qquad c_{x,y}(i) = \begin{cases} x & \text{if } i = 1, \\ y & \text{if } i = 2 \end{cases}$$

for each $(x,y) \in X \times X$. In short, $X \times X$ can be represented as a set of functions.

Let us generalize this representation to the case of the product $X \times Y$ of X with a set Y not necessarily equal to X. For $(x,y) \in X \times Y$, define the function $c_{x,y}\colon \{1,2\} \to X \cup Y$ just as in (2), and let

$$P = \{c_{x,y} \mid (x,y) \in X \times Y\}$$

be the set of all functions so obtained. Formula (1) defines a map

$$\varphi\colon X \times Y \to P$$

which is surjective by definition of P. This map is also injective, for if (x,y), $(a,b) \in X \times Y$ with $(x,y) \neq (a,b)$, then either $x \neq a$ whence $c_{x,y}(1) \neq c_{a,b}(1)$, or $y \neq b$ whence $c_{x,y}(2) \neq c_{a,b}(2)$. Thus φ is a one-to-one correspondence between $X \times Y$ and the set P of functions.

Let us describe P in different terms. Each element c of P is a function with domain $I = \{1,2\}$ and codomain $X \cup Y$ such that $c(1) \in X$ and $c(2) \in Y$. If we define a family $(X_i \mid i \in I)$ by

$$X_1 = X, \qquad X_2 = Y,$$

then an element of P is just a family $c = (c_i \mid i \in I)$ in $\bigcup_{i \in I} X_i$ such that $c_i \in X_i$ for each $i \in I$. Thus, the elements of P are nothing but choice functions for $(X_i \mid i \in I)$.

This description suggests a definition.

12.4. DEFINITION

Let $(X_i \mid i \in I)$ be a family of sets. The *(Cartesian) product of* $(X_i \mid i \in I)$, denoted by

$$\times_{i\in I} X_i,$$

is defined to be the class

$$\{x \mid x \text{ is a choice function for } (X_i \mid i \in I)\}$$

(the notation $\prod_{i\in I} X_i$ also appears in the literature). In other words, $\times_{i\in I} X_i$ is the class of all families $x = (x_i \mid i \in I)$ in $\bigcup_{i\in I} X_i$ and indexed by I such that

$$x_i \in X_i \qquad (i \in I).$$

If $j \in I$, the jth coordinate X_j of the given family $(X_i \mid i \in I)$ is also called the *jth factor of* $\times_{i\in I} X_i$, and the map

$$pr_j\colon \times_{i\in I} X_i \to X_j$$
$$x \mapsto x_j$$

is called the *jth projection.*

If $x \in \times_{i\in I} X_i$, then for each $i \in I$ the ith coordinate x_i of x is given by

$$x_i = pr_i(x),$$

so that

$$x = (pr_i(x) \mid i \in I).$$

The product of $(X_i \mid i \in I)$ may be visualized as follows. Represent the index class I by a horizontal line segment, and above each point $i \in I$, draw a vertical segment representing X_i (see Figure 12.1). Then each element $x = (x_i \mid i \in I)$ of the product is represented by the "curve" obtained by marking the point x_i on the vertical segment X_i for each $i \in I$.

A word on notation. When $I = \omega$, $\times_{i\in I} X_i$ is also denoted by $\times_{i=0}^{\infty} X_i$ or informally by

$$X_0 \times X_1 \times \ldots .$$

If $I = n + 1 = \{0,1,\ldots,n\}$ for some $n \in \omega$, then $\times_{i\in I} X_i$ is also denoted by $\times_{i=0}^{n} X_i$ or informally by

$$X_0 \times X_1 \times \ldots \times X_n.$$

Other notations such as

$$X_3 \times X_4 \times \ldots \times X_9$$

should now be self-explanatory.

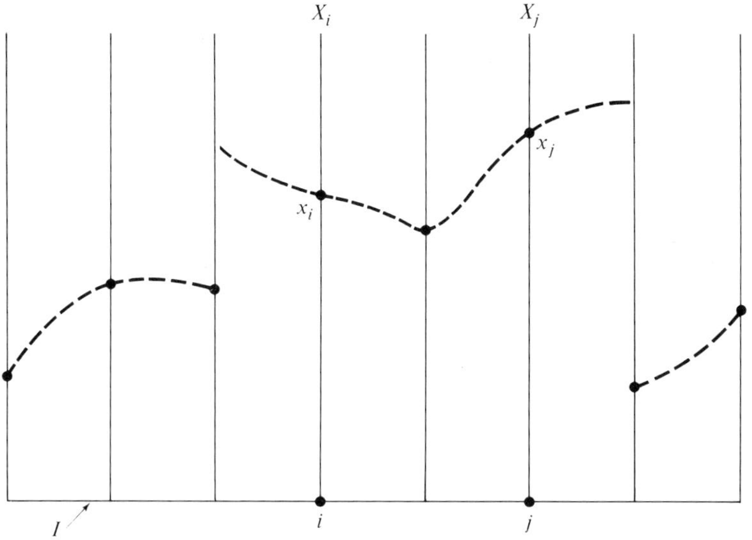

Figure 12.1

12.5. EXAMPLES

(1) Let X and Y be sets. In the discussion preceding 12.4, we constructed a bijection from $X \times Y$ to the product $\times_{i \in I} X_i$ of the family $(X_i \mid i \in I)$, where $I = \{1,2\}$ and $X_1 = X$, $X_2 = Y$. The inverse of this bijection is the bijection

$$f: \times_{i \in I} X_i \to X \times Y$$

$$x \mapsto (x_1,x_2).$$

(2) Let X be a set and I be a class. Define $(X_i \mid i \in I)$ to be the constant family such that

$$X_i = X \qquad (i \in I).$$

Then $\times_{i \in I} X_i$ is the class of all functions

$$x: I \to \bigcup_{i \in I} X_i = X$$

such that

$$x_i = x(i) \in X_i = X \qquad (i \in I),$$

that is,

$$\times_{i \in I} X_i = \text{Map }(I,X).$$

Moreover, for each $j \in I$ the evaluation map

$$e_j: \text{Map }(I,X) \to X$$

$$f \mapsto f(j) \qquad (f \in \text{Map }(I,X))$$

is nothing but the jth projection

$$pr_j: \times_{i\epsilon I} X_i \to X_j.$$

(3) Suppose $I = \{i\}$ for some set i. Then a family $(X_i \mid i \in I)$ indexed by I is just a constant family with value some set $X = X_i$, and the product $\times_{i\epsilon I} X_i$ is the set of all functions from $\{i\}$ to X. The map

$$f: \times_{i\epsilon I} X_i \to X$$

$$x \mapsto x_i$$

is a bijection. Thus any set can be regarded as a product.

(4) Let $(X_i \mid i \in I)$ be a family of sets such that each X_i is a singleton, say $X_i = \{c_i\}$, $i \in I$. Then

$$\times_{i\epsilon I} X_i = \{(c_i \mid i \in I)\}$$

is also a singleton.

12.6. LEMMA

Let X and Y be sets. Then Map (X,Y) is a set.

Proof

Let $\mathfrak{F}$ be the class of all functional relations f such that $\operatorname{dmn} f = X$ and $\operatorname{rng} f \subset Y$. The map

$$\mathfrak{F} \to \text{Map }(X,Y)$$

$$f \mapsto \langle X, f, Y\rangle \qquad (f \in \mathfrak{F})$$

is surjective (in fact, bijective), so by the axiom of replacement it suffices to show that $\mathfrak{F}$ is a set.

If $f \in \mathfrak{F}$, then

$$f \subset X \times Y,$$

f is a set since its domain X is, and so

$$f \in \mathcal{P}(X \times Y).$$

Thus

$$\mathfrak{F} \subset \mathcal{P}(X \times Y).$$

Because X and Y are sets, so is $\mathcal{P}(X \times Y)$. Hence $\mathfrak{F}$ is a set. $\square$

12.7. THEOREM

If $(X_i \mid i \in I)$ is a family of sets whose index class I is also a set, then $\times_{i\epsilon I} X_i$ is a set.

Proof

We have

$$\mathsf{X}_{i\epsilon I}\, X_i \subset \text{Map}\,(I, \bigcup_{i\epsilon I} X_i),$$

and $\bigcup_{i\epsilon I} X_i$ is a set. Now apply the lemma. □

In view of Example 12.5(2), Lemma 12.6 is actually a special case of Theorem 12.7.

The axiom of choice (more precisely, Theorem 9.24) has the following product version.

12.8. THEOREM (MULTIPLICATIVE PRINCIPLE)

If $(X_i \mid i \in I)$ is a family of nonempty sets, then $\mathsf{X}_{i\epsilon I}\, X_i \neq \varnothing$.

12.9. COROLLARY

Let $(X_i \mid i \in I)$ be a family of nonempty sets. Then for each $j \in I$ the jth projection

$$pr_j\colon \mathsf{X}_{i\epsilon I}\, X_i \to X_j$$

is surjective.

Proof

Fix $j \in I$. By 12.8, there exists some

$$c \in \mathsf{X}_{i\epsilon I\setminus\{j\}}\, X_i.$$

Let $y \in X_j$. If $x \in \mathsf{X}_{i\epsilon I}\, X_i$ is defined by

$$x_i = \begin{cases} c_i & \text{if } i \in I \setminus \{j\}, \\ y & \text{if } i = j, \end{cases}$$

then $pr_j(x) = y$. □

PRODUCTS AND MAPS

Probably the most important feature of products is their use in representing a family of maps with a common domain as a single map (compare 8.19).

12.10. THEOREM

Let X be a class, and let $(f_i \mid i \in I)$ be a nonempty family of maps each having X as its domain. For each $i \in I$ let Y_i be the codomain of f_i, so that

$$f_i\colon X \to Y_i.$$

Then there exists a *unique* map

$$f: X \to \times_{i \in I} Y_i$$

making the following diagrams commute for all $j \in I$:

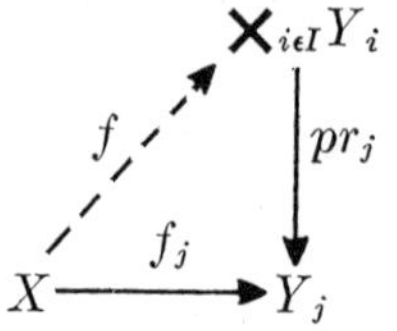

Proof

If $f_i = pr_i \circ f$ for all $i \in I$, then for each $x \in X$ we have

$$f(x) = (pr_i(f(x)) \mid i \in I) = (f_i(x) \mid i \in I).$$

This establishes the uniqueness of f. To prove existence, take f to be the map defined by

$$x \mapsto (f_i(x) \mid i \in I) \qquad (x \in X). \quad \square$$

In the notation of 12.10, one says that f is *induced by* $(f_i \mid i \in I)$.

Theorem 12.10 even allows any family of maps to be represented by a single map.

12.11. COROLLARY

Let $(f_i \mid i \in I)$ be a nonempty family of maps. For each $i \in I$ let X_i be the domain and Y_i be the codomain of f_i, so that

$$f_i: X_i \to Y_i.$$

Then there exists a *unique* map f making the following diagrams commute for all $j \in I$:

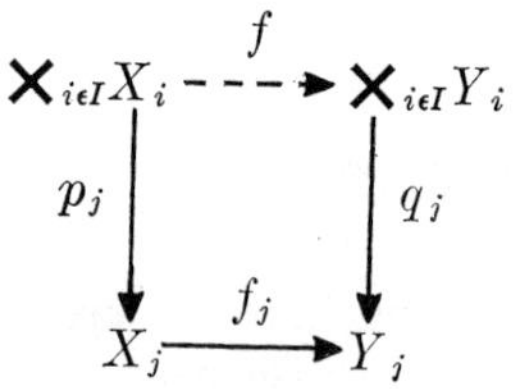

(here p_j and q_j are the projections).

Proof

Let

$$X = \times_{i \in I} X_i,$$

and for each $i \in I$ let

$$g_i = f_i \circ p_i: X \to Y_i.$$

Now apply 12.10 to the family $(g_i \mid i \in I)$. $\square$

In the notation of 12.11, one calls f the *extension of* $(f_i \mid i \in I)$ *to products.*

The representation given by 12.10 actually establishes a one-to-one correspondence between families $(f_i \mid i \in I)$ and maps f.

12.12. COROLLARY

Let X be a set and $(Y_i \mid i \in I)$ be a family of sets. Let

$$\psi\colon \times_{i\in I} \text{Map}\,(X,Y_i) \to \text{Map}\,(X, \times_{i\in I} Y_i)$$

be the map such that for each family

$$(f_i \mid i \in I) \in \times_{i\in I} \text{Map}\,(X,Y_i),$$
$$\psi((f_i \mid i \in I))\colon X \to \times_{i\in I} Y_i$$

is the map induced by $(f_i \mid i \in I)$. Then ψ is bijective, and the inverse is the map

$$\varphi\colon \text{Map}\,(X, \times_{i\in I} Y_i) \to \times_{i\in I} \text{Map}\,(X,Y_i)$$
$$f \mapsto (pr_i \circ f \mid i \in I).$$

Proof

If

$$(f_i \mid i \in I) \in \times_{i\in I} \text{Map}\,(X,Y_i)$$

and if f is the map induced by this family, then

$$(\varphi \circ \psi)((f_i \mid i \in I)) = \varphi(f) = (pr_i \circ f \mid i \in I) = (f_i \mid i \in I).$$

Hence $\varphi \circ \psi$ is the identity map on dmn ψ.

To see that $\psi \circ \varphi$ is the identity map on dmn φ, let

$$f\colon X \to \times_{i\in I} Y_i$$

be a map belonging to dmn φ, and let

$$g = (\psi \circ \varphi)(f),$$

so that

$$g = \psi((pr_i \circ f \mid i \in I)).$$

We must show $g = f$. Consider the diagrams below for all $j \in I$.

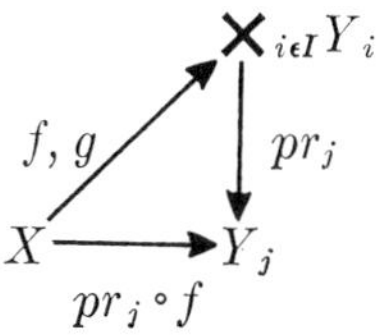

Since g is induced by $(pr_i \circ f \mid i \in I)$, g makes these diagrams commute. By the uniqueness assertion of 12.10, $f = g$. ◻

DISTRIBUTIVITY AND COMMUTATIVITY

By using products, a union of intersections can be written as an intersection of unions, and vice versa.

12.13. THEOREM (DISTRIBUTIVE LAWS)

Let $(J_i \mid i \in I)$ be a nonempty family of nonempty sets, and for each $i \in I$ let $(X_{i,j} \mid j \in J_i)$ be a family of sets indexed by J_i. Define

$$J = \times_{i \in I} J_i.$$

Then

$$\bigcup_{i \in I} \left(\bigcap_{j \in J_i} X_{i,j}\right) = \bigcap_{c \in J} \left(\bigcup_{i \in I} X_{i,c(i)}\right)$$

and

$$\bigcap_{i \in I} \left(\bigcup_{j \in J_i} X_{i,j}\right) = \bigcup_{c \in J} \left(\bigcap_{i \in I} X_{i,c(i)}\right).$$

Proof

Let

$$x \in \bigcup_{i \in I} \left(\bigcap_{j \in J_i} X_{i,j}\right).$$

If $c \in J$, then since

$$x \in \bigcap_{j \in J_i} X_{i,j}$$

for some $i \in I$,

$$x \in X_{i,c(i)}$$

for such an i, and hence

$$x \in \bigcup_{i \in I} X_{i,c(i)}.$$

Thus

$$x \in \bigcap_{c \in J} \left(\bigcup_{i \in I} X_{i,c(i)}\right).$$

This proves the inclusion

$$\bigcup_{i \in I} \left(\bigcap_{j \in J_i} X_{i,j}\right) \subset \bigcap_{c \in J} \left(\bigcup_{i \in I} X_{i,c(i)}\right).$$

To prove the opposite inclusion, suppose

$$x \notin \bigcup_{i \in I} \left(\bigcap_{j \in J_i} X_{i,j}\right).$$

Then

$$(\forall i \in I)(\exists j \in J_i)(x \notin X_{i,j}).$$

By the axiom of choice, there exists a choice function c for the family

$$(\{j \mid j \in J_i \;\&\; x \notin X_{i,j}\} \mid i \in I).$$

Then $c \in J$, $x \notin X_{i,c(i)}$ for each $i \in I$, and

$$x \notin \bigcup_{i \in I} X_{i,c(i)}.$$

Hence

$$x \notin \bigcap_{c \in J} (\bigcup_{i \in I} X_{i,c(i)}).$$

This completes the proof of the first stated equality. The proof of the second is similar. □

The commutative law for products takes a form different from the commutative laws for union and intersection. One would hardly expect a product to be left unchanged by permutations of its factors, for $X \times Y$ would not be equal to $Y \times X$ in general. For a permutation σ of the index class I of a family $(X_i \mid i \in I)$, one gets only a bijection

$$\varphi: \mathsf{X}_{i \in I} X_i \simeq \mathsf{X}_{i \in I} X_{\sigma(i)}.$$

This result is a special case of the next proposition.

12.14. PROPOSITION

Let $(X_i \mid i \in I)$ and $(Y_j \mid j \in J)$ be families of sets, let

$$\sigma: J \simeq I,$$

and let $(f_j \mid j \in J)$ be a family of maps with

$$f_j: X_{\sigma(j)} \simeq Y_j \qquad (j \in J).$$

Then

$$x \mapsto (f_j(x_{\sigma(j)}) \mid j \in J)$$

defines a bijection

$$\varphi: \mathsf{X}_{i \in I} X_i \simeq \mathsf{X}_{j \in J} Y_j.$$

Proof

Observe that

$$(f_{\sigma^{-1}(i)})^{-1}: Y_{\sigma^{-1}(i)} \simeq X_i$$

for each $i \in I$. Hence we may define a map

$$\psi: \mathsf{X}_{j \in J} Y_j \to \mathsf{X}_{i \in I} X_i$$

by

$$\psi(y) = ((f_{\sigma^{-1}(i)})^{-1}(y_{\sigma^{-1}(i)}) \mid i \in I).$$

An easy computation shows that $\psi \circ \varphi$ and $\varphi \circ \psi$ are the identity maps on their respective domains. □

In addition to the commutative law mentioned above, two other consequences of 12.14 are of note. For the first, take $J = I$, $\sigma = 1_I$; from bijections

$$f_i: X_i \simeq Y_i \qquad (i \in I)$$

one gets a bijection

$$\varphi: \times_{i\in I} X_i \simeq \times_{i\in I} Y_i.$$

For the second, take $X_i = X$ for all $i \in I$ and $Y_j = Y$ for all $j \in J$:

12.15. COROLLARY

Let I and J be classes and X and Y be sets. If

$$\sigma: J \simeq I$$

and

$$f: X \simeq Y,$$

then

$$\varphi: \text{Map}\ (I,X) \simeq \text{Map}\ (J,Y),$$

where

$$\varphi(x) = f \circ x \circ \sigma \qquad (x \in \text{Map}\ (I,X)).$$

For another proof of 12.15, see Chapter 9, Exercise Q.

EXERCISES

A. Prove the converse of 12.8.

B. Let X be a set and I be a nonempty set. For each $i \in I$, let e_i: Map $(I,X) \to X$ be the evaluation map at i. Use the results of this section to prove there is a unique map $\delta: X \to$ Map (I,X) such that $e_i \circ \delta = 1_X$ for every $i \in I$.

C. Let $(X_i \mid i \in I)$ be a family of sets.
(a) If $j \in I$ and $A \subset X_j$, show that $pr_j^{-1}\langle A \rangle \simeq \times_{i\in I} Y_i$, where $Y_i = X_i$ if $i \neq j$ and $Y_j = A$.
(b) If $J \subset I$ and $(A_j \mid j \in J)$ is a family with $A_j \subset X_j$ for each $j \in J$, represent $\bigcap_{j\in J} pr_j^{-1}\langle A \rangle$ as a product of subsets of the X_i's.

D. Let $A \subset \times_{i\in I} X_i$. Show that

$$A \subset \times_{i\in I}\, pr_i\langle A \rangle$$

but that this inclusion may be strict.

E. In the notation of 12.11, show:
(a) f is injective if each f_i is injective.
(b) f is surjective if each f_i is surjective.

F. Let X and Y be sets. Let

$$\alpha: X \times Y \simeq \times_{i\in I} X_i, \qquad \beta: Y \times X \simeq \times_{i\in I} Y_i$$

be the representations of $X \times Y$, $Y \times X$ as products defined as in 12.5(1). If

$$\varphi: \times_{i\in I} X_i \simeq \times_{i\in I} Y_i$$

is the bijection given by 12.14, which map ψ makes the following diagram commute?

$$\begin{array}{ccc} X \times Y & \xrightarrow{\psi} & Y \times X \\ \Big\downarrow{\scriptstyle \alpha} & & \Big\downarrow{\scriptstyle \beta} \\ \mathsf{X}_{i\in I} X_i & \xrightarrow{\varphi} & \mathsf{X}_{i\in I} Y_i \end{array}$$

G. For a permutation σ of I, describe explicitly the bijection

$$\varphi\colon \mathsf{X}_{i\in I}\, X_i \simeq \mathsf{X}_{i\in I}\, X_{\sigma(i)}$$

given by the commutative law.

H. Let X, Y, Z be sets. Construct bijections

$$\varphi\colon \mathrm{Map}\,(Y, \mathrm{Map}\,(X,Z)) \simeq \mathrm{Map}\,(X, \mathrm{Map}\,(Y,Z))$$

and

$$\psi\colon \mathrm{Map}\,(X \times Y, Z) \simeq \mathrm{Map}\,(X, \mathrm{Map}\,(Y,Z)).$$

I. Let $(X_i \mid i \in I)$ and $(Y_j \mid j \in J)$ be nonempty families of sets. Prove:

(a) $(\bigcap_{i\in I} X_i) \cup (\bigcap_{j\in J} Y_j) = \bigcap_{(i,j)\in I\times J} (X_i \cup Y_j)$.

(b) $(\bigcup_{i\in I} X_i) \cap (\bigcup_{j\in J} Y_j) = \bigcup_{(i,j)\in I\times J} (X_i \cap Y_j)$.

(c) $(\bigcup_{i\in I} X_i) \times (\bigcup_{j\in J} Y_j) = \bigcup_{(i,j)\in I\times J} (X_i \times Y_j)$.

(d) $(\bigcap_{i\in I} X_i) \times (\bigcap_{j\in J} Y_j) = \bigcap_{(i,j)\in I\times J} (X_i \times Y_j)$.

J. Let $(X_i \mid i \in I)$ and $(Y_i \mid i \in I)$ be two families of sets indexed by the same nonempty index class I. Prove:

$$(\textstyle\bigcap_{i\in I} X_i) \times (\bigcap_{i\in I} Y_i) = \bigcap_{i\in I} (X_i \times Y_i).$$

Does the analog of this equation for union also hold?

K. Let $(X_i \mid i \in I)$, $(Y_i \mid i \in I)$ be as in Exercise J. Explain why the equation

$$\mathsf{X}_{i\in I}\, X_i \cap \mathsf{X}_{i\in I}\, Y_i = \mathsf{X}_{i\in I}\, (X_i \cap Y_i)$$

is technically incorrect, and render it correct through a suitable modification. [*Hint*: Each $x \in \mathsf{X}_{i\in I}\, X_i$ is a map with domain I and codomain $\bigcup_{i\in I} X_i$.]

L. Let $(Y_i \mid i \in I)$ be a nonempty family of sets. Let Y be a class, and let $(q_i \mid i \in I)$ be a family of maps with

$$q_i\colon Y \to Y_i \qquad (i \in I).$$

Assume the following: If $(f_i \mid i \in I)$ is a family of maps and if X is a class with

$$f_i\colon X \to Y_i \qquad (i \in I),$$

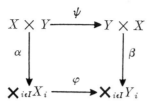

then there exists a unique map f making the diagram

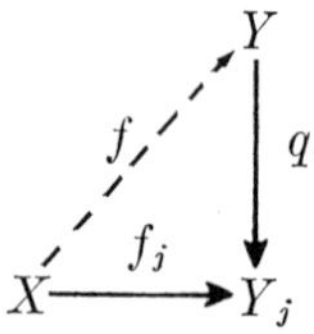

commute for all $j \in I$. Under this assumption, show that there is a *unique* bijection

$$h\colon \times_{i\in I} Y_i \simeq Y$$

making the diagrams

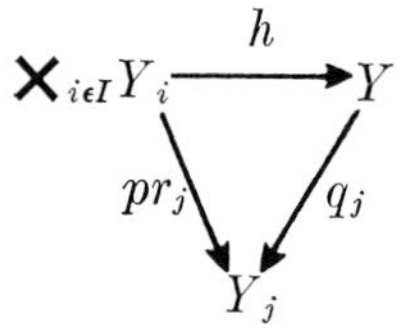

commute for all $j \in I$. (Thus, the property expressed in 12.10 determines $\times_{i\in I} Y_i$ and its projections uniquely "up to a one-to-one correspondence.")

M. Let $(X_i \mid i \in I)$ be a nonempty family of nonempty sets, and for each $i \in I$ let R_i be an equivalence relation on X_i. Construct an equivalence relation R on

$$X = \times_{i\in I} X_i$$

and a bijection f making the diagrams below commute for all $j \in I$, where p, p_j, r_j, q_j are the projections.

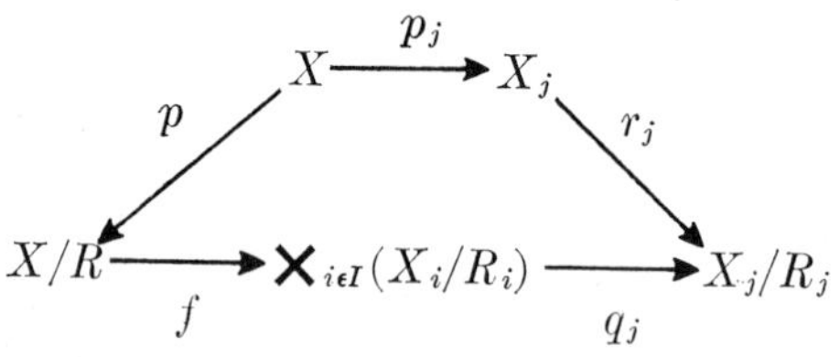

("The product of quotients is a quotient of the product.")

N. Let $(X_i \mid i \in I)$ be a nonempty, disjoint family of sets, and let Y be a set. Construct a one-to-one correspondence between $\times_{i\in I} \operatorname{Map}(X_i, Y)$ and $\operatorname{Map}(\bigcup_{i\in I} X_i, Y)$.

P. Let $(X_i \mid i \in I)$ be a family of nonempty sets, and let $j \in I$. Determine all sections of the jth projection.

Q. Let $(I, \leq)$ be a directed set (see Chapter 11, Exercise P). Let $X = (X_i \mid i \in I)$ be a family of sets and $f = (f_j{}^i \mid i, j \in I \;\&\; i \leq j)$

be a family of maps with

$$f_j{}^i\colon X_j \to X_i \qquad (i \leq j)$$

such that

$$f_i{}^i = \text{identity map of } X_i \qquad (i \in I),$$

$$f_j{}^i \circ f_k{}^j = f_k{}^i \qquad (i \leq j \leq k);$$

then one calls (X, f) an *inverse* (or *projective*) *system on* $(I, \leq)$.
(a) Let $(X_i \mid i \in \omega)$ be a sequence of sets and $(f_i \mid i \in \omega)$ a sequence of maps with $f_i\colon X_{i+1} \to X_i$ for all $i \in \omega$. Construct an inverse system (X, f) on $(\omega, \leq)$ such that $f_{i+1}{}^i = f_i$ for all $i \in \omega$.
(b) Repeat (a) starting with families indexed by $\mathbf{Z}$, where $\mathbf{Z}$ is directed by its usual ordering $\leq$.
(c) Let $(Y_\alpha \mid \alpha \in A)$ be a nonempty family of nonempty sets. Let $I = \mathcal{P}(A)$. Define a suitable relation $\leq$ in I which directs I and an inverse system (X, f) on $(I, \leq)$ for which

$$X_i = \mathsf{X}_{\alpha \in i}\, Y_\alpha \qquad (i \in I).$$

R. (Continuation of Q.) Define

$$X_\infty = \{x \mid x \in \mathsf{X}_{i \in I}\, X_i \,\&\, (\forall i, j \in I)(i \leq j \Rightarrow x_i = f_j{}^i(x_j))\}.$$

One calls X_∞ the *inverse limit of* (X, f) and denotes it by inv lim (X, f).
(a) If (X, f) is the inverse system of Exercise Q(c), construct a bijection between X_∞ and $\mathsf{X}_{\alpha \in A}\, Y_\alpha$.
(b) Suppose Z is a set with $X_i = Z$ for all $i \in I$ and $f_j{}^i = 1_Z$ whenever $i \leq j$. Compute X_∞ in this case.
For each $i \in I$, let

$$f_\infty{}^i = pr_i \mid X_\infty\colon X_\infty \to X_i.$$

(c) Show that

$$f_j{}^i \circ f_\infty{}^j = f_\infty{}^i \qquad (i, j \in I,\ i \leq j).$$

(d) Suppose $I = \omega$ and $f_j{}^i$ is surjective for all i, j with $i \leq j$. Suppose also $X_i \neq \varnothing$ for all $i \in I$. Prove that $f_\infty{}^i$ is then surjective for every $i \in I$, and hence $X_\infty \neq \varnothing$.

S. (Continuation of R.) Let Y be a set and $(h^i \mid i \in I)$ be a family of maps, with $h^i\colon Y \to X_i$ for each $i \in I$, satisfying

$$f_j{}^i \circ h^j = h^i \qquad (i \leq j)$$

(the superscript i on h^i is just an index). Prove:
(a) There is a unique map $h\colon Y \to X_\infty$ such that

$$h^i = f_\infty{}^i \circ h \qquad (i \in I).$$

(b) The map h is injective if and only if $(h^i \mid i \in I)$ "separates points of Y," that is,

$$y, z \in Y \,\&\, y \neq z \Rightarrow (\exists i \in I)(h^i(y) \neq h^i(z)).$$

T. (Continuation of S.) Let (Y,g) be a second inverse system on $(I,\leq)$. Let $\alpha = (\alpha^i \mid i \in I)$ be a family of maps with

$$\alpha^i \colon X_i \to Y_i \qquad (i \in I)$$

such that

$$g_j{}^i \circ \alpha^j = \alpha^i \circ f_j{}^i \qquad (i \leq j);$$

one calls α an *inverse system of maps from* (X,f) *to* (Y,g).

(a) Prove that there is a unique map

$$\alpha^\infty \colon X_\infty \to Y_\infty$$

such that

$$g_\infty{}^i \circ \alpha^\infty = \alpha^i \circ f_\infty{}^i.$$

(b) Prove that the map α^∞ is injective (respectively, bijective) if each α^i is.

(c) Let $\varphi \colon A \to A$ be a surjection, where A is a nonempty set. Let (X,f) be the inverse system on $(\mathbb{Z},\leq)$ such that $f_{i+1}{}^i = \varphi$ and $X_i = A$ for all $i \in \mathbb{Z}$ [see Exercise Q(b)]. Define $\alpha = (\alpha^i \mid i \in \mathbb{Z})$ by $\alpha^i = \varphi$ for all $i \in \mathbb{Z}$. Verify that α is an inverse system of maps from (X,f) to (X,f). Compute α^∞ explicitly and show that it is bijective.

(d) State and prove the "dual" of Chapter 11, Exercise S(c) for inverse systems.

13. Equipollence

We begin in this chapter our treatment of Cantor's theory of cardinality, which concerns the size of sets.

One way to determine the size of a given collection of objects is to count it: Arrange the objects in some order, match each object in turn with one of a standard collection of "numbers" (already arranged in a standard order), and measure the size of the given collection by the last number matched with an object of that collection. Through such a process we shall in Chapter 19 assign to each set a measure of its size. (The necessary machinery will be developed in Chapters 15–18.) Before that, in Chapter 14, we shall measure the size of those sets, the finite sets, that correspond to the collections actually encountered in concrete experience.

Underlying the idea of counting is another method for obtaining information about the size of a given collection without determining exactly how large it is: Find whether the collection has the same size as another collection by seeing whether the members of the given collection can be paired one-by-one with the members of the other. This method is more primitive than counting. Even the child who cannot yet recite the names of numbers can see he has the same number of hands as feet by placing a hand on each foot. It is this simple idea of comparison, whose systematic exploitation was initiated by Cantor, that we now study.

EQUIPOLLENCE

13.1. DEFINITION

Let X and Y be classes. One says that X is *equipollent to* Y and writes

$$X \simeq Y$$

to mean

$$(\exists f)(f\colon X \simeq Y).$$

Thus one class is equipollent to a second if the elements of the first can be put into some one-to-one correspondence with the elements of the second.

13.2. EXAMPLES

(1) If X is a class, then $X \simeq \varnothing$ if and only if $X = \varnothing$.
(2) Since there is no surjection $f\colon \{0\} \to \{0,1\}$, 1 is not equipollent to 2.
(3) Let

$$E = \{n \mid n \in \omega \;\&\; 2 \text{ divides } n\}.$$

The map

$$\begin{aligned} &\omega \to E \\ &k \mapsto 2k \qquad (k \in \omega) \end{aligned}$$

is bijective, so $\omega \simeq E$. However, $E \subsetneq \omega$. Thus a set can be equipollent to a strict subset of itself! This phenomenon may seem contrary to intuition, for in fact it does not occur with the sets modeled on concretely experienced objects in the "real world." It is simply forced upon us when the formal analog of "has the same number of elements as" is applied to the abstract constructs of set theory which transcend concrete experience.

(4) We have

$$\omega \simeq \omega \setminus \{0\},$$

for the map

$$\begin{aligned} &\omega \to \omega \setminus \{0\} \\ &n \mapsto n+1 \qquad (n \in \omega) \end{aligned}$$

is bijective. Of course there are many bijections from ω to $\omega \setminus \{0\}$, for example, the map f defined by

$$f(n) = \begin{cases} 9 & \text{if } n = 0, \\ n & \text{if } 1 \leq n \leq 8, \\ n+1 & \text{if } n > 8. \end{cases}$$

(5) If X is a class, then $X \times \{y\} \simeq X$ for each set y.
(6) If $X \sim Y$, then $\mathcal{P}(X) \simeq \mathcal{P}(Y)$.
(7) If

$$X_1 \simeq X_2, \qquad Y_1 \simeq Y_2,$$

then

$$X_1 \times Y_1 \simeq X_2 \times Y_2,$$

$$\text{Map }(X_1,Y_1) \simeq \text{Map }(X_2,Y_2),$$

and

$$X_1 \cap Y_1 = \varnothing = X_2 \cap Y_2 \Rightarrow X_1 \cup Y_1 \simeq X_2 \cup Y_2.$$

Examples (8)–(12) below concern sets of real numbers. If $a, b \in \mathbb{R}$ with $a < b$, then the open interval $\{x \mid x \in \mathbb{R} \,\&\, a < x < b\}$ with endpoints a and b will be denoted by $]a,b[$, and the corresponding left-closed, right-open interval $\{x \mid x \in \mathbb{R} \,\&\, a \leq x < b\}$ by $[a,b[$.

(8) Any open interval in $\mathbb{R}$ is equipollent to any other. In fact, let $a, b, c, d \in \mathbb{R}$ with $a < b$ and $c < d$. Then the map f defined by

$$(*) \qquad f(x) = \frac{d - c}{b - a} \cdot (x - a) + c$$

is a bijection from $]a,b[$ to $]c,d[$. This map may be visualized in the plane as a projection through a point (Figure 13.1). Note that the same formula (*) defines a bijection from $[a,b[$ to $[c,d[$.

(9) The restriction of $x \mapsto \tan x$ to the interval $]-\pi/2, \pi/2[$ defines a bijection from this interval to $\mathbb{R}$. Composing this with a bijection from $]0,1[$ to $]-\pi/2, \pi/2[$ we obtain a bijection from $]0,1[$ to $\mathbb{R}$, so

$$]0,1[\simeq \mathbb{R}.$$

(10) We claim that also

$$[0,1[\simeq \mathbb{R}.$$

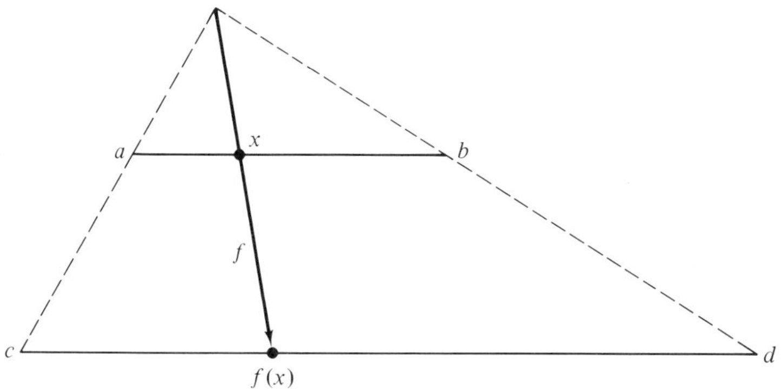

Figure 13.1

We prove this by constructing a bijection from $[0,1[$ to $]0,1[$. For each $n \in \omega$ define

$$I_n = [1 - 1/2^n, 1 - 1/2^{n+1}[, \qquad J_n = [1/2^{n+1}, 1/2^n[.$$

Then $(I_n \mid n \in \omega)$ partitions $[0,1[$, and $(J_n \mid n \in \omega)$ partitions $]0,1[$ (Figure 13.2).

By (1), we can construct for each $n \in \omega$ a bijection

$$f_n\colon I_n \simeq J_n.$$

Then the map $f\colon [0,1[\to]0,1[$ such that $f \mid I_n = f_n$ for each $n \in \omega$ is the desired bijection.

(11) Fix a natural number $b \geq 2$. Define S to be the subset

$$\{s \mid s \in \text{Map}\,(\omega,b)\ \&\ (\forall n \in \omega)(\exists m \in \omega)(m > n\ \&\ s_m \neq b - 1)\}$$

of Map (ω,b) consisting of all those sequences in $\{0,1,\ldots,b-1\}$ which are not constantly $b - 1$ from some point on. We shall show

$$[0,1[\simeq S.$$

From this and (10) it will follow that

$$\mathbb{R} \simeq S.$$

We recall some facts about the expansion of real numbers to the base b (the reader may wish to keep in mind the cases of $b = 2$, binary expansion, and $b = 10$, decimal expansion). If $s \in S$, then by comparison with the convergent geometric series $\sum_{n=0}^{\infty} b^{-n}$, the series $\sum_{n=0}^{\infty} s_n b^{-(n+1)}$ converges to some number in $[0,1[$.

Now let $x \in [0,1[$. There is a unique sequence

$$s(x) = (s_0(x), s_1(x), \ldots) \in S$$

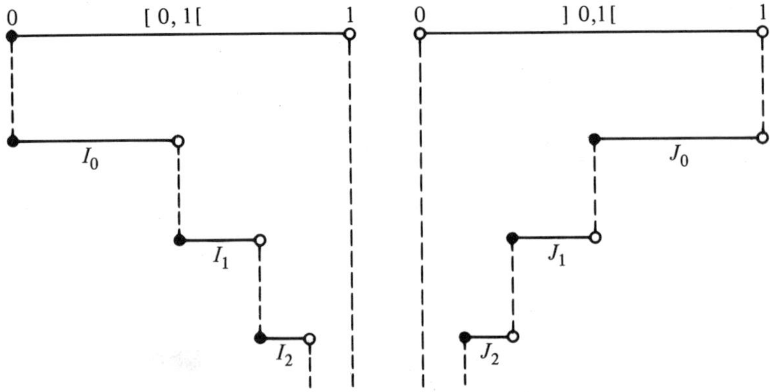

Figure 13.2

such that

$$x = \sum_{n=0}^{\infty} s_n(x) b^{-(n+1)}.$$

This sequence is defined recursively (using Chapter 10, Exercise L) as follows: Take $s_0(x)$ to be the unique $k \in \{0,1,\ldots,b-1\}$ such that

$$kb^{-1} \le x < (k+1)b^{-1}.$$

Once $s_0(x), \ldots, s_n(x)$ have been determined, take $s_{n+1}(x)$ to be the unique $k \in \{0,1,\ldots,b-1\}$ such that

$$kb^{-(n+2)} \le x - \sum_{i=0}^{n} s_i(x) b^{-(i+1)} < (k+1)b^{-(n+2)}.$$

The map $x \mapsto s(x)$ from $[0,1[$ to S is then the desired bijection.

(12) Let S be as in (11). The map

$$(s,t) \mapsto (s_0,t_0,s_1,t_1,\ldots)$$

from $S \times S$ into S is bijective, for it has the inverse

$$u \mapsto ((u_0,u_2,u_4,\ldots),\ (u_1,u_3,u_5,\ldots))$$

from S to $S \times S$. Then $S \times S \simeq S$. From (11), $\mathbb{R} \simeq S$. Hence by (7),

$$\mathbb{R} \times \mathbb{R} \simeq \mathbb{R}.$$

Thus there are just as many points on the real line as there are in the entire plane! (Of course there are essential differences between the line and the plane. There are precise notions of algebraic and topological dimension in terms of which the line is one-dimensional and the plane is two-dimensional. These notions involve structure in addition to the purely set-theoretic structure with which equipollence is concerned.)

From results in Chapter 9 we obtain:

13.3. PROPOSITION

(1) For any class X, $X \simeq X$.
(2) If $X \simeq Y$, then $Y \simeq X$.
(3) If $X \simeq Y$ and $Y \simeq Z$, then $X \simeq Z$.

In particular, the relation R in $\mathcal{U}$ such that

$$XRY \Leftrightarrow X \simeq Y$$

is an equivalence relation on $\mathcal{U}$.

Here is a useful example of equipollence.

13.4. THEOREM

Let X be a class. Then

$$\mathcal{P}(X) \simeq \text{Map}\,(X,2).$$

Proof

For each $E \in \mathcal{P}(X)$, let

$$\chi_E: X \to \{0,1\}$$

be the characteristic function of E in X, so that

$$\chi_E(x) = \begin{cases} 1 & \text{if } x \in E, \\ 0 & \text{if } x \in X \setminus E. \end{cases}$$

Consider the map

$$\varphi: \mathcal{P}(X) \to \text{Map }(X,2)$$
$$E \mapsto \chi_E \qquad (E \in \mathcal{P}(X))$$

and the map

$$\psi: \text{Map }(X,2) \to \mathcal{P}(X)$$
$$f \mapsto f^{-1}\langle\{1\}\rangle \qquad (f \in \text{Map }(X,2)).$$

If $E \in \mathcal{P}(X)$, then

$$(\psi \circ \varphi)(E) = \psi(\chi_E) = \chi_E{}^{-1}\langle\{1\}\rangle = E.$$

Hence $\psi \circ \varphi$ is the identity map of $\mathcal{P}(X)$.

If $f \in \text{Map }(X,2)$ and if $E = f^{-1}\langle\{1\}\rangle$, then since

$$f(x) = 1 \Leftrightarrow x \in f^{-1}\langle\{1\}\rangle \Leftrightarrow \chi_E(x) = 1,$$

we have

$$(\varphi \circ \psi)(f) = \varphi(E) = \chi_E = f.$$

Hence $\varphi \circ \psi$ is the identity map of Map $(X,2)$.

We conclude that

$$\varphi: \mathcal{P}(X) \simeq \text{Map }(X,2). \quad \square$$

DOMINANCE

The interpretation of '$X \simeq Y$' as "X has the same number of elements as Y" suggests the following formal analogs to "X has at most as many elements as Y" and "X has strictly fewer elements than Y."

13.5. DEFINITION

Let X and Y be classes. One says that X is *dominated by* Y and writes

$$X \preceq Y$$

to mean $X \simeq Z$ for some subclass Z of Y. One says that X is *strictly dominated by* Y and writes

$$X \prec Y$$

to mean $X \preceq Y$ & $X \not\simeq Y$.

Clearly $X \preceq Y$ if and only if there exists an injection $f: X \to Y$. Moreover,

$$X \preceq Y \Leftrightarrow X \prec Y \vee X \simeq Y.$$

13.6. EXAMPLES

(1) If $X \subset Y$, then $X \preceq Y$.
(2) Since $\omega \setminus \{0\} \subset \omega$,

$$\omega \setminus \{0\} \preceq \omega.$$

Even though $\omega \setminus \{0\}$ is strictly contained in ω,

$$\omega \setminus \{0\} \not\prec \omega$$

since $\omega \setminus \{0\} \simeq \omega$ by 13.2(4).

Additional examples of strict dominance are furnished by

13.7. THEOREM (CANTOR)

Let X be a set. Then

$$X \prec \mathcal{P}(X).$$

Proof

The map

$$X \to \mathcal{P}(X)$$

$$x \mapsto \{x\} \qquad (x \in X)$$

is an injection, so $X \preceq \mathcal{P}(X)$.

Suppose $X \simeq \mathcal{P}(X)$. Let

$$f: X \simeq \mathcal{P}(X).$$

Note that $f(x) \subset X$ for each $x \in X$. Define

$$Y = \{x \mid x \in X \,\&\, x \notin f(x)\}.$$

Then $Y \subset X$, so $Y \in \mathcal{P}(X)$. Since f is surjective, we may choose $x \in X$ with $f(x) = Y$. If $x \in Y$, then $x \notin f(x)$ by definition of Y, that is, $x \notin Y$. Conversely, if $x \notin Y$, then $x \in f(x)$, that is, $x \in Y$. Hence $x \in Y$ if and only if $x \notin Y$, a contradiction. □

The argument in the first paragraph of the proof does not use the assumption that X be a set, so

$$X \preceq \mathcal{P}(X)$$

for any class X.

Cantor's theorem indicates that given any set, there is a still larger set.

From 9.4 together with 9.12 and from 9.22 we obtain the following criterion.

13.8. THEOREM

Let X and Y be nonempty classes. A necessary condition for $X \preceq Y$ to hold is that there exist some surjection $f: Y \to X$. In case Y is a set, this condition is also sufficient.

13.9. COROLLARY

If Y is a set and if $X \preceq Y$ (in particular, if $X \simeq Y$), then X is a set.

Proof

Use 13.8 and the axiom of replacement. □

The following facts are evident:

(1) $X \preceq X$.
(2) If $X \preceq Y$ and $Y \preceq Z$, then $X \preceq Z$.
(3) $X \nprec X$.
(4) If $X' \simeq X$ and $Y' \simeq Y$, then

$$X \preceq Y \Rightarrow X' \preceq Y', \qquad X \prec Y \Rightarrow X' \prec Y'.$$

Notice the similarities of properties (1) and (2) with the properties

$$X \subset X, \qquad X \subset Y \,\&\, Y \subset Z \Rightarrow X \subset Z$$

of the inclusion relation on the one hand and with the properties

$$n \leq n, \qquad m \leq n \,\&\, n \leq k \Rightarrow m \leq k$$

of the usual order relation in ω on the other.

SCHRÖDER-BERNSTEIN THEOREM

In addition to being reflexive and transitive, the relations $\subset$ and $\preceq$ are also "antisymmetric" in the sense that

$$(*) \qquad X \subset Y \,\&\, Y \subset X \Rightarrow X = Y,$$

$$m \leq n \,\&\, n \leq m \Rightarrow m = n.$$

It is decidedly *false*, however, that $X \preceq Y \,\&\, Y \preceq X$ implies $X = Y$, for if $X \simeq Y$ and $X \neq Y$, then $X \preceq Y$ and $Y \preceq X$ (consider, for example, ω and $\omega \setminus \{0\}$). The most that can be said by way of analogy with (*) is the following famous theorem, which had already been conjectured by Cantor.

13.10. THEOREM (SCHRÖDER-BERNSTEIN)

Let X and Y be sets such that

$$X \preceq Y, \qquad Y \preceq X.$$

Then

$$X \simeq Y.$$

Proof

There exist injections

$$f: X \to Y, \qquad g: Y \to X.$$

Define recursively a sequence $(E_n \mid n \in \omega)$ of subsets of X by

$$E_0 = X \setminus g\langle Y \rangle,$$

$$E_{n+1} = (g \circ f)\langle E_n \rangle \qquad (n \in \omega),$$

and let

$$E = \bigcup_{n \in \omega} E_n.$$

Since g is injective, its restriction $g_0: Y \to \operatorname{rng} g$ is bijective and so has the inverse

$$g_0^{-1}: \operatorname{rng} g \to Y.$$

Now

$$X \setminus E \subset \operatorname{dmn} g_0^{-1} = \operatorname{rng} g$$

since $E_0 \subset E$, so we may form the map

$$h: X \to Y$$

such that

$$h \mid E = f \mid E, \qquad h \mid X \setminus E = g_0^{-1} \mid (X \setminus E).$$

Since f and g_0^{-1} are both injective, h is injective. To show that $h: X \simeq Y$, it remains only to show that $Y \subset h\langle X \rangle$.

Let $y \in Y$. Just suppose

$$y \notin h\langle X \setminus E \rangle = g^{-1}\langle X \setminus E \rangle.$$

We show that $y \in f\langle E \rangle = h\langle E \rangle$. By supposition, $g(y) \notin X \setminus E$, that is, $g(y) \in E$. Then $g(y) \in E_n$ for some $n \in \omega$. We must have $n > 0$, so $g(y) = g(f(x))$ for some $x \in E_{n-1}$. But g is injective, so $y = f(x)$ with $x \in E$. Hence $y \in f\langle E \rangle$. $\square$

How intuitively appealing this theorem is! If each of two sets has at most as many elements as the other, then the two have the same number of elements.

One application of 13.10 is the proof that

$$\mathbb{R} \simeq \operatorname{Map}(\omega, 2).$$

Take first $b = 2$ in 13.2(11) to see that $\mathbb{R}$ is equipollent to a subset of Map $(\omega,2)$, so that

$$\mathbb{R} \preceq \text{Map}\ (\omega,2).$$

Now take $b = 10$ and consider the set $S \subset$ Map $(\omega,10)$ defined in 13.2(11) for this choice of b. Since $b - 1 = 9 > 2$, if $s \in$ Map $(\omega,2)$, then the element $\bar{s} \in$ Map $(\omega,10)$ having the same graph as s belongs to S, and $s \mapsto \bar{s}$ defines an injection of Map $(\omega,2)$ into S. Thus

$$\text{Map}\ (\omega,2) \preceq S \simeq \mathbb{R},$$

and

$$\text{Map}\ (\omega,2) \preceq \mathbb{R}.$$

The Schröder-Bernstein theorem now gives $\mathbb{R} \simeq$ Map $(\omega,2)$.

If $f: X \to Y$ and $g: Y \to X$ are given injections, the Schröder-Bernstein theorem does *not* promise that $f: X \simeq Y$ or $g: Y \simeq X$; consider, for example, the inclusion map of $\omega \setminus \{0\}$ into ω and the map $n \mapsto 2n + 1$ from ω into $\omega \setminus \{0\}$. However, our proof of the Schröder-Bernstein theorem does furnish the following corollary: *If X and Y are sets and if $f: X \to Y$ and $g: Y \to X$ are injections, then there are partitions $\{X_1,X_2\}$ of X and $\{Y_1,Y_2\}$ of Y such that $f\langle X_1\rangle = Y_1$ and $g\langle Y_2\rangle = X_2$.*

As analogs to familiar properties of the strict order relation in ω, we have:

13.11. COROLLARY

Let X, Y, and Z be sets. Then:

(1) If $X \prec Y$, then $Y \not\preceq X$.
(2) If $X \prec Y$ and $Y \prec Z$, then $X \prec Z$.

Having observed the similarities in behavior of the relations $\simeq$, $\preceq$, $\prec$ in $\mathfrak{U}$ to the relations $=$, $\leq$, $<$ in ω, let us examine the connection between them more closely.

13.12. LEMMA

Let X and Y be classes, and let x_0 and y_0 be sets with $x_0 \notin X$ and $y_0 \notin Y$. Then

$$X \simeq Y \Leftrightarrow X \cup \{x_0\} \simeq Y \cup \{y_0\}.$$

Proof

If $f: X \simeq Y$, then the map

$$g: X \cup \{x_0\} \to Y \cup \{y_0\}$$

such that $g \mid X = f$ and $g(x_0) = y_0$ is bijective.

Assume

$$g: X \cup \{x_0\} \simeq Y \cup \{y_0\}.$$

Let

$$y_1 = g(x_0).$$

If $y_1 = y_0$, then of course the restriction of g to X and Y is a bijection from X into Y. In the general case, we first interchange y_0 and y_1. More precisely, define a permutation σ of $Y \cup \{y_0\}$ by

$$\sigma(y_1) = y_0,$$

$$\sigma(y_0) = y_1,$$

$$\sigma(y) = y \qquad (y \in Y \setminus \{y_0, y_1\}).$$

Then $(\sigma \circ g) \mid X$ is an injection with range Y. □

13.13. PROPOSITION

Let $m, n \in \omega$. Then:

(1) $m = n \Leftrightarrow m \simeq n$.
(2) $m < n \Leftrightarrow m \prec n$.
(3) $m \leq n \Leftrightarrow m \preceq n$.

Proof

(1) If $m = n$, then of course $m \simeq n$. To prove the converse, define

$$E = \{n \mid n \in \omega \,\&\, (\forall m \in \omega)(n \simeq m \Rightarrow n = m)\}.$$

We use induction to show $E = \omega$. Clearly $0 \in E$.

Assume $n \in E$. We show $n + 1 \in E$. Let $m \in \omega$ with $n + 1 \simeq m$. We must prove $n + 1 = m$. Now $m > 0$, and

$$n \cup \{n\} = n + 1 \simeq m = (m - 1) \cup \{m - 1\}.$$

By 13.12, $n \simeq m - 1$. Since $n \in E$, $n = m - 1$. Then $n + 1 = (m - 1) + 1 = m$.

(2) Assume $m < n$. Then $m \in n$, $m \subset n$ since n is full, and $m \preceq n$. Since $m \neq n$, $m \not\simeq n$ by (1). Hence $m \prec n$.

Conversely, assume $m \prec n$ and just suppose $m \geq n$. By (1), $m \neq n$, so $m > n$. By what was proved in the preceding paragraph, $n \prec m$, in contradiction to 13.11.

(3) follows from (1) and (2). □

It is both plausible and true that of any two given sets, one of them has at least as many elements as the other.

13.14. THEOREM (COMPARABILITY THEOREM)

If X and Y are sets, then

$$X \preceq Y \vee Y \preceq X.$$

Surprisingly, without considerable preparation a proof of this theorem would be very difficult to construct. Therefore the proof is deferred until much more machinery is at our disposal. Once this machinery has been developed, the proof will be a triviality (see 17.19 and Chapter 16, Exercise S).

13.15. THEOREM (TRICHOTOMY LAW)

If X and Y are sets, then exactly one of the following three alternatives holds:

$$X \prec Y, \qquad X \simeq Y, \qquad Y \prec X.$$

Proof

That at least one alternative holds follows from 13.14. That at most one holds follows from 13.11. □

Theorems 13.10, 13.11, 13.14, and 13.15 are actually true for arbitrary classes, not just sets. In fact, in case some of the classes involved there are proper classes, we have the following more precise results:

(1) If X and Y are proper classes, then $X \simeq Y$.
(2) If X is a set and Y is a proper class, then $X \prec Y$.

These facts will be proved later (see 18.27).

EXERCISES

A. Prove or disprove:
 (a) If $X \simeq A$ and $Y \simeq B$, then $X \cap Y \simeq A \cap B$.
 (b) If $X \preceq A$ and $Y \preceq B$, then Map $(X,Y) \preceq$ Map (A,B).

B. Call any subset of $\mathbb{R}$ of one of the following kinds an "interval": a bounded interval, whether open, closed, or half-open and half-closed; a ray, to the right or to the left, whether or not it includes its endpoint; $\mathbb{R}$ itself. Prove that any two intervals in $\mathbb{R}$ are equipollent to one another.

C. Show that $\mathbb{R} \setminus]0,1[\simeq \mathbb{R}$.

D. Prove that $\mathbb{R}^n \simeq \mathbb{R}$ for every $n \geq 2$.

E. Generalize 13.12 to the case where $n \geq 1$ distinct elements $x_0, \ldots, x_{n-1}$ are adjoined to X and n distinct elements $y_0, \ldots, y_{n-1}$ are adjoined to Y.

F. If $(E_n \mid n \in \omega)$ is a disjoint sequence of nonempty sets, is $E_i \prec \bigcup_{n \in \omega} E_n$ for every $i \in \omega$? for some $i \in \omega$?

G. Where does the proof of 13.7 break down if X happens to be a proper class? Is the restriction in 13.7 that X be a set really needed?

H. Use Cantor's theorem to prove anew that $\mathcal{U}$ is not a set.

I. Let $f\colon X \to E$ be an injection from a set X into a subset E of X. Define

$$D = \{y \mid y \in X \,\&\, (\exists n \in \omega)(\exists x \in X \setminus E)(y = f^n(x))\}.$$

Let $h\colon X \to E$ be the map such that

$$h \mid D = f \mid D, \qquad h \mid X \setminus D = 1_{X \setminus D}.$$

Show that h is bijective.

J. Use Exercise I to obtain a new proof of the Schröder-Bernstein theorem.

K. Show that the map

$$f\colon \mathcal{P}(\omega) \to \mathbb{R}$$

$$E \mapsto \textstyle\sum_{n=0}^{\infty} \chi_E(n)/2^{2n} \qquad (E \in \mathcal{P}(\omega))$$

is injective; here χ_E denotes the characteristic function of E in ω.

L. Compare the sizes of the following subsets of the Euclidean plane $\mathbb{R} \times \mathbb{R}$:

(a) $\mathbb{R} \times \mathbb{R}$.
(b) $[-1,1] \times [-1,1]$.
(c) $\{(x,y) \mid (x,y) \in \mathbb{R} \times \mathbb{R} \,\&\, x^2 + y^2 \leq 1\}$.
(d) $\{(x,y) \mid (x,y) \in \mathbb{R} \times \mathbb{R} \,\&\, 0 \leq x \,\&\, 0 \leq y \,\&\, x + y \leq 1\}$.
(e) $]-1,1[\times]-1,1[$.
(f) $\{(x,y) \mid (x,y) \in \mathbb{R} \times \mathbb{R} \,\&\, x^2 + y^2 < 1\}$.
(g) $]-1,1[\times \{0\}$.

M. Compare the sizes of the following sets of real-valued sequences:

(a) $\{x \mid x \in \text{Map}(\omega,\mathbb{R}) \,\&\, \lim_{n\to\infty} x_n \text{ exists}\}$.
(b) $\{x \mid x \in \text{Map}(\omega,\mathbb{R}) \,\&\, \lim_{n\to\infty} x_n = 0\}$.
(c) $\{x \mid x \in \text{Map}(\omega,\mathbb{R}) \,\&\, \sum_{n=0}^{\infty} x_n \text{ converges}\}$.

N. For a closed interval $[a,b] \subset \mathbb{R}$, let

$$[a,b]^* = [a,\, a + (b-a)/3] \cup [a + 2(b-a)/3,\, b]$$

be the set obtained by deleting the open middle-third interval $]a + (b-a)/3,\, a + 2(b-a)/3[$ from $[a,b]$. For a union $E = \bigcup_{i=1}^{k} [a_i,b_i]$ of a disjoint family $([a_i,b_i] \mid 1 \leq i \leq k)$ of closed intervals, let

$$E^* = \textstyle\bigcup_{i=1}^{k} [a_i,b_i]^*.$$

Define the sequence $(K_n \mid n \in \omega)$ recursively as follows:

$$K_0 = [0,1],$$

$$K_{n+1} = K_n^* \qquad (n \in \omega).$$

Let

$$K = \bigcap_{n\in\omega} K_n.$$

The set K is called the *Cantor discontinuum* (or *middle third*, or *ternary, set*). Prove:

(a) $K_{n+1} \subset K_n$ for all $n \in \omega$.

(b) For each $n \in \omega$, K_n is the union of a disjoint family of 2^n closed intervals each of length 3^{-n}.

(c) K contains no nonempty open interval.

(d) $K \simeq \mathbb{R}$. [*Hint*: Determine which elements of [0,1] belong to K in terms of their ternary (base 3) expansions.]

P. A set $A \subset \mathbb{R}^2$ is said to be "similar and similarly situated" to a set $B \subset \mathbb{R}^2$ in case $B = p + \lambda A$ for some $p \in \mathbb{R}^2$ and some $\lambda > 0$; here

$$p + \lambda A = \{p + \lambda a \mid a \in A\}$$

and for $p = (p_1,p_2)$, $a = (a_1,a_2)$,

$$p + \lambda a = (p_1 + \lambda a_1,\, p_2 + \lambda a_2),$$

as is usual in vector algebra.

(a) A "ball" in $\mathbb{R}^2$ is the set of all points in $\mathbb{R}^2$ whose distance from some fixed point in $\mathbb{R}^2$ is less than some fixed positive number. Show that any ball in $\mathbb{R}^2$ is similar and similarly situated to any other ball in $\mathbb{R}^2$.

(b) Let X and Y be any two subsets of $\mathbb{R}^2$ which are bounded (each is contained in some ball) and have nonempty interior (each contains a ball). Prove that there are partitions $\{X_1,X_2\}$ of X and $\{Y_1,Y_2\}$ of Y such that X_1 is similar and similarly situated to Y_1 and X_2 to Y_2. [*Hint*: For $p \in \mathbb{R}^2$ and $\lambda > 0$, the map $x \mapsto p + \lambda x$ is injective.]

14. Finite and Countable Sets

In this chapter we consider sets which are "small" in the sense of being equipollent to subsets of ω, and to some of these, the finite sets, we assign standard measures of their sizes. The discussion, of interest in its own right, provides a model for the later generalization to cardinal numbers of arbitrary sets.

FINITE SETS

Recall (13.13) that two equipollent natural numbers are actually equal. Hence if a set is equipollent to both a natural number m and a natural number n, then $m = n$. This justifies part of the following definition.

14.1. DEFINITION

A class X is said to be *finite* if $X \simeq n$ for some $n \in \omega$, and *infinite* otherwise. If X is finite, the natural number

$$(\iota n)\,(n \in \omega \,\&\, X \simeq n)$$

is denoted by $\#\,(X)$ and is called the *number of elements in* X.

14.2. EXAMPLES

(1) The empty set is finite, and $\#\,(\varnothing) = 0$.

(2) If x is a set, then $\{x\}$ is finite, and $\#\,(\{x\}) = 1$. If y is another set distinct from x, then $\{x,y\}$ is finite, and $\#\,(\{x,y\}) = 2$.

(3) Let $n \in \omega$. Then n is finite, and $\#(n) = n$.

(4) *The set ω is infinite.* In fact, suppose to the contrary that $\omega \simeq n$ for some $n \in \omega$. Then $n > 0$, $n = m \cup \{m\}$ where $m = n - 1$, $m \simeq \omega \setminus \{0\}$ by 13.12, $m \simeq \omega$ since $\omega \setminus \{0\} \simeq \omega$, and hence $m \simeq n$, an impossibility since $m < n$.

Let $n \in \omega$ with $n > 0$. Then

$$n = \{0,\ldots,n-1\} \simeq \{1,\ldots,n\}.$$

Hence if X is a finite class, then $\#(X) = n$ if and only if there is a one-to-one correspondence from $\{1,\ldots,n\}$ to X, in other words, X has the form

$$X = \{x_1,\ldots,x_n\}$$

for a family $(x_1,\ldots,x_n)$ of distinct elements of X.

Obviously a finite class X is a set, and if Y is another class with $Y \simeq X$, then Y is also finite and $\#(Y) = \#(X)$.

14.3. THEOREM

If Y is a finite set and $X \subsetneq Y$, then X is also finite, and $\#(X) < \#(Y)$.

Proof

If $n \in \omega$ and $f\colon Y \simeq n$, then a subset X of Y is finite if and only if $f\langle X\rangle$ is a finite subset of n, and in this case $\#(X) = \#(f\langle X\rangle)$. Hence it suffices to prove the theorem for the case $Y \in \omega$.

We use induction on n to show

$$(\forall X \subsetneq n)(X \text{ is finite } \& \ \#(X) < n).$$

There is nothing to prove for $n = 0$.

Let $n \in \omega$ and assume each $X \subsetneq n$ is finite and satisfies $\#(X) < n$. Let $X \subsetneq n + 1$. We must show X is finite and $\#(X) < n + 1$. If $X \subset n$, then this follows from the inductive hypothesis. Now suppose $X \not\subset n$. Then $n \in X$. Define

$$Z = X \setminus \{n\}.$$

Then $Z \subset n$, and since $X \subsetneq n + 1 = n \cup \{n\}$, $Z \subsetneq n$. By the inductive hypothesis, Z is finite and $\#(Z) < n$. Letting $m = \#(Z)$, we have

$$X = Z \cup \{n\} \simeq m \cup \{m\} = m + 1,$$

so X is finite and

$$\#(X) = m + 1 \leq n < n + 1. \quad \square$$

14.4. COROLLARY

Let Y be a set and X be a finite set. Then:

(1) $Y < X$ if and only if Y is finite and $\#(Y) < \#(X)$.

(2) $Y \preceq X$ if and only if Y is finite and $\#(Y) \leq \#(X)$.

The next three theorems assert the finiteness of sets constructed from finite sets.

14.5. THEOREM

Let X and Y be *disjoint* finite sets. Then $X \cup Y$ is finite, and

$$\#(X \cup Y) = \#(X) + \#(Y).$$

Proof

Let $m = \#(X)$ and $n = \#(Y)$, and define

$$E = \{i \mid i \in \omega \,\&\, m \leq i < m + n\}.$$

The map $i \mapsto i + m$ is a bijection from n to E, so

$$Y \simeq n \simeq E.$$

Then

$$\begin{aligned} X \cup Y \simeq m \cup E &= \{i \mid i \in \omega \,\&\, 0 \leq i < m\} \cup E \\ &= \{i \mid i \in \omega \,\&\, 0 \leq i < m + n\} = m + n. \quad \square \end{aligned}$$

14.6. COROLLARY

Let X be a finite set and let $Y \subset X$. Then

$$\#(X \setminus Y) = \#(X) - \#(Y).$$

Proof

We have

$$X = (X \setminus Y) \cup Y, \qquad (X \setminus Y) \cap Y = \varnothing.$$

By 14.5,

$$\#(X) = \#(X \setminus Y) + \#(Y). \quad \square$$

14.7. COROLLARY

Let X and Y be finite sets (not necessarily disjoint). Then $X \cup Y$ is finite, and

$$\#(X \cup Y) = \#(X) + \#(Y) - \#(X \cap Y).$$

Proof

We have

$$X \cup Y = (X \setminus (X \cap Y)) \cup (Y \setminus (X \cap Y)) \cup (X \cap Y).$$

Applying 14.5 twice, we see that $X \cup Y$ is finite, and

$$\#(X \cup Y) = \#(X \setminus (X \cap Y)) + \#(Y \setminus (X \cap Y)) + \#(X \cap Y).$$

The computation of $\#(X \cup Y)$ is completed by using 14.6 twice. $\square$

When we speak of a "finite family" below, we mean, of course, a family whose index class is finite.

14.8. COROLLARY

The union of a finite family of finite sets is finite.

Proof

Use 14.7 and induction on the number of elements in the index set. □

14.9. THEOREM

Let X and Y be finite sets. Then $X \times Y$ is finite, and

$$\#(X \times Y) = \#(X) \cdot \#(Y).$$

Proof

If $m = \#(X)$ and $n = \#(Y)$, then $m \times n \simeq X \times Y$. Hence it suffices to prove

$$m \times n \simeq m \cdot n \qquad\qquad (m, n \in \omega).$$

We use induction on n. If $n = 0$, then $m \times n = \varnothing = 0 = m \cdot n$ for all $m \in \omega$.

Let $n \in \omega$, and assume $m \times n \simeq m \cdot n$ for all $m \in \omega$. Let $m \in \omega$. Then

$$m \times (n + 1) = m \times (n \cup \{n\}) = (m \times n) \cup (m \times \{n\}).$$

Now

$$m \times n \simeq m \cdot n, \qquad m \times \{n\} \simeq m,$$

$$(m \times n) \cap (m \times \{n\}) = \varnothing.$$

By 14.5,

$$m \times (n + 1) \simeq (m \cdot n) + m = m(n + 1). \quad □$$

Theorem 14.9 has the interesting interpretation that if one thing can be done in m ways and another thing can be done in n ways, then there are in all $m \cdot n$ ways of doing the two things in order. (Represent the ways of doing the first as the elements of a set X, the ways of doing the second as the elements of a set Y; each element of $X \times Y$ represents a way of doing the first, then the second thing.)

14.10. COROLLARY

The product of a finite family of finite sets is finite.

14.11. THEOREM

If X and Y are finite sets, then Map (X,Y) is finite, and

$$\#(\text{Map }(X,Y)) = \#(Y)^{\#(X)}.$$

Proof

As usual it is enough to consider the case $X, Y \in \omega$. Fix $m \in \omega$. For each $n \in \omega$, Map (n,m) is the product of a family of n sets equal to m and is finite by 14.10. We shall show $\#(\text{Map }(n,m)) = m^n$ for all $n \in \omega$. This is evident if $m = 0$. Suppose $m > 0$.

If $n = 0$, then Map (n,m) consists of the empty map alone, and $\#(\text{Map }(n,m)) = 1 = m^n$.

Now let $n \in \omega$ and assume

$$\#(\text{Map }(n,m)) = m^n.$$

The map

$$\text{Map }(n+1, m) \to \text{Map }(n,m) \times m$$

$$f \mapsto (f \mid n, f(n))$$

is bijective, so

$$\text{Map }(n+1, m) \simeq \text{Map }(n,m) \times m.$$

From 14.9,

$$\#(\text{Map }(n+1, m)) = \#(\text{Map }(n,m)) \cdot \#(m) = m^n \cdot m = m^{n+1}. \quad \square$$

14.12. COROLLARY

If X is a finite set, then $\mathcal{P}(X)$ is finite, and

$$\#(\mathcal{P}(X)) = 2^{\#(X)}.$$

Proof

Use 13.4. $\square$

COUNTABLE SETS

We now pass to sets which are larger than the finite sets.

14.13. DEFINITION

A class is said to be *denumerable* if it is equipollent to ω, *countable* if it is either finite or denumerable, and *uncountable* if it is not countable.

Caution: The usage of these terms in the literature is far from uniform. "Countable" sometimes means what we have called "denumerable," and then "at most countable" is used for what we have called "countable." The term "countably infinite" is sometimes used to mean "denumerable" in our sense.

Note that a class cannot be both finite and denumerable, for ω is infinite.

A denumerable class is an infinite set; a countable class is a set, either finite or infinite.

If X is denumerable, then Y is denumerable if and only if $X \simeq Y$; if X is countable and if $X \simeq Y$, then Y is countable.

Evidently a necessary condition for X to be countable is that $X \preceq \omega$. This condition is also sufficient, as will be shown presently.

14.14. EXAMPLE

The class $\mathcal{P}(\omega)$ is uncountable, for $\omega \prec \mathcal{P}(\omega)$ by Cantor's theorem (13.7).

In Chapter 13 it was shown that $\mathbb{R} \simeq \text{Map}\ (\omega,2)$. Since $\text{Map}\ (\omega,2) \simeq \mathcal{P}(\omega)$ by 13.4, it follows that *the set* $\mathbb{R}$ *of all real numbers is uncountable.*

The uncountability of $\mathbb{R}$ can also be proved directly, without the use of Cantor's theorem. It suffices to show that the interval $[0,1[$ is not denumerable. We sketch the proof of this, following Cantor's original diagonal procedure.

Just suppose $[0,1[$ is denumerable. Then

$$[0,1[\ = \{x_n \mid n \in \omega\}$$

for a sequence $(x_0,x_1,\ldots)$ of real numbers. Each x_n has some decimal expansion

$$x_n = \textstyle\sum_{i=0}^{\infty} x_{ni}10^{-(i+1)}$$

with the x_{ni}'s natural numbers between 0 and 9 inclusive, but not all equal to 9 from some point on [see 13.2(11)]. The digits x_{ni} used in the expansions of all the x_n's may be displayed as follows in an "infinite array."

$$\begin{array}{cccc} x_{00} & x_{01} & x_{02} & \cdots \\ \\ x_{10} & x_{11} & x_{12} & \cdots \\ \\ x_{20} & x_{21} & x_{22} & \cdots \\ \cdot & \cdot & \cdot & \\ \cdot & \cdot & \cdot & \\ \cdot & \cdot & \cdot & \end{array}$$

The entries on the nth row (counting the top row as the 0th) are, in order, the digits used in the expansion of x_n.

Consider the entries

$$x_{00},\ x_{11},\ x_{22},\ \ldots$$

on the principal diagonal of this array. Construct a sequence $(y_n \mid n \in \omega)$

by letting

$$y_n = \begin{cases} 0 & \text{if } x_{nn} \neq 0, \\ 1 & \text{if } x_{nn} = 0. \end{cases}$$

Then

$$y = \textstyle\sum_{i=0}^{\infty} y_i 10^{-(i+1)}$$

is a number in $[0,1[$. However, for each $n \in \omega$ the nth digit y_n in the expansion of y differs from the nth digit x_{nn} in the expansion of x_n, so $y \neq x_n$. This contradicts the assumption that each element of $[0,1[$ is some x_n.

We are going to show next that each infinite set X contains some denumerable set. The idea of the proof is simple enough: Since $X \neq \varnothing$, we may choose some $x_0 \in X$. Now $\{x_0\}$ is finite, so $X \setminus \{x_0\}$ is infinite, and we may choose some $x_1 \in X \setminus \{x_0\}$. Since $X \setminus \{x_0,x_1\}$ is still infinite, we may choose some $x_2 \in X \setminus \{x_0,x_1\}$. Repeating this procedure again and again, we obtain a sequence $(x_0,x_1,x_2,\ldots)$ of distinct elements of X.

14.15. THEOREM

Let X be an infinite set. Then X has some denumerable subset.

Proof

There exists a choice function c for $\mathcal{P}(X) \setminus \{\varnothing\}$. Define

$$G\colon \mathcal{P}(X) \to \mathcal{P}(X)$$

by

$$G(E) = \begin{cases} E \cup \{c(X \setminus E)\} & \text{if } E \neq X, \\ \varnothing & \text{if } E = X. \end{cases}$$

By recursion there exists a sequence $(E_n \mid n \in \omega)$ of subsets of X such that

$$E_0 = \varnothing,$$

$$E_{n+1} = G(E_n) \qquad (n \in \omega).$$

Note that if E is a finite subset of X, then $E \neq X$ and so $G(E) = E \cup \{c(X \setminus E)\}$. A trivial induction shows that E_n is a finite subset of X for each $n \in \omega$, so

$$c(X \setminus E_n) \in X \setminus E_n \subset X \qquad (n \in \omega).$$

Hence we may define a map

$$h\colon \omega \to X$$

by

$$h(n) = c(X \setminus E_n) \qquad (n \in \omega).$$

We need only show h is injective, for then $h\langle\omega\rangle$ is a denumerable subset of X.

We first use induction on n to show

$$m < n \Rightarrow h(m) \in E_n \qquad (m, n \in \omega).$$

There is nothing to prove for $n = 0$. Let $n \in \omega$ and assume $k < n$ implies $h(k) \in E_n$. Let $m \in \omega$ with $m < n + 1$. Since E_n is finite,

$$E_{n+1} = G(E_n) = E_n \cup \{c(X \setminus E_n)\} = E_n \cup \{h(n)\}.$$

If $m = n$, then certainly $h(m) \in E_{n+1}$. If $m < n$, then $h(m) \in E_n \subset E_{n+1}$ by the inductive hypothesis. Hence $h(m) \in E_{n+1}$ in either case.

Finally, we show h is injective. Let $m, n \in \omega$ with $m < n$. By the preceding paragraph,

$$h(m) \in E_n.$$

However, $h(n) \notin E_n$ since

$$h(n) = c(X \setminus E_n) \in X \setminus E_n.$$

Hence $h(m) \neq h(n)$. ▯

14.16. COROLLARY

Let X be a set. Then:

(1) X is infinite if and only if $\omega \preceq X$.
(2) X is finite if and only if $X \prec \omega$.

Proof

(1) Assume X is infinite. By 14.15, there exists $E \subset X$ with $\omega \simeq E$. Then $\omega \preceq X$.

Conversely, assume $\omega \preceq X$. If X is finite, then $X \simeq n \subset \omega$ for some $n \in \omega$, $n \prec \omega$ since ω is infinite, and hence $X \prec \omega$, a contradiction.

(2) If $X \prec \omega$, then it is not the case that $\omega \preceq X$, so X is finite by (1). Conversely, if X is finite, then $X \simeq n \prec \omega$ for some $n \in \omega$, so $X \prec \omega$. ▯

Notice that the proof (2) did not use the as yet unproved trichotomy law (13.15). Of course, (2) is a trivial consequence of (1) with the aid of the trichotomy law.

Part (1) of 14.16 says that the smallest infinite sets are the denumerable sets.

14.17. THEOREM

An infinite subset of a countable set is denumerable. Hence any subset of a countable set is countable.

Proof

Let X be countable and Y be infinite with $Y \subset X$. By 14.16, $\omega \preceq Y$. But $Y \preceq X \preceq \omega$, so $Y \simeq \omega$ by the Schröder-Bernstein theorem. ▯

14.18. COROLLARY

Let X be a set. Then:

(1) X is countable if and only if $X \preceq \omega$.
(2) X is uncountable if and only if $\omega \prec X$.

Proof

(1) It has already been observed that $X \preceq \omega$ if X is countable. The converse follows from 14.17.

(2) If $\omega \prec X$, then it is not the case that $X \preceq \omega$, so X is uncountable by (1). Conversely, if X is uncountable, then X is infinite, $\omega \preceq X$ by 14.16, and hence $\omega \prec X$. $\square$

Part (1) of 14.18 implies that a nonempty set X is countable if and only if there is a map of ω onto X, in other words, there is a sequence $(x_0,x_1,\ldots)$ of elements of X with $X = \{x_n \mid n \in \omega\}$. In contrast, X is denumerable if and only if there is a sequence $(x_0,x_1,\ldots)$ of *distinct* elements of X with $X = \{x_n \mid n \in \omega\}$.

Recall that the infinite set ω strictly contains the subset $\omega \setminus \{0\}$ to which it is equipollent. With the aid of 14.15 we can prove that such behavior is characteristic of all infinite classes.

14.19. THEOREM (DEDEKIND)

Let X be a set. Then the following two conditions are equivalent:

(1) X is infinite.
(2) There exists some $Y \subsetneq X$ such that $Y \simeq X$.

Proof

Assume (1). We show (2). By 14.15, there exists an injection

$$j\colon \omega \to X.$$

Define

$$f\colon X \to X$$

by

$$f(x) = \begin{cases} x & \text{if } x \notin j\langle\omega\rangle, \\ j(n+1) & \text{if } n \in \omega \,\&\, x = j(n). \end{cases}$$

Let $Y = f\langle X\rangle$. Clearly f is injective, so $Y \simeq X$. Moreover, $Y = X \setminus \{j(0)\}$.

Conversely, assume (2) and just suppose X is finite. There exists $n \in \omega$ and a bijection

$$f\colon X \simeq n.$$

Since X has some element not belonging to Y, $n > 0$. Let $E = f\langle Y\rangle$. Then $E \subsetneq n$. We shall derive a contradiction by showing that $\#(X) < n$. If $n - 1 \notin E$, then $E \subset n - 1$ and hence $\#(E) \leq n - 1 < n$. If $n - 1 \in E$, there is some j with $0 \leq j < n - 1$ and $j \notin E$, and the reasoning in the preceding sentence may be applied to the set

$$(E \setminus \{n - 1\}) \cup \{j\}$$

which is equipollent to E. $\square$

A set X is sometimes said to be "Dedekind infinite" if it satisfies condition (2) above, and then X is said to be "ordinary infinite" if it is infinite in our sense.

COUNTABLE UNIONS AND PRODUCTS

We now discuss the countability and denumerability of sets constructed from countable and denumerable families of countable sets. Here "countable family" means a family whose index class is countable, and "denumerable family" means one whose index class is denumerable.

14.20. PROPOSITION

Let X be a finite and Y be a denumerable set. Then $X \cup Y$ is denumerable.

Proof

There exists $n \in \omega$ and a bijection

$$f: n \simeq X.$$

There also exists a bijection

$$g: \omega \simeq Y.$$

The map

$$\begin{aligned} s: \omega \setminus n &\to \omega \\ i &\mapsto i - n \qquad (i \in \omega \setminus n) \end{aligned}$$

is a bijection. Then the map

$$h: \omega \to X \cup Y$$

such that

$$h \mid n = f, \qquad h \mid (\omega \setminus n) = g \circ s$$

is bijective. $\square$

14.21. LEMMA

There is a partition $(A_n \mid n \in \omega)$ of ω such that A_n is denumerable for each $n \in \omega$.

Proof

The construction of the partition is based on the following "diagonal procedure." Arrange the elements of ω in an "infinite array" extending indefinitely to the right and downwards by putting 0 in the leftmost and top position, next putting 1 and 2 on the diagonal just below 0, and so on.

$$\begin{array}{ccccccccc}
0 & & 2 & & 5 & & 9 & & 14 \quad \ldots \\
& \nearrow & & \nearrow & & \nearrow & & \nearrow & \\
1 & & 4 & & 8 & & 13 & & \ldots \\
& \nearrow & & \nearrow & & \nearrow & & & \\
3 & & 7 & & 12 & & \ldots & & \\
& \nearrow & & \nearrow & & & & & \\
6 & & 11 & & \ldots & & & & \\
& \nearrow & & & & & & & \\
10 & & \ldots & & & & & & \\
\cdot & & & & & & & & \\
\cdot & & & & & & & & \\
\cdot & & & & & & & &
\end{array}$$

The direction of writing the entries on each diagonal is upwards and to the right, as indicated by the arrows.

Calling the top row the 0th, the next row the 1st, and so on, we shall define A_n to be the set of entries on the nth row for each n. More precisely, for n, $i \in \omega$, let $f(n,i)$ denote the entry on the nth row and ith column, where the leftmost column is called the 0th. Then

$$(1) \qquad A_n = \{ f(n,i) \mid i \in \omega\} \qquad (n \in \omega).$$

To make the definition of the A_n's rigorous, and to prove they have the desired properties, we need an explicit expression for $f(n,i)$ in terms of n and i.

For each $k \geq 1$ there are $k + 1$ elements on the kth diagonal, where the diagonal consisting of 1 and 2 is called the 1st. Then for each $m \geq 1$ there are in all

$$\textstyle\sum_{k=0}^{m-1} (k + 1) = \sum_{k=1}^{m} k = \dfrac{m(m + 1)}{2}$$

elements above the mth diagonal. Hence $f(m,0)$, the initial entry on the mth diagonal, is given by

$$(2) \qquad f(m,0) = \frac{m(m + 1)}{2} .$$

This holds for $m = 0$, also.

Now let n, $i \in \omega$ with $n + i > 0$. Then $f(n,i)$ is on the $(n + i)$th diagonal and is the ith element on this diagonal following $f(n + i, 0)$, the initial element on this diagonal. Hence

$$f(n,i) = f(n + i, 0) + i.$$

Using (2) with $m = n + i$, we obtain

$$(3) \qquad f(n,i) = \frac{(n+i)(n+i+1)}{2} + i.$$

This also holds for $n = i = 0$.

Reversing our standpoint, we now *define* a map

$$f\colon \omega \times \omega \to \omega$$

by formula (3), and define a sequence $(A_n \mid n \in \omega)$ by (1).

To prove, firstly, that each A_n is denumerable, it suffices to show that for each fixed n the map $i \mapsto f(n,i)$ is injective. To prove, secondly, that the A_n's are pairwise disjoint, it suffices to show that if $m \neq n$, then $f(m,j) \neq f(n,i)$ for all i, j. Hence to prove both these things it suffices to show that f is injective. The details of this are left to the reader (consider the cases $m + j = n + i$, $m + j \neq n + i$ separately).

To prove that $\omega = \bigcup_{n\in\omega} A_n$, we must show that f is surjective. We use induction on k to show that $k \in \operatorname{rng} f$ for all $k \in \omega$. First, $f(0,0) = 0$.

Let $k \in \omega$ and assume $k \in \operatorname{rng} f$. We show $k + 1 \in \operatorname{rng} f$. Let $(n,i) \in \omega \times \omega$ with

$$f(n,i) = k.$$

An easy computation shows

$$k + 1 = \begin{cases} f(i+1, 0) & \text{if } n = 0, \\ f(n-1, i+1) & \text{if } n > 0. \end{cases}$$

Hence $k + 1 \in \operatorname{rng} f$. □

14.22. LEMMA

Let $(X_n \mid n \in \omega)$ be a disjoint sequence of denumerable sets. Then $\bigcup_{n\in\omega} X_n$ is denumerable.

Proof

Let $(A_n \mid n \in \omega)$ be as in 14.21. If $n \in \omega$, then $A_n \simeq \omega \simeq X_n$, and there exists a bijection from A_n to X_n. By the axiom of choice, we may choose simultaneously a sequence $(f_n \mid n \in \omega)$ of maps such that

$$f_n\colon A_n \simeq X_n \qquad (n \in \omega).$$

Then the map

$$f\colon \omega \to \bigcup_{n\in\omega} X_n$$

such that

$$f(x) = f_n(x) \qquad (n \in \omega,\ x \in A_n)$$

is a bijection. □

14.23. THEOREM

The union X of a nonempty countable family $(X_i \mid i \in I)$ of denumerable sets is denumerable.

Proof

Since $I \preceq \omega$, we may assume without loss of generality that $I \subset \omega$. We may further assume that $I = \omega$, for if not we choose some $j \in I$, define $X_n = X_j$ for all $n \in \omega \setminus I$, and observe that $X = \bigcup_{n\in\omega} X_n$.

The set X is infinite since it contains the denumerable set X_0. Hence we need only show that X is countable. Now $(X_n \times \{n\} \mid n \in \omega)$ is a *disjoint* sequence with $X_n \times \{n\} \simeq X_n$ denumerable for each $n \in \omega$, so

$$Y = \bigcup_{n\in\omega} X_n \times \{n\}$$

is denumerable by 14.22. But the map

$$f: Y \to X$$
$$(x,n) \mapsto x$$

is surjective. Hence $X \preceq Y$, and X is countable by 14.18. □

As an application of 14.23, we show that *the set* $\mathbb{Z}$ *of all integers is denumerable.* Let

$$N = \{n \mid n \in \mathbb{Z} \,\&\, n < 0\}.$$

Then $\mathbb{Z} = \omega \cup N$. Now N is denumerable since

$$\omega \to N$$
$$n \mapsto -n - 1 \qquad (n \in \omega)$$

is bijective.

14.24. COROLLARY

The union of a countable family $(X_i \mid i \in I)$ of countable sets is countable.

Proof

We may assume $I \neq \varnothing$. We enlarge those X_i's which are finite by defining

$$Y_i = \begin{cases} X_i & \text{if } X_i \text{ is denumerable,} \\ X_i \cup \omega & \text{if } X_i \text{ is finite.} \end{cases}$$

By 14.20, Y_i is denumerable for each $i \in I$. Now

$$\bigcup_{i\in I} X_i \subset \bigcup_{i\in I} Y_i,$$

by 14.23 the set on the right is denumerable, and hence by 14.17 the set on the left is countable. □

14.25. THEOREM

Let X and Y be denumerable sets. Then $X \times Y$ is denumerable.

Proof

For each $y \in Y$, $X \times \{y\} \simeq X$ and so is denumerable. By 14.23 (or by 14.22),

$$X \times Y = \bigcup_{y \in Y} X \times \{y\}$$

is also denumerable. □

Theorem 14.25 also follows directly from the construction used to prove 14.21, without any intervention of the axiom of choice as was used for 14.22. In fact, $X \simeq \omega \simeq Y$ implies $X \times Y \simeq \omega \times \omega$, and in the proof of 14.21 we constructed explicitly a bijection $f: \omega \times \omega \simeq \omega$.

As an application of 14.25, we show that *the set* $\mathbb{Q}$ *of all rational numbers is denumerable.* The sets $\mathbb{Z}$ and $\mathbb{Z} \setminus \{0\}$ are both denumerable, so $\mathbb{Z} \times (\mathbb{Z} \setminus \{0\})$ is also denumerable. Now the map

$$\mathbb{Z} \times (\mathbb{Z} \setminus \{0\}) \to \mathbb{Q}$$

$$(m,n) \mapsto m/n$$

is surjective, so $\mathbb{Q}$ is countable. Finally, $\mathbb{Q}$ is denumerable since it contains the infinite set $\mathbb{Z}$.

Since $\mathbb{Q}$ is denumerable and $\mathbb{R} = \mathbb{Q} \cup (\mathbb{R} \setminus \mathbb{Q})$ is uncountable, it follows that *the set* $\mathbb{R} \setminus \mathbb{Q}$ *of all irrational numbers is uncountable.*

14.26. COROLLARY

The product of a nonempty finite family of denumerable sets is denumerable. The product of a finite family of countable sets is countable.

Proof

The first assertion follows from 14.25 by use of induction on the number of elements in the index set. The second assertion follows from the first. □

14.27. COROLLARY

If X is a nonempty finite set and Y is a denumerable set, then Map (X,Y) is denumerable. If X is finite and Y is countable, then Map (X,Y) is countable.

Although ω is denumerable and 2 is finite, the set Map $(\omega,2)$ is uncountable. Hence *the product of a countable family of countable sets need not be countable.*

EXERCISES

A. Give interesting interpretations of 14.5 and 14.7 concerning numbers of ways of doing things (compare the discussion following 14.9).

B. Show that a necessary and sufficient condition for a map f from a finite set X into X to be a bijection (in other words, a permutation of X) is that f be an injection, or equivalently, that f be a surjection.

C. If $(m_i \mid i \in I)$ is a nonempty finite family of natural numbers, we define

$$\sum_{i \in I} m_i = \sum_{i=1}^{n} m_{f(i)}$$

where $n = \#(I)$ and

$$f: \{1,2,\ldots,n\} \simeq I$$

is some bijection. In a similar way we define $\prod_{i \in I} m_i$.

(a) Show that the above definition of $\sum_{i \in I} m_i$ is independent of the choice of f.

Now let $(X_i \mid i \in I)$ be a nonempty finite family of finite sets.

(b) Generalize 14.5 by showing that if $(X_i \mid i \in I)$ is disjoint, then

$$\#(\bigcup_{i \in I} X_i) = \sum_{i \in I} \#(X_i).$$

(c) Sharpen 14.8 by proving that

$$\#(\bigcup_{i \in I} X_i) \leq \sum_{i \in I} \#(X_i).$$

(d) Sharpen 14.10 by proving that

$$\#(\times_{i \in I} X_i) = \prod_{i \in I} \#(X_i).$$

D. The *Dirichlet principle* asserts that if n objects are distributed into $m < n$ piles, then at least one pile will contain more than a single object. Formulate and prove a set-theoretic statement expressing this principle. (The Dirichlet principle is also known as the "pigeon-hole" principle: if n letters are distributed into m compartments—pigeon-holes—of a desk, then. . . .)

E. Generalize 14.7 to n finite sets.

F. Let X be a finite set. Prove that the set $\mathfrak{S}(X)$ of all permutations of X is finite and

$$\#(\mathfrak{S}(X)) = (\#(X))!.$$

G. Let X be a finite set with $\#(X) = n$. Let $m \in \omega$ with $m \leq n$. Define

$$\mathcal{P}_m(X) = \{E \mid E \subset X \;\&\; \#(E) = m\}.$$

Show that $\mathcal{P}_m(X)$ is finite and

$$\#(\mathcal{P}_m(X)) = \binom{n}{m}.$$

[*Hint*: One may assume $X = n$. Use induction on n. Note that if $0 < m < n + 1$, then

$$\mathcal{P}_m(n+1) = \mathcal{A} \cup \mathcal{P}_m(n), \qquad \mathcal{A} \cap \mathcal{P}_m(n) = \varnothing,$$

where

$$\mathcal{A} = \{E \mid E \subset n + 1 \,\&\, \#(E) = m \,\&\, n \in E\}.]$$

H. Let X be a finite set with $n = \#(X)$, and let $m \in \omega$ with $m \leq n$. Define

$$\mathcal{S} = \{f \mid f\colon m \to X \,\&\, f \text{ is injective}\}.$$

The set $\mathcal{S}$ may be interpreted as the totality of all ways of selecting in some order exactly m elements from X. Prove that

$$\#(\mathcal{S}) = \frac{n!}{(n-m)!}.$$

$$\left[\textit{Hint}: \quad \frac{n!}{(n-m)!} = m!\binom{n}{m}.\right]$$

I. Use 14.12 and Exercise G to prove the special case

$$\textstyle\sum_{m=0}^{n} \binom{n}{m} = 2^n$$

of the binomial theorem.

J. Let X be a set having a subset Y with $Y \neq X$ and $Y \simeq X$. Without using 14.15 or 14.19, prove directly that X has some denumerable subset. (Thus, had we defined "infinite" to mean "Dedekind infinite," we could still have proved 14.15 without first proving 14.19.)

K. Furnish another proof of 14.21 by taking

$$A_0 = \{0\} \cup \{k \mid k \in \omega \,\&\, k \text{ is odd}\}$$

$$A_n = \{k \mid k \in \omega \,\&\, 2^n \text{ divides } k \,\&\, 2^{n+1} \text{ does not divide } k\} \qquad (n > 0).$$

L. Make precise the use of the axiom of choice in the proof of 14.22.

M. A real number is said to be *algebraic* if it is a root of some polynomial

$$(*) \quad p = a_0x^n + a_1x^{n-1} + \ldots + a_{n-1}x + a_n$$

having as coefficients $a_0, a_1, \ldots, a_n$ integers. Show that the set A of all algebraic real numbers is denumerable and hence that the set $\mathbb{R} \setminus A$ of all *transcendental* numbers is uncountable. [*Hint*: You may assume that a polynomial of degree n has at most n real roots. If p is a polynomial given by (*), define the "height" $h(p)$ by

$$h(p) = n + |a_0| + |a_1| + \ldots + |a_n|.$$

How many polynomials p are there with $h(p) = m$?]

N. Let X be a set having at least two distinct elements. Show that X is infinite if and only if $X \simeq X \setminus F$ for each finite subset F of X.

P. Let X be denumerable. Prove:
(a) The set of all finite subsets of X is denumerable.
(b) The set of all denumerable subsets of X is uncountable.

Q. In the notation of 13.2(11), prove that Map $(\omega, b) \setminus S$ is denumerable.

R. Give another proof of 14.25 by showing that the map

$$f: \omega \times \omega \to \omega$$
$$(m,n) \mapsto 2^m \cdot 3^n$$

is injective.

S. (a) Prove that the plane is not the union of countably many lines.
(b) Prove that three-dimensional space is not the union of countably many planes.

T. For a natural number $n > 0$, an "n-class" is a class $\mathcal{A}$ of nonempty finite sets with $\#(A) = n$ for each $A \in \mathcal{A}$. Without using the axiom of choice, prove:
(a) Every 1-class has a choice function.
(b) If for some $k > 0$ every kn-class has a choice function, then every n-class has a choice function.
(c) If every 2-class has a choice function, then every 4-class has a choice function.

15. Order Relations

The technical apparatus developed in Chapters 16–18 for measuring the sizes of sets concerns the notion of ordering a class by a relation. The specific relations needed—well-orderings—are defined in Chapter 17. In the present chapter the general features of relations which impose orderings are discussed. The discussion here is very general and quite simple and deals largely with terminology. Results worthy of the title "theorem" first appear only in Chapter 16, where the axiom of choice is brought to bear upon ordering relations.

ORDERING BY RELATIONS

To prescribe an "ordering" of a class X is to give a rule determining for any given $x, y \in X$ whether "x precedes y" in some sense. Then such an ordering is described by giving a relation R in X. If $x, y \in X$, then we interpret "xRy" as "x precedes y." Of course, such a relation should have properties reflecting the connotations of "precedes." It is reasonable to require that x precede z in case x precedes y and y precedes z, in other words, that R be transitive. (The real world is not always so reasonable. It has been observed that voters may prefer candidate x to candidate y, and y to candidate z, yet prefer z to x. Such perversity in human behavior must always be kept in mind when building mathematical models of it.)

The prototypical examples which follow indicate that before saying a relation determines an ordering, we may reasonably require it to have one of the following two properties: (i) each $x \in X$ precedes itself; (ii) no $x \in X$ precedes itself. Property (i) simply requires the relation to be reflexive on X; (ii) requires the relation to be "irreflexive" in accordance with the next definition.

15.1. DEFINITION

Let S be a relation in a class X. One says that S is *irreflexive* if $\neg(xSx)$ for each $x \in X$. (In contrast, S is *not* reflexive if and only if $\neg(xSx)$ for *some* $x \in X$.)

15.2. EXAMPLES

(1) The relation R in ω such that

$$xRy \Leftrightarrow x \leq y$$

is the *usual order relation in* ω. The relation S in ω such that

$$xSy \Leftrightarrow x < y$$

is the *usual strict order relation in* ω.

(2) Replacing ω by $\mathbb{Z}$ in (1), we obtain the *usual order relation in* $\mathbb{Z}$ and the *usual strict order relation in* $\mathbb{Z}$.

(3) Let X be a class. The relation R in $\mathcal{P}(X)$ such that

$$ARB \Leftrightarrow A \subset B$$

is the *inclusion relation in* $\mathcal{P}(X)$. The relation S in $\mathcal{P}(X)$ such that

$$ASB \Leftrightarrow A \subsetneq B$$

is the *strict inclusion relation in* $\mathcal{P}(X)$. (By analogy with Examples (1) and (2), it would appear desirable to denote '$\subset$' by '$\subseteq$' and '$\subsetneq$' by '$\subset$', but we follow instead the more common usage.)

In each of these three examples we have two transitive relations in a class X—a reflexive relation R and an irreflexive relation S, the relation R determining one ordering of X, S a second. In a sense the two orderings are the same, for R is completely determined by S, and vice versa. The way in which the two relations determine each other is made precise by the next definition.

15.3. DEFINITION

Let X be a class.

(1) If S is a relation in X, then the *weak relation in X corresponding to S* is the relation R in X such that

$$xRy \Leftrightarrow (xSy \lor x = y).$$

(2) If R is a relation in X, then the *strict relation in X corresponding to R* is the relation S in X such that

$$xSy \Leftrightarrow (xRy \,\&\, x \neq y).$$

15.4. PROPOSITION

Let S be a relation in a class X. Denote by R the weak relation in X corresponding to S. Then:

(1) R is reflexive and dmn $R = X$.
(2) If S is reflexive and dmn $S = X$, then $R = S$.
(3) If S is transitive, then R is transitive.
(4) If S is irreflexive, then S is the strict relation in X corresponding to R.

Proof

(1) Obvious.

(2) Assume S is reflexive and dmn $S = X$. If xSy, then of course xRy. Conversely, if xRy, then either xSy or $x = y$, so xSy in either case.

(3) Assume S is transitive. Suppose xRy and yRz. If $x = y$ or $y = z$, then xRz. So suppose $x \neq y$ and $y \neq z$. Then xSy and ySz, so xSz. Hence xRz.

(4) Assume S is irreflexive. Let $\bar{S}$ denote the strict relation corresponding to R. Then

$$x\bar{S}y \Leftrightarrow xRy \,\&\, x \neq y \Leftrightarrow (xSy \vee x = y) \,\&\, x \neq y$$

$$\Leftrightarrow xSy \,\&\, x \neq y \Leftrightarrow xSy.$$

Hence $\bar{S} = S$. □

Proposition 15.4 says: Suppose a transitive relation S describes an ordering of a class X. From S we obtain a transitive relation R which is reflexive on X. Now for S to describe an ordering of X, either S is reflexive or else it is irreflexive. If S is reflexive, then $R = S$, so nothing new is obtained. If S is irreflexive, then S may be recovered from R. Thus in describing orderings, nothing is lost by restricting our attention to reflexive, transitive relations.

PREORDERINGS

15.5. DEFINITION

A transitive, reflexive relation in a class X whose domain is X is said to *preorder* X and is called a *preordering* of X.

Except when ambiguity may result, the symbol '$\leq$' is used to denote a preordering, and then '$<$' is used to denote the corresponding strict relation. With this notation, we use the following terminology.

If $x \leq y$, we call x a *predecessor of* y and y a *successor* of x, and we say that x *precedes* y (or x is *less than or equal to* y) and that y *follows* x (or y is *greater than or equal to* x). If moreover $x < y$, then we call x a *strict predecessor of* y and y a *strict successor of* x, and we say that x *strictly precedes* y (or x is *less than* y) and that y *strictly follows* x (or y is *greater than* x). If $x \leq y$ and $y \leq z$, we say y is *between* x *and* z; if moreover $x < y$ and $y < z$, we say that y is *strictly between* x *and* z.

If $\leq$ is a preordering of a class X, we call $\langle X, \leq \rangle$ a *preordered class*; when X is actually a set, we speak of a *preordered set.* Frequently we call the class X itself a preordered class when we really mean that $\langle X, \leq \rangle$ is a preordered class for a certain preordering of X. Such an ellipsis may lead to ambiguity, for a given class may admit many preorderings of itself. Nevertheless, we follow the practice of suppressing mention of the specific preordering in question so long as no ambiguity is likely to arise.

We give next two ways of obtaining new preorderings from old.

15.6. DEFINITION

Let the relation $\leq$ preorder the class X, and let $Y \subset X$. Then

$$R = (\leq) \cap (Y \times Y)$$

is a preordering of Y, said to be *induced by* $\leq$. If $y, z \in X$, then

$$yRz \Leftrightarrow y \in Y \,\&\, z \in Y \,\&\, y \leq z.$$

Following our usual practice, we indiscriminately denote again by $\leq$ the preordering of Y induced by $\leq$.

If X is a preordered class whose preordering is $\leq$, then to call a subclass Y of X a "preordered subclass of X" means we are considering the preordering of Y induced by $\leq$.

15.7. DEFINITION

Let R preorder the class X. Then the inverse R^{-1} of R preorders X and is called the *reverse* (or *opposite*) *of* R.

Clearly the strict relation corresponding to R^{-1} is just the inverse of the strict relation corresponding to R. When the preordering of X is denoted $\leq$, then the opposite preordering is denoted $\geq$, and the strict relation corresponding to $\geq$ is denoted $>$. Then

$$x \geq y \Leftrightarrow y \leq x,$$

$$x > y \Leftrightarrow y < x.$$

Thus to pass from a preordering to its opposite is to consider one thing smaller than another when originally is was considered larger.

Note that the opposite of the opposite of a preordering is the original preordering: $(R^{-1})^{-1} = R$.

To every statement about preorderings there is a corresponding "dual" statement obtained by replacing each preordering in the statement by its opposite. Then to each definition concerning a preordering there is a dual definition. Hence to each theorem concerning preorderings there is a corresponding dual statement which is automatically true. We illustrate this principle of duality.

15.8. DEFINITION

Let X be a preordered class and let $Y \subset X$. By an *upper bound of Y in X* is meant an $x \in X$ such that $y \leq x$ for every $y \in Y$; one says that Y is *bounded above in X* if there exists some upper bound of Y in X. Dually, by a *lower bound of Y in X* is meant an element $x \in X$ such that $x \leq y$ for every $y \in Y$; one says that Y is *bounded below in X* if there exists some lower bound of Y in X. Thus x is a lower bound of Y in X (for the preordering $\leq$) if and only if x is an upper bound of Y in X for the preordering $\geq$ opposite to $\leq$, and so Y is bounded below in X (for $\leq$) if and only if Y is bounded above in X for $\geq$.

By a *supremum of Y in X* is meant an $x \in X$ which is an upper bound of Y in X and which precedes every upper bound of Y in X. Dually, an *infimum of Y in X* is a supremum of Y in X for the opposite preordering.

15.9. EXAMPLES

(1) Consider ω with its usual order relation. Then the set of all natural numbers is not bounded above in ω.

(2) Let X be a nonempty set, and preorder $\mathcal{P}(X)$ by inclusion. Then X is the only upper bound of $\mathcal{P}(X) \setminus \{X\}$ in $\mathcal{P}(X)$, but $X \notin \mathcal{P}(X) \setminus \{X\}$.

(3) Consider the usual order relation in ω, and let $Y = \{1,2\}$. Then each natural number $n \geq 2$ is an upper bound of Y in ω, and 2 is the only supremum of Y in ω.

(4) Let $X = \omega \cup \{-1,-2\}$. Define $\leq'$ to be the preordering of X such that

$$m \leq' n \Leftrightarrow m \leq n \qquad (m, n \in \omega),$$

$$n \leq' -1 \;\&\; n \leq' -2 \qquad (n \in \omega),$$

$$-1 \leq' -1,\ -1 \leq' -2,\ -2 \leq' -1,\ -2 \leq' -2.$$

Thus $\leq'$ induces on ω its usual order relation, places -1 and -2 after each $n \in \omega$, and places -1 and -2 after themselves and after each other. The numbers -1 and -2 are distinct suprema of ω in X.

(5) Let $\leq'$ be the preordering of $\mathbb{Z}$ such that

$$\begin{aligned} m \leq' n \Leftrightarrow & (m \geq 0 \,\&\, n \geq 0 \,\&\, m \leq n) \\ & \vee\ (m < 0 \,\&\, n < 0 \,\&\, m \leq n) \\ & \vee\ (m \geq 0 \,\&\, n < 0). \end{aligned}$$

Thus $\leq'$ places all negative integers, in their usual order, after all nonnegative integers. Let $Y = \{m \mid m \in \mathbb{Z} \,\&\, m \geq 0\}$. Then with respect to $\leq'$ each negative integer is an upper bound of Y in $\mathbb{Z}$, but Y does not have a supremum in $\mathbb{Z}$.

(6) Let X be any preordered class. Then every element of X is both a lower bound and an upper bound of $\varnothing$ in X.

(7) Let X be a set, and preorder $\mathcal{P}(X)$ by inclusion. If $\varnothing \neq \mathcal{A} \subset \mathcal{P}(X)$, then $\bigcap \mathcal{A}$ is an infimum and $\bigcup \mathcal{A}$ a supremum of $\mathcal{A}$ in $\mathcal{P}(X)$.

15.10. PROPOSITION

Let X be a preordered class in which each nonempty subclass that is bounded above has a supremum. Then each nonempty subclass of X that is bounded below in X has an infimum in X.

Proof

Let $\varnothing \neq Y \subset X$ such that Y is bounded below in X. Define

$$L = \{x \mid x \text{ is a lower bound of } Y \text{ in } X\}.$$

Then $\varnothing \neq L \subset X$, and L is bounded above in X since each element of the nonempty class Y is an upper bound of L in X. By hypothesis, L has a supremum b in X.

We show that b is an infimum of Y in X. If $y \in Y$, then y is an upper bound of L in X, so $b \leq y$ since b is a supremum of L in X. Thus b is a lower bound of Y in X. Since b is an upper bound of L in X, $b \geq x$ for each lower bound x of Y in X. Thus b is an infimum of Y in X. $\square$

From 15.10 we obtain automatically the dual proposition: Let X be a preordered class in which each nonempty subclass that is bounded below has an infimum. Then each nonempty subclass of X that is bounded above in X has a supremum in X.

PARTIAL ORDERINGS

In each of the examples in 15.2, the relations R and S respectively have the properties named in the next definition.

15.11. DEFINITION

A relation R in a class X is called *antisymmetric* if

$$xRy \,\&\, yRx \Rightarrow x = y.$$

A relation S in X is called *asymmetric* if

$$xSy \Rightarrow \neg (ySx).$$

If R is a reflexive relation in a class X, and if S is the corresponding strict relation, then R is antisymmetric if and only if S is asymmetric.

15.12. DEFINITION

An antisymmetric preordering of a class X is called a *partial ordering of* X and is said to *partially order* X. A preordered class $\langle X, \leq \rangle$ is called a *partially ordered class* if the preordering $\leq$ is actually a partial ordering of X; following the usual convention, we frequently call X itself a partially ordered class.

Suppose X is a partially ordered class. If $Y \subset X$, then the induced preordering of Y is actually a partial ordering of Y. The preordering of X opposite to the given partial ordering of X is actually a partial ordering of X.

With a single exception all specific preorderings described above in this chapter are actually partial orderings. The exception is 15.9(4): $-1 \leq' -2$ and $-2 \leq' -1$. This exceptional example is somewhat artificial. Here are some more natural exceptional examples.

15.13. EXAMPLES

(1) The universe is preordered, but not partially ordered, by $\leq$.

(2) For $n \in \mathbf{Z}$, let

$$|n| = \begin{cases} n & \text{if } n \geq 0, \\ -n & \text{if } n < 0. \end{cases}$$

Define a relation $\leq'$ in $\mathbf{Z}$ by

$$m \leq' n \Leftrightarrow |m| \leq |n|.$$

Then $\leq'$ preorders $\mathbf{Z}$ but does not partially order $\mathbf{Z}$ ($-1 \leq' 1$ and $1 \leq' -1$).

15.14. PROPOSITION

Let the relation R preorder the set X. Define a relation $\sim$ in X by

$$x \sim y \Leftrightarrow xRy \;\&\; yRx.$$

Then $\sim$ is an equivalence relation on X with which R is compatible.

Denote $X/\sim$ by $\tilde{X}$, for each $x \in X$ let $\tilde{x} \in \tilde{X}$ denote the equivalence class of x modulo $\sim$, and let $\tilde{R}$ be the relation in $\tilde{X}$ obtained from R by passing to quotients, so that

$$xRy \Leftrightarrow \tilde{x}\tilde{R}\tilde{y}.$$

Then $\tilde{R}$ partially orders $\tilde{X}$. If R partially orders X, then

$$x \sim y \Leftrightarrow x = y.$$

The proof is trivial.

Proposition 15.14 says that we can obtain from a preordering of X a partial ordering by "dividing out" a certain equivalence relation (and nothing essentially new is obtained in case X is already partially ordered). It is tempting to conclude that nothing is lost in the study of "orderings" by restricting attention to partial orderings. This conclusion is ill-founded, for there is no way of recovering X and R from $\tilde{X}$ and $\tilde{R}$. The fact is that many theorems frequently stated just for partial orderings are true more generally for preorderings, and no substantial difficulties arise by proving them, in greater generality, for arbitrary preorderings.

15.15. DEFINITION

Let R be a relation in a class X. One calls an $x \in X$ a *greatest* (or *largest*, or *last*) *element of* X if $y \in X$ and $y \neq x$ implies yRx. Dually, one calls an $x \in X$ a *least* (or *smallest*, or *first*) element of X if $y \in X$ and $y \neq x$ implies xRy.

Suppose R is a preordering of X. Then x is a greatest (respectively, least) element of X if and only if x is an upper (respectively, a lower) bound of X in X.

Suppose, in addition, that R partially orders X. By antisymmetry, if X has a greatest (respectively, least) element it has only one such, and then we may speak of *the* greatest (respectively, *the* least) element of X.

Let X be a partially ordered class, let $Y \subset X$, and suppose Y has a supremum (respectively, an infimum) b in X. Denote by B the class of all upper (respectively, lower) bounds of Y in X. Then b is the least (respectively, greatest) element of B, is unique, may be called *the* supremum (respectively, *the* infimum) *of* Y *in* X, and is frequently denoted by $\sup Y$ (respectively, $\inf Y$). For obvious reasons b is also called *the least upper bound* (respectively, *the greatest lower bound*) *of* Y *in* X and is denoted by lub Y (respectively, glb Y).

TOTAL ORDERINGS

The following considerations account for the appearance of the adjective "partial" in the term "partial ordering."

15.16. DEFINITION

Let R be a relation in a class X. If $x, y \in X$, then one says that x and y are *comparable* (each *to* the other) if xRy or yRx. One says that R *connects*

X and that R is *connected* to mean that each two distinct elements of X are comparable.

15.17. DEFINITION

A relation $\leq$ in a class X is called a *total ordering of* X and is said to *totally order* X if $\leq$ is a connected partial ordering of X. (Synonyms for "total" are *simple* and *linear*.) We may then speak of a *totally ordered class*. A totally ordered set is called a *chain*.

Suppose X is a preordered class and $Y \subset X$. To call Y a "totally ordered subclass of X" means, of course, that Y is totally ordered by the preordering induced on Y by the given preordering of X (even though the latter need not be a total or even a partial ordering). A totally ordered subset of a preordered class X is also called a *chain in* X.

15.18. EXAMPLES

(1) The usual order relation in ω (or in $\mathbb{Z}$) is a total ordering.

(2) Let X be a class having at least two elements. Then the inclusion relation in $\mathcal{P}(X)$ is a partial ordering of $\mathcal{P}(X)$ which does not totally order $\mathcal{P}(X)$. In fact, if $x, y \in X$ with $x \neq y$, then $\{x\}$ is not comparable to $\{y\}$.

Evidently certain subclasses of $\mathcal{P}(X)$ are totally ordered by inclusion. A chain in $\mathcal{P}(X)$ is sometimes called a *nest*.

(3) Let $X = \omega \setminus \{0\}$, and let $|$ be the relation in X such that $x \mid y$ if and only if x divides y. Then $|$ is a partial ordering of X which does not totally order X.

(4) If X is a totally ordered class, and if $Y \subset X$, then the induced preordering of Y totally orders Y.

(5) The opposite of a total ordering is again a total ordering.

(6) Let $(X_i \mid i \in I)$ be a family of partially ordered sets. Let $X = \bigtimes_{i \in I} X_i$, and let $\leq$ be the relation in X such that

$$x \leq y \Leftrightarrow (\forall i \in I)(x_i \leq y_i).$$

Then $\leq$ is a partial ordering of X, said to be defined *coordinatewise*. This partial ordering need not totally order X even when the partial ordering of each X_i totally orders X_i: see the next example.

(7) Let X and Y be sets, and let Y be partially ordered by $\leq$. From (6) we obtain a partial ordering $\leq$ of the set Map (X,Y) of all maps from X to Y defined by

$$f \leq g \Leftrightarrow (\forall x \in X)(f(x) \leq g(x)).$$

This partial ordering of Map (X,Y) is said to be defined *pointwise*.

Suppose X and Y each have at least two elements. Then $\leq$ does not totally order Map (X,Y), even if the partial ordering of Y totally orders

Y. In fact, let $x_1, x_2 \in X$ with $x_1 \neq x_2$ and let $y_1, y_2 \in Y$ with $y_1 \neq y_2$. Let f and g be any two maps from X to Y such that

$$f(x_1) = y_1, \qquad f(x_2) = y_2,$$
$$g(x_1) = y_2, \qquad g(x_2) = y_1.$$

Then f is not comparable to g.

(8) Let X and Y be sets. Define $\mathfrak{M}$ to be the set of all maps f having Y as codomain such that $\operatorname{dmn} f \subset X$. Define a relation $\leq$ in $\mathfrak{M}$ as follows:

$$f \leq g \Leftrightarrow \operatorname{dmn} f \subset \operatorname{dmn} g \;\&\; g \mid \operatorname{dmn} f = f,$$

in other words, $f \leq g$ if and only if g is an extension of f to $\operatorname{dmn} g$. Then $\leq$ partially orders $\mathfrak{M}$. In fact, if $f, g \in \mathfrak{M}$, then $f \leq g$ if and only if the graph of f is contained in the graph of g.

The following result is completely trivial, but it is used so frequently that it is worth stating it explicitly just once.

15.19. LEMMA

Let X be a *totally* ordered class, and let $x, y \in X$. Then

$$x \leq y \Leftrightarrow (\forall z \in X)(z < x \Rightarrow z < y).$$

EXERCISES

A. (a) Let R be a reflexive relation in a class X with $\operatorname{dmn} R = X$, and let S be the strict relation corresponding to R. Show that S is irreflexive and that R is the weak relation in X corresponding to S.

(b) Suppose X is a set. Construct a one-to-one correspondence between the set of all reflexive relations in X with domain X and the set of all irreflexive relations in X.

(c) Let R be a reflexive relation in X with $\operatorname{dmn} R = X$, and let S be the corresponding strict relation. By 15.4, if S is transitive, then R is transitive. Give a counterexample to the converse of this implication. Under what additional conditions on R is the converse true?

B. Determine all possible preorderings (respectively, partial orderings, total orderings) of $\{1,2,3\}$ and of $\{1,2,3,4\}$.

C. Let R be a relation in a class X. Prove:

(a) If R is asymmetric, then R is irreflexive.

(b) If R is asymmetric and $R \neq \varnothing$, then R is not symmetric.

(c) If R is antisymmetric and if $R \not\subset \Delta_X$, then R is not symmetric.

D. Is the product of a finite family of totally ordered sets totally ordered by its partial ordering defined coordinatewise?

E. Let X be a partially ordered class. A subclass D of X is said to be *order-dense in* X if strictly between any two distinct elements of X there is some element of D, that is,

$$x, y \in X \;\&\; x \neq y \Rightarrow (\exists z \in D)(x < z < y \vee y < z < x).$$

(a) Prove that X is totally ordered if some $D \subset X$ is order-dense in X.

(b) Suppose there exist disjoint $A, B \subset X$ with $A \cup B = X$ such that A has a greatest element, B has a least element, and $a < b$ whenever $a \in A$ and $b \in B$ (then one calls $\langle A, B \rangle$ a *jump in* X). Prove that no $D \subset X$ is order-dense in X.

F. A preordered class X is said to be *conditionally complete* if each nonempty subclass of X which is bounded above in X has a supremum in X.

(a) Show that $\mathbb{Z}$ is conditionally complete for its usual order relation.

(b) Let $\mathfrak{F}$ be the set of all finite subsets of an infinite set. Is $\mathfrak{F}$ conditionally complete when partially ordered by inclusion?

G. A preordered class X is said to be *complete* if every subclass of X has a supremum in X.

(a) If X is a set, show that $\mathcal{P}(X)$ is complete when partially ordered by inclusion.

(b) Prove that a complete preordered class has both a least and a greatest element.

(c) Does every subclass of a complete preordered class X have an infimum in X?

(d) Let X be an infinite set, and let $\mathcal{E}$ be the set of all $E \subset X$ such that either E is finite or $X \setminus E$ is finite. Is $\mathcal{E}$ complete when partially ordered by inclusion?

H. Let X be the product of a family $(X_i \mid i \in I)$ of partially ordered sets whose index class I is a nonempty subset of ω. The *dictionary* (or *lexicographic*) *ordering of* X is the relation $\leq$ in X defined as follows: $x \leq y$ if and only if either $x = y$ or else $x_n < y_n$ where n is the least $i \in I$ such that $x_i \neq y_i$.

(a) Verify that this relation partially orders X.

(b) Suppose each X_i is totally ordered. Must the dictionary ordering of X totally order X?

I. Let $\sim$ be an equivalence relation on a set X, and let $\leq$ be a preordering (respectively, partial ordering, total ordering) of X which is compatible with $\sim$. Show that the relation in $X/\sim$ induced by $\leq$ is again a preordering (respectively, partial ordering, total ordering).

J. A *lattice* is a partially ordered set in which each doubleton has both a supremum and an infimum. For example, if X is any set, then $\mathcal{P}(X)$ is a lattice when partially ordered by inclusion.
(a) Show that each totally ordered set is a lattice.
(b) Must a finite partially ordered set be a lattice?
(c) Show that the partially ordered set of Example 15.18(3) is a lattice, and determine $\sup \{x,y\}$ and $\inf \{x,y\}$ for each doubleton $\{x,y\}$.
(d) Let X be a nonempty set and let L be a lattice. Is Map (X,L) a lattice when it is partially ordered pointwise?
(e) Show that the partially ordered set $\mathcal{E}$ of Exercise G(d) is a lattice.
(f) Let X be an infinite set, and let $\mathcal{C}$ be the set of all $E \subset X$ such that E is countable or $X \setminus E$ is countable. Is $\mathcal{C}$ a lattice when partially ordered by inclusion?

K. Let $\mathcal{L}$ be the set of all partial orderings of a nonempty set X, and partially order $\mathcal{L}$ by inclusion. Is $\mathcal{L}$ a lattice?

L. Let L be a lattice. For $x, y \in L$, define

$$x \vee y = \sup \{x,y\}, \qquad x \wedge y = \inf \{x,y\};$$

one calls $x \vee y$ the *join* and $x \wedge y$ the *meet of x and y*. Establish the following for all $x, y, z \in L$:
(a) $x \vee x = x$ and $x \wedge x = x$.
(b) $x \vee y = y \vee x$ and $x \wedge y = y \wedge x$.
(c) $x \vee (y \vee z) = (x \vee y) \vee z$ and
$x \wedge (y \wedge z) = (x \wedge y) \wedge z$.
(d) $x \vee (x \wedge y) = x$ and $x \wedge (x \vee y) = x$.
[*Hint*: L is also a lattice when partially ordered by the opposite of the given partial ordering. Use duality to reduce your work.]

M. (Continuation of L.) Let L be a set in which there are given two binary operations

$$(x,y) \mapsto x \vee y, \qquad (x,y) \mapsto x \wedge y$$

having properties (a)–(d) of Exercise L. Define a relation $\leq$ in L by

$$x \leq y \Leftrightarrow x \vee y = y.$$

Prove that $\langle L, \leq \rangle$ is a lattice in which

$$\sup \{x,y\} = x \vee y, \qquad \inf \{x,y\} = x \wedge y$$

for all $x, y \in L$. [*Hint*: Show that

$$x \vee y = y \Leftrightarrow x \wedge y = x$$

for all $x, y \in L$.]

N. A lattice L is said to be *distributive* if the following two conditions hold:

(i) $x \wedge (y \vee z) = (x \wedge y) \vee (x \wedge z) \qquad (x, y, z \in L),$

(ii) $x \vee (y \wedge z) = (x \vee y) \wedge (x \vee z) \qquad (x, y, z \in L).$

Prove that in a lattice L the conditions (i) and (ii) are actually equivalent to one another.

P. Let L be a lattice. Suppose L has a least element 0 and a greatest element 1 (the symbols '0' and '1' here do not stand for natural numbers but are merely conventional notations). Suppose also

$$(\forall x \in L)(\exists y \in L)(x \wedge y = 0 \,\&\, x \vee y = 1).$$

Then the lattice L is said to be *complemented*.

(a) Give an example of a complemented lattice.

(b) If L is a complemented, distributive lattice, show that for each $x \in L$, there is a unique $x' \in L$ such that $x \wedge x' = 0$ and $x \vee x' = 1$.

Q. A complemented, distributive lattice is said to be *Boolean*. Which of the lattices mentioned in Exercise J are Boolean?

Additional information about Boolean lattices (and the related Boolean algebras) can be found in Halmos [14].

R. A subclass J of a partially ordered class I is said to be *cofinal in* I if $(\forall i \in I)(\exists j \in J)(j \geq i)$.

(a) Which subsets of ω are cofinal in ω?

(b) Let I be a partially ordered set whose partial ordering directs I (Chapter 11, Exercise P). Let J be cofinal in I. Show that the induced partial ordering of J directs J.

(c) With I, J as in (b), let (X, f) be an inverse system over I (Chapter 12, Exercise Q). Let (Y, g) be the inverse system over J such that

$$Y = (Y_j \mid j \in J), \qquad g = (g_k{}^j \mid j, k \in J \,\&\, j \leq k)$$

where $Y_j = X_j$ for all $j \in J$ and $g_k{}^j = f_k{}^j$ for all $j, k \in J$ with $j \leq k$. Prove that

$$\text{inv lim } (Y, g) \simeq \text{inv lim } (X, f).$$

16. Zorn's Lemma

This chapter is devoted to deducing from the axiom of choice an important consequence asserting that certain preordered sets have members which are "almost" least and greatest elements.

MAXIMAL AND MINIMAL ELEMENTS

For motivation, consider the set $\mathcal{P}(\omega)$ partially ordered by inclusion. Then $\varnothing$ is the least and ω the greatest element of $\mathcal{P}(\omega)$. However, the subset

$$X = \mathcal{P}(\omega) \setminus \{\varnothing, \omega\}$$

of $\mathcal{P}(\omega)$, with the induced partial ordering, has neither a least nor a greatest element. In fact, let $x \in X$. For some $n \in \omega$, $n \notin x$. Then $x \not\subset \{n\}$ so that x is not a least element of X, and $\{n\} \not\subset x$ so that x is not a greatest element of X.

This set X has many elements which are "almost" least and greatest elements. In fact, take any $n \in \omega$. Then $\{n\} \leq y$ for every $y \in X$ comparable to $\{n\}$, and $y \leq \omega \setminus \{n\}$ for every $y \in X$ comparable to $\omega \setminus \{n\}$. The only thing preventing $\{n\}$ from being a least element, and $\omega \setminus \{n\}$ from being a greatest element, of X is that these elements are not comparable to certain elements of X. Even though X has no least or greatest element, it thus has the next best things!

16.1. DEFINITION

Let X be a preordered class. A member $m \in X$ is said to be *maximal* if

$$x \in X \;\&\; x \text{ comparable to } m \Rightarrow x \leq m.$$

Dually, a member $m \in X$ is said to be *minimal* if

$$x \in X \;\&\; x \text{ comparable to } m \Rightarrow m \leq x,$$

that is, if m is maximal with respect to the preordering of X opposite to the given one.

Before giving some examples, we note the following simple criterion.

16.2. LEMMA

Let X be a *partially* ordered class, and let $m \in X$. Then the following statements are equivalent:

(1) m is maximal.
(2) If $x \in X$ and $m \leq x$, then $x = m$.
(3) If $x \in X$, then $m \nless x$.

16.3. EXAMPLES

(1) If $X = \mathcal{P}(\omega) \setminus \{\varnothing,\omega\}$, then by the discussion preceding 16.1, X has infinitely many maximal and infinitely many minimal elements.

(2) If X is a partially ordered class and if X has a greatest (respectively, least) element, then this element is the unique maximal (respectively, minimal) element of X.

(3) Partially order $\omega \setminus \{0,1\}$ by the relation R given by

$$xRy \Leftrightarrow y \mid x$$

($y \mid x$ means y divides x). Then $m \in \omega \setminus \{0,1\}$ is maximal if and only if m is a "prime" number, that is, $m \neq xy$ for all $x, y \in \omega \setminus \{0,1\}$.

(4) Let X and Y be sets, and let $\mathfrak{F}$ be a set of maps such that $\operatorname{dmn} f \subset X$ and $\operatorname{rng} f \subset Y$ for each $f \in \mathfrak{F}$. Assume $\mathfrak{F}$ has the following extension property: If $f \in \mathfrak{F}$ and $x \in X \setminus \operatorname{dmn} f$, there exists $g \in \mathfrak{F}$ such that

$$\{x\} \cup \operatorname{dmn} f \subset \operatorname{dmn} g, \qquad g \mid \operatorname{dmn} f = f.$$

Partially order $\mathfrak{F}$ as in 15.18(8):

$$f \leq g \Leftrightarrow \operatorname{dmn} f \subset \operatorname{dmn} g \;\&\; g \mid \operatorname{dmn} f = f.$$

Then $h \in \mathfrak{F}$ is maximal if and only if $\operatorname{dmn} h = X$.

(5) This example is from linear algebra. Let V be a vector space (over some field, say the field of real numbers). Define

$$\mathfrak{F} = \{E \mid E \subset V \,\&\, E \text{ is linearly independent}\},$$

$$\mathcal{G} = \{S \mid S \subset V \,\&\, S \text{ spans } V\}.$$

Partially order $\mathfrak{F}$ and $\mathcal{G}$ by inclusion. Then a subset B of V is a basis for V if and only if B is a maximal member of $\mathfrak{F}$, or equivalently, if and only if B is a minimal member of $\mathcal{G}$.

Below we consider classes $\mathcal{S} \subset \mathcal{P}(X)$ for some class X. *To call a subset M of X a maximal element of* $\mathcal{S}$ without further qualification *means that M is maximal with respect to the inclusion relation in* $\mathcal{S}$, in other words, that M is not strictly contained in any member of $\mathcal{S}$. A dual remark holds for "minimal."

HAUSDORFF CHAIN THEOREM

We now proceed toward the main result of this chapter, Zorn's lemma (16.6). The crucial step is the following technical lemma. Despite the formidable length of its proof, the guiding idea is quite simple: to reduce the existence of a maximal member of the given class $\mathcal{S}$ to the existence of a "fixed point" of a certain map $f: \mathcal{S} \to \mathcal{S}$, that is, an element $E \in \mathcal{S}$ for which $f(E) = E$.

16.4. LEMMA

Let X be a nonempty set. Let $\mathcal{S} \subset \mathcal{P}(X)$, and partially order $\mathcal{S}$ by inclusion. Suppose:

(i) $\varnothing \in \mathcal{S}$.
(ii) If $E \in \mathcal{S}$ and $F \subset E$, then $F \in \mathcal{S}$.
(iii) If $\mathcal{C}$ is a chain in $\mathcal{S}$, then $\bigcup \mathcal{C} \in \mathcal{S}$.

Then $\mathcal{S}$ has a maximal member.

Proof

By the axiom of choice, there exists a choice function c for $\mathcal{P}(X) \setminus \{\varnothing\}$. Define a map

$$g: \mathcal{S} \to \mathcal{P}(X)$$

by

$$g(E) = \{x \mid x \in X \,\&\, E \cup \{x\} \in \mathcal{S}\} \qquad (E \in \mathcal{S}).$$

In view of (ii), a set $E \in \mathcal{S}$ is maximal in $\mathcal{S}$ if and only if $g(E) = E$. Hence it suffices to prove that $g(E) = E$ for some $E \in \mathcal{S}$. Now if $E \in \mathcal{S}$, then of course $E \subset g(E)$, but $g(E)$ may conceivably contain many more elements than does E.

If $E \in \mathcal{S}$ with $g(E) \neq E$, then $g(E) \setminus E \neq \varnothing$, $c(g(E) \setminus E) \in g(E)$, and so

$$E \cup \{c(g(E) \setminus E)\} \in \mathcal{S}.$$

Hence we may define a map

$$f\colon \mathcal{S} \to \mathcal{S}$$

by

$$f(E) = \begin{cases} E \cup \{c(g(E) \setminus E)\} & \text{if } g(E) \neq E, \\ E & \text{if } g(E) = E. \end{cases}$$

If $E \in \mathcal{S}$, then

$$E \subset f(E),$$

and $f(E)$ contains exactly one element not in E or else $f(E) = E$. In fact,

$$g(E) = E \Leftrightarrow f(E) = E.$$

Hence it is enough to show that $f(E) = E$ for some $E \in \mathcal{S}$.

In the remainder of the proof, we shall mean by a *tower* any collection $\mathfrak{I} \subset \mathcal{S}$ such that:

(a) $\varnothing \in \mathfrak{I}$.
(b) If $E \in \mathfrak{I}$, then $f(E) \in \mathfrak{I}$.
(c) If $\mathcal{C}$ is a chain in $\mathfrak{I}$, then $\bigcup \mathcal{C} \in \mathfrak{I}$.

For example, $\mathcal{S}$ is itself a tower.

Our aim is to produce a tower that is a chain in $\mathcal{S}$. For suppose that the tower $\mathfrak{M}$ is a chain in $\mathcal{S}$. By (c),

$$M = \bigcup \mathfrak{M} \in \mathfrak{M}.$$

By (b), $f(M) \in \mathfrak{M}$. Then $f(M) \subset M$, and so $f(M) = M$, which is enough to complete the proof.

We now proceed to show that some tower is a chain in $\mathcal{S}$. Since there is at least one tower, namely $\mathcal{S}$ itself, then

$$\mathfrak{M} = \bigcap \{\mathfrak{I} \mid \mathfrak{I} \text{ is a tower}\}$$

is a subset of $\mathcal{S}$. It is easy to see that $\mathfrak{M}$ is in fact a tower. The proof will be completed if we can show that $\mathfrak{M}$ is a chain. To this end, define

$$\mathcal{E} = \{E \mid E \in \mathcal{S} \,\&\, (\forall M \in \mathfrak{M})(E \subset M \vee M \subset E)\}.$$

Then $\mathfrak{M}$ is a chain if $\mathfrak{M} \subset \mathcal{E}$. Since $\mathfrak{M}$ is contained in every tower, we need only show that $\mathcal{E}$ is a tower.

Obviously $\varnothing \in \mathcal{E}$, so $\mathcal{E}$ satisfies condition (a) in the definition of a tower.

We show next that $\mathcal{E}$ satisfies condition (c). Let $\mathcal{C}$ be a chain in $\mathcal{E}$, and let $E = \bigcup \mathcal{C}$. By hypothesis (iii), $E \in \mathcal{S}$. To show that $E \in \mathcal{E}$, let $M \in \mathfrak{M}$ and just suppose $M \not\subset E$. Then for each $C \in \mathcal{E}$ we have $M \not\subset C$ and so $C \subset M$. Hence $E \subset M$.

Finally, we show that $\mathcal{E}$ satisfies (b). Let $E \in \mathcal{E}$. We must show that $f(E) \in \mathcal{E}$, that is,

$$(\forall M \in \mathfrak{M})\,(f(E) \subset M \vee M \subset f(E)).$$

Define

$$\mathfrak{J} = \{M \mid M \in \mathfrak{M} \,\&\, (f(E) \subset M \vee M \subset E)\}.$$

Since $E \subset f(E)$, it suffices to show that $\mathfrak{M} \subset \mathfrak{J}$. Hence it suffices to show that $\mathfrak{J}$ is a tower. Obviously $\varnothing \in \mathfrak{J}$. The argument that the union of a chain in $\mathfrak{J}$ is itself an element of $\mathfrak{J}$ is similar to the argument given in the preceding paragraph.

To complete the proof that $\mathfrak{J}$ is a tower, let $M \in \mathfrak{J}$; we must show that $f(M) \in \mathfrak{J}$. Since $M \in \mathfrak{M}$ and $\mathfrak{M}$ is a tower, $f(M) \in \mathfrak{M}$. It remains to show that

$$f(E) \subset f(M) \vee f(M) \subset E.$$

Since $M \in \mathfrak{J}$, we have

$$f(E) \subset M \vee M \subset E.$$

We can distinguish three cases.

Case (1)

$f(E) \subset M$. Then $f(E) \subset f(M)$ since $M \subset f(M)$.

Case (2)

$M = E$. Then $f(E) = f(M)$.

Case (3)

$M \subsetneq E$. Now $E \in \mathcal{E}$ and $f(M) \in \mathfrak{M}$, so by definition of $\mathcal{E}$,

$$f(M) \subset E \vee E \subsetneq f(M).$$

If $E \subsetneq f(M)$, then $M \subsetneq E \subsetneq f(M)$ which is impossible since $f(M)$ has at most one element not belonging to M. Hence $f(M) \subset E$. □

The next theorem refers to a "maximal chain" M in a preordered set X, and this reference is understood to mean a maximal member M of the set $\mathcal{S}$ of all chains in X. Now $\mathcal{S} \subset \mathcal{P}(X)$, so by our convention such an M is simply a chain in X which is not strictly contained in any chain in X.

16.5. THEOREM (HAUSDORFF CHAIN THEOREM)

Let X be a preordered set. Then there exists a maximal chain M in X.

Proof

If $X = \varnothing$, we may take $M = \varnothing$, so we suppose $X \neq \varnothing$. Define

$$\mathcal{S} = \{C \mid C \text{ is a chain in } X\}.$$

It suffices to show that $\mathcal{S}$ satisfies the three hypotheses of Lemma 16.4. Clearly (i) and (ii) are satisfied.

To prove that $\mathcal{S}$ satisfies (iii), let $\mathcal{C}$ be a chain in $\mathcal{S}$. (One should not confuse $\mathcal{C}$ with a chain in X; although $\mathcal{C} \subset \mathcal{S} \subset \mathcal{P}(X)$, each chain in X is a subset of X.) We must show that $C = \bigcup \mathcal{C}$ is a chain in X. Let $x, y \in C$. There exist $E, F \in \mathcal{C}$ with $x \in E$, $y \in F$. Since $\mathcal{C}$ is totally ordered by inclusion, $E \subset F$ or $F \subset E$. Without loss of generality, we may assume $E \subset F$. Then $x \in F$ and $y \in F$. But F is a chain in X, so x is comparable to y for the given preordering of X. ▯

ZORN'S LEMMA

With the use of the Hausdorff chain theorem, we can prove the main result of this chapter.

16.6. THEOREM (ZORN'S LEMMA)

Let X be a nonempty preordered set in which each chain is bounded above. Then X has a maximal member.

Proof

By 16.5, there exists a maximal chain M in X. By hypothesis, M has some upper bound m in X. We claim that m is a maximal member of X.

Let $x \in X$ with x comparable to m. Just suppose $x \nleq m$. Then $m \leq x$. Now $y \in M$ implies $y \leq m$ and so $y \leq x$. Hence

$$C = M \cup \{x\}$$

is another chain in X. Now $x \notin M$ since $x \nleq m$. Then $M \subsetneq C$, and this contradicts the maximality of M among all chains in X. ▯

Although it is now the customary practice, followed here, to call Theorem 16.6 "Zorn's lemma," this attribution is hardly appropriate. Quite a few maximal principles, which could serve in proofs as substitutes for the well-ordering theorem and transfinite induction (see Chapter 17), appeared in the literature prior to Zorn's. Apparently the first was Theorem 16.5, which F. Hausdorff deduced from the well-ordering theorem in 1914. In 1922,

C. Kuratowski deduced forms of 16.8 and 16.9 from the well-ordering theorem. In 1935, M. Zorn published a paper in which he formulated 16.8, used it to prove several theorems in algebra, and stated without proof that it is equivalent to the axiom of choice. In 1940, J. W. Tukey published the generalization 16.6 of Zorn's principle. (For bibliographic citations and further historical comments, see Rubin and Rubin [33].)

By applying 16.6 to the opposite of a given preordering, one obtains immediately a dual result.

16.7. COROLLARY

Let X be a nonempty preordered set in which each chain is bounded below. Then X has a minimal member.

As a concrete instance of Zorn's lemma, where the preordering is actually the inclusion relation, we have:

16.8. COROLLARY (MAXIMAL PRINCIPLE)

Let $\mathcal{S}$ be a nonempty set of sets. Suppose for each nest $\mathfrak{N}$ in $\mathcal{S}$, some member of $\mathcal{S}$ contains every element of $\mathfrak{N}$. Then $\mathcal{S}$ has a maximal member.

The hypothesis of 16.8 is satisfied, in particular, if $\bigcup\mathfrak{N} \in \mathcal{S}$ for each nest $\mathfrak{N}$ in $\mathcal{S}$. Then the technical lemma 16.4 is just a very special case of Zorn's lemma, and the statement of 16.4 need no longer be remembered.

From 16.7 one obtains the concrete instance:

16.9. COROLLARY (MINIMAL PRINCIPLE)

Let $\mathcal{S}$ be a nonempty set of sets. Suppose for each nest $\mathfrak{N}$ in $\mathcal{S}$, some member of $\mathcal{S}$ is contained in every element of $\mathfrak{N}$. Then $\mathcal{S}$ has a minimal member.

The hypothesis of 16.9 is satisfied if $\bigcap\mathfrak{N} \in \mathcal{S}$ for each nonempty nest $\mathfrak{N}$ in $\mathcal{S}$.

Additional well-known versions of Zorn's lemma appear in Exercises G–I.

The reader should not be misled by the word "lemma" as part of the name of 16.6 into believing that Zorn's lemma is just an auxiliary result which may be forgotten once it has been applied. To the contrary, Zorn's lemma is one of the most powerful tools the mathematician has at his disposal, and significant applications occur in nearly every branch of contemporary mathematics.

Zorn's lemma will be used in the remainder of this book to prove a number of fundamental theorems in set theory. The commentary upon Exercises M–P indicates an application to each of the fields of analysis, topology, and algebra.

AN APPLICATION

We give here just one application outside set theory proper.

16.10. EXAMPLE

Let X be a real vector space, and let

$$f_0: W \to \mathbb{R}$$

be a given linear map on a linear subspace W of X. Then there exists a linear map

$$f_1: X \to \mathbb{R}$$

such that

$$f_1 \mid W = f_0.$$

To establish this result, define $\mathfrak{F}$ to be the set of all maps

$$f: \operatorname{dmn} f \to \mathbb{R}$$

such that $\operatorname{dmn} f$ is a linear subspace of X, f is linear, $W \subset \operatorname{dmn} f$, and

$$f \mid W = f_0.$$

We must show there is some $f_1 \in \mathfrak{F}$ for which $\operatorname{dmn} f_1 = X$.

Partially order $\mathfrak{F}$ as in 16.3(4) by letting

$$f \leq g \Leftrightarrow \operatorname{dmn} f \subset \operatorname{dmn} g \;\&\; g \mid \operatorname{dmn} f = f.$$

We are going to show that $\mathfrak{F}$ satisfies the hypothesis of Zorn's lemma and has the following extension property:

$$(*) \quad f \in \mathfrak{F} \;\&\; x \in X \setminus \operatorname{dmn} f \Rightarrow$$
$$(\exists g \in \mathfrak{F})(\{x\} \cup \operatorname{dmn} f \subset \operatorname{dmn} g \;\&\; g \mid \operatorname{dmn} f = f).$$

This will establish the desired result. In fact, Zorn's lemma will yield a maximal element f_1 of $\mathfrak{F}$, and $\operatorname{dmn} f_1 = X$ according to 16.3(4).

We show that $\mathfrak{F}$ has property (*). Let $f \in \mathfrak{F}$ and $x \in X \setminus \operatorname{dmn} f$. Let V be the linear subspace of X spanned by $\{x\} \cup \operatorname{dmn} f$, that is,

$$V = \{\alpha x + y \mid \alpha \in \mathbb{R} \;\&\; y \in \operatorname{dmn} f\};$$

it is readily checked that V is indeed a linear subspace of X. Moreover, $x = 1 \cdot x + 0 \in V$, and $y = 0 \cdot x + y \in V$ for each $y \in \operatorname{dmn} f$. Thus $\{x\} \cup \operatorname{dmn} f \subset V$. Since $W \subset \operatorname{dmn} f$, $W \subset V$.

Note that if $v \in V$, then there are *unique* $\alpha \in \mathbb{R}$, $y \in \operatorname{dmn} f$ with $v = \alpha x + y$. In fact, suppose $y, y' \in \operatorname{dmn} f$ and $\alpha, \alpha' \in \mathbb{R}$ with

$$\alpha x + y = v = \alpha' x + y'.$$

If $\alpha \neq \alpha'$, then

$$x - (\alpha' - \alpha)^{-1}(y - y') \in \operatorname{dmn} f$$

since dmn f is a vector space, but $x \notin \operatorname{dmn} f$ by assumption. Hence $\alpha = \alpha'$, and a similar argument shows $y = y'$.

To complete the verification of (*), we construct a linear map $g: V \to \mathbb{R}$ such that $g \mid \operatorname{dmn} f = f$. For $v \in V$ define

$$g(v) = \alpha + f(y),$$

where $\alpha \in \mathbb{R}$, $y \in \operatorname{dmn} f$ are the unique elements for which

$$v = \alpha x + y.$$

If $v \in \operatorname{dmn} f$, then $v = 0 + v$ and $g(v) = 0 + f(v) = f(v)$. Hence g extends f. The verification that g is linear is left to the reader. This completes the proof of (*).

We show that $\mathfrak{F}$ satisfies the hypothesis of Zorn's lemma. Of course $\mathfrak{F} \neq \varnothing$, for $f_0 \in \mathfrak{F}$. Let $\mathfrak{C}$ be a nonempty chain in $\mathfrak{F}$. We must show $\mathfrak{C}$ has an upper bound in $\mathfrak{F}$. Define

$$V = \bigcup\{\operatorname{dmn} g \mid g \in \mathfrak{C}\}.$$

It suffices to show that V is a linear subspace of X, for then the unique map $f: V \to \mathbb{R}$ such that

$$f \mid \operatorname{dmn} g = g \qquad (g \in \mathfrak{C})$$

evidently belongs to $\mathfrak{F}$ and is an upper bound of $\mathfrak{C}$ in $\mathfrak{F}$.

We show that V is a linear subspace of X. Let $x, y \in V$ and $\alpha, \beta \in \mathbb{R}$. For some $g, h \in \mathfrak{C}$, $x \in \operatorname{dmn} g$ and $y \in \operatorname{dmn} h$. Since $\mathfrak{C}$ is totally ordered, we may assume without loss of generality that $g \leq h$. Then $x, y \in \operatorname{dmn} h$, and since dmn h is a linear subspace of X, $\alpha x + \beta y \in \operatorname{dmn} h \subset V$. □

This example is quite typical of applications of Zorn's lemma to extend a map. Notice in the construction of an upper bound of the chain $\mathfrak{C}$ the implicit use of 8.28 to obtain a common extension to $\bigcup\{\operatorname{dmn} g \mid g \in \mathfrak{C}\}$ of every $g \in \mathfrak{C}$.

EXERCISES

A. Let X be a totally ordered class. If m is a maximal element of X, show that m is a (and hence the) greatest element of X. Dualize.

B. (a) Determine all maximal and minimal members of $\mathcal{P}(\omega) \setminus \{\varnothing, \omega\}$.
(b) Does $\mathcal{P}(\omega) \setminus \{\varnothing, \omega\}$ satisfy the hypothesis of Zorn's lemma?
(c) Find all maximal and minimal members of the class of all infinite strict subsets of ω.

C. A set X is said to be *Tarski-finite* if each nonempty set of subsets of X has a minimal member (under inclusion).

(a) Prove that a set X is Tarski-finite if and only if each nonempty set of subsets of X has a maximal member.

(b) Prove that a Tarski-finite set X is finite. [*Hint*: If X is infinite, it is Dedekind infinite and so has a strict subset Y with $X \simeq Y$, say $f: X \simeq Y$. Define

$$\mathcal{S} = \{E \mid E \subset X \,\&\, f\langle E\rangle \subset E\}.$$

Show that $\mathcal{S}$ has no minimal member.]

The converse of (b) is true. See Suppes [33, pages 100–102].

D. Let X be a partially ordered set. One says that X satisfies the *ascending chain condition* if there exists no sequence $(x_n \mid n \in \omega)$ in X with $x_n < x_{n+1}$ for all $n \in \omega$.

(a) Show that X satisfies the ascending chain condition if and only if every nonempty subset of X has a maximal element.

(b) Dualize to a "descending chain condition."

E. Find all maximal chains in the partially ordered set $\langle \omega \setminus \{0,1\}, R\rangle$ of Example 16.3(3). [*Hint*: Let M be such a maximal chain. If p is the least element of ω (for $\leq$) belonging to M, use the method of proof of 16.6 to show that p is a prime number.]

F. Is Zorn's lemma still true if "set" is replaced by "class"?

G. (a) Deduce from Zorn's lemma: Let X be a nonempty preordered set in which each chain is bounded above, and let $x \in X$. Then there exists a maximal element m of X such that $m \geq x$.

(b) State and prove the dual of (a).

H. (a) Deduce from Exercise G: Let X be a preordered set, and let C be a chain in X. Then there exists a maximal chain M in X such that $C \subset M$. [*Remark*: Taking $C = \varnothing$ here, one obtains 16.5 as a consequence of 16.6. Since 16.6 was deduced from 16.5, it follows that 16.5 and 16.6 are equivalent statements.]

(b) Deduce from (a): If $\mathfrak{N}$ is a nest in a set $\mathcal{S}$ of sets, then there exists a maximal nest $\mathfrak{M}$ in $\mathcal{S}$ with $\mathfrak{N} \subset \mathfrak{M}$.

I. A class $\mathfrak{F}$ of sets is said to be *of finite character* if it has the following property: A set E belongs to $\mathfrak{F}$ precisely when each finite subset of E belongs to $\mathfrak{F}$.

(a) Let x be a given set. Show that $\{X \mid x \notin X\}$ is of finite character.

(b) Let X be a preordered set. Show that the set of all chains in X is of finite character.

(c) Give an example of a class $\mathfrak{F}$ not of finite character.

(d) Suppose $\mathfrak{F}$ is of finite character and $X \in \mathfrak{F}$. Prove that $Y \in \mathfrak{F}$ for each $Y \subset X$. (Hence $\mathfrak{F} \neq \varnothing \Rightarrow \varnothing \in \mathfrak{F}$.)

J. Prove the *Teichmüller-Tukey lemma*: If a nonempty set $\mathfrak{F}$ of sets is of finite character, then $\mathfrak{F}$ has a maximal member.

K. Let R be a relation which is a set. Prove there exists a maximal set Y such that $Y \times Y \subset R$.

L. Let L be a lattice (Chapter 15, Exercise J). A subset J of L is called an *ideal in L* if

$$x \in J \,\&\, y \in J \Rightarrow \sup\{x,y\} \in J,$$

$$x \in L \,\&\, y \in J \Rightarrow \inf\{x,y\} \in J.$$

An ideal J in L is said to be *proper* if $J \neq L$. Suppose L has a greatest element and at least one other element. Show that then L has a maximal proper ideal.

M. Let $\perp$ be a symmetric relation in a set X and let $z \in X$ with $X \neq \{z\}$. Call a subset N of X "orthogonal" if $z \notin N$ and if $x \perp y$ whenever $x, y \in N$ with $x \neq y$. Show there exists some orthogonal subset M of X with the property that $x \in X$ and $x \perp y$ for all $y \in M$ implies $x = z$.

[*Comment*: Take for X a Hilbert space (or even an inner product space), for z the zero vector in X, and for $\perp$ the usual orthogonality relation in X. Then the above result guarantees the existence of a complete orthonormal system in X. Of course, this can be proved using the Gram-Schmidt process in case X is finite-dimensional or separable.]

N. Let X be a set. A nonempty subset $\mathfrak{F}$ of $\mathcal{P}(X)$ is called a *filter on X* if $\varnothing \notin \mathfrak{F}$, $E \cap F \in \mathfrak{F}$ whenever $E \in \mathfrak{F}$ and $F \in \mathfrak{F}$, and $E \in \mathfrak{F}$ whenever $E \subset X$ and $E \supset F$ for some $F \in \mathfrak{F}$. Let φ be the set of all filters on X, and partially order φ by inclusion. A maximal element of φ is called an *ultrafilter on X*.

(a) If $x \in X$, show that

$$\{B \mid x \in B \subset X\}$$

is an ultrafilter on X.

(b) Show that

$$\{F \mid F \subset X \,\&\, X \setminus F \text{ is finite}\}$$

is a filter on X which is not an ultrafilter if X is infinite.

(c) Prove that each filter on X is contained in some ultrafilter on X.

(d) If $p: X \rightarrow Y$ is a surjection and $\mathfrak{F}$ is an ultrafilter on X, show that

$$\{p\langle E\rangle \mid E \in \mathfrak{F}\}$$

is an ultrafilter on Y.

[*Comment*: Part (c) is used to prove that a topological space X is compact if and only if each ultrafilter on X "converges." When applied to the projections of a product onto its factors, part (d) provides an elegant proof of "Tychonoff's theorem" which asserts that the product of any nonempty family of compact spaces is itself compact.]

P. Let X be a set and

$$s: \mathcal{P}(X) \to \mathcal{P}(X)$$

be a map. Assume the following hold for all $A, B \in \mathcal{P}(X)$ and all $x, y \in X$:

(i) If $A \subset B$, then $s(A) \subset s(B)$.
(ii) $A \subset s(A)$.
(iii) If $x \in s(A)$, then $x \in s(F)$ for some finite $F \subset A$.
(iv) If $y \in s(A \cup \{x\})$ and $y \notin s(A)$, then $x \in s(A \cup \{y\})$.

Call a subset A of X "free" if

$$x \in A \Rightarrow x \notin s(A \setminus \{x\}).$$

Show that $s(B) = X$ for some free set B. [*Hint*: Consider the class of all free sets.]

[*Comment*: Take X to be a vector space over some field, and for each $A \in \mathcal{P}(X)$ define $s(A)$ to be the span of A, that is, the linear subspace of X consisting of all finite linear combinations of elements of A. Then (i)–(iii) are clearly satisfied, and (iv) is a restatement of the "exchange theorem." A subset of X is free if and only if it is linearly independent. The conclusion then says that some linearly independent set spans X, in other words, that X has a basis.]

Q. Deduce from Zorn's lemma: If $\mathcal{A}$ is a *set* of nonempty sets, then there exists a choice function for $\mathcal{A}$. [*Hint*: Consider the set of all functions f such that $\operatorname{dmn} f \subset \mathcal{A}$ and f is a choice function for $\operatorname{dmn} f$.]

R. This exercise refers to Example 16.10.
(a) Why could we not obtain the extension $f_1: X \to \mathbb{R}$ of f_0 simply by taking $f_1 \mid W = f_0$ and $f_1(y) = 0$ for all $y \in X \setminus W$?
(b) Suppose $x_0 \in X \setminus W$ and $0 \neq t \in \mathbb{R}$. Prove there is a linear extension $f_2: X \to \mathbb{R}$ of f_0 such that $f_2(x_0) = t$.

S. Use Zorn's lemma to prove the comparability theorem (13.14): If X and Y are sets, then $X \preceq Y$ or $Y \preceq X$. [*Hint*: Consider all injections $f: E \to Y$ for all subsets E of X.]

17. Well-Ordering

According to the well-ordering principle (6.20), the usual order relation in ω has the property that each nonempty subset of ω has a unique first element. In this chapter other relations having this property are studied abstractly. The theory developed here permits at last a proof of the comparability theorem (13.14) and provides much of the machinery needed for the subsequent discussion of ordinal numbers.

WELL-ORDERINGS

17.1. DEFINITION

A relation R in a class X is called a *well-ordering of* X, and is said to *well-order* X, if each nonempty subclass of X has a *unique* first element.

17.2. EXAMPLES

(1) The set ω is well-ordered by its usual order relation $\leq$. In the sequel, any reference to a well-ordering of ω is understood to mean this one.

(2) If X is well-ordered by a relation R and if $Y \subset X$, then the relation in Y induced by R well-orders Y. In particular, if $n \in \omega$, then the relation in n induced by the usual order relation in ω well-orders n (recall that $n \subset \omega$), and any reference to a well-ordering of n is to this one.

(3) The total ordering $\geq$ of ω opposite to the usual order relation in ω is not a well-ordering, for ω has no first element for $\geq$.

(4) Let X and Y be disjoint classes and let R, S well-order X, Y, respectively. Then there is a unique well-ordering T of $X \cup Y$ which induces R on X and S on Y and which satisfies

$$xTy \qquad (x \in X, y \in Y),$$

namely, $T = R \cup S \cup (X \times Y)$. One calls T the *ordinal sum of R and S*, and $\langle X \cup Y, T\rangle$ the *ordinal sum of* $\langle X,R\rangle$ and $\langle Y,S\rangle$.

(5) Let R well-order X, and let z be a set with $z \notin X$. Then the ordinal sum T of R and the unique well-ordering of $\{z\}$ is the unique well-ordering T of $X \cup \{z\}$ which induces R on X and which satisfies xTz for all $x \in X$; one calls T the *extension of R making z the last element.*

In particular, a well-ordering of a set X provides through this construction a well-ordering of the successor $X^+ = X \cup \{X\}$ of X; given a well-ordering R of a set X, any reference to a well-ordering of X^+ will be understood to mean the extension of R making X the last element.

Use of the term "ordering" in 17.1 is justified by:

17.3. PROPOSITION

Let the relation R well-order the class X. Then R totally orders X.

Proof

Let x, $y \in X$. Since $\{x,y\}$ has a first element for R, x is comparable to y. Taking $y = x$, it follows that xRx. If xRy and yRx, uniqueness of the first element of $\{x,y\}$ implies $x = y$. Thus R connects X, is reflexive on X, and is antisymmetric.

To prove the transitivity of R, let x, y, $z \in X$ with xRy and yRz. The set $\{x,y,z\}$ has a unique first element u. If $u = x$, then xRz. If $u = y$, then yRx, $x = y$ by antisymmetry, and xRz. Similarly, xRz if $u = z$. ☐

In view of 17.3, the usual conventions for preordered classes may be applied to well-orderings. Thus one calls X itself a well-ordered class to mean that $\langle X,\leq\rangle$ is a preordered class with the particular $\leq$ actually being a well-ordering of X.

If a relation $\leq$ in a class X is already known to be antisymmetric, then to verify that $\leq$ well-orders X it suffices to check that each nonempty subclass of X has *some* first element.

THE WELL-ORDERING THEOREM

The "well-ordering theorem" which follows says that any set can be well-ordered (it will be proved in Chapter 18 that any proper class can be well-ordered). Cantor had considered it self-evident and accordingly used it to prove other theorems. In 1904, E. Zermelo deduced it from the axiom of choice.

The well-ordering theorem implies that the class $\mathcal{W}$ of all well-ordered sets is a proper class, for it says that the map

$$\mathcal{W} \to \mathcal{U}$$
$$\langle X, \leq \rangle \mapsto X$$

is surjective. Thus the examples of 17.2 are but a meager sample of the totality of well-ordered sets.

17.4. THEOREM (WELL-ORDERING THEOREM)

Let X be a set. Then there exists a well-ordering of X.

Proof

Define

$$\mathcal{E} = \{(E,R) \mid E \subset X \;\&\; R \text{ is a well-ordering of } E\}.$$

We must prove $(X,R) \in \mathcal{E}$ for some R. Note that $\mathcal{E} \neq \emptyset$ since $(\emptyset,\emptyset) \in \mathcal{E}$, and that $\mathcal{E}$ is a set.

Define a relation $\leq$ in $\mathcal{E}$ as follows: For (E,R), $(F,S) \in \mathcal{E}$, $(E,R) \leq (F,S)$ means each of the following three conditions holds:

(i) $E \subset F$;
(ii) S induces R on E, that is, $R \subset S$;
(iii) if $x \in E$ and $y \in F \setminus E$, then xSy.

It is easy to see that $\leq$ partially orders $\mathcal{E}$. We are going to apply Zorn's lemma to the partially ordered set $\langle \mathcal{E}, \leq \rangle$.

Let $\mathcal{A}$ be a chain in $\mathcal{E}$; we want to find an upper bound of $\mathcal{A}$ in $\mathcal{E}$. We may write

$$\mathcal{A} = \{(A_i,R_i) \mid i \in I\}$$

for some family $(A_i \mid i \in I)$ of subsets of X and some family $(R_i \mid i \in I)$ of relations. Note that $\{A_i \mid i \in I\}$ is a chain in $\mathcal{P}(X)$ in view of (i), and $\{R_i \mid i \in I\}$ is a chain in $\mathcal{P}(X \times X)$ in view of (ii). Define

$$B = \bigcup_{i \in I} A_i, \qquad S = \bigcup_{i \in I} R_i.$$

We claim (B,S) is an upper bound of $\mathcal{A}$ in $\mathcal{E}$. First we must show S well-orders B.

Let $x, y \in B$. Then $x, y \in A_i$ for some $i \in I$, and for any such i,

$$xSy \Leftrightarrow xR_iy.$$

In fact, xR_iy implies xSy since $R_i \subset S$. Conversely, suppose xSy with $x, y \in A_i$. For some $j \in I$, xR_jy. From (ii) and the comparability of (A_i,R_i) with (A_j,R_j), we conclude xR_iy.

By the preceding paragraph and the antisymmetry of each R_i, S is antisymmetric.

Let E be a nonempty subset of B. We show that E has a first element for S. There exists $i \in I$ such that $E \cap A_i \neq \varnothing$, and then $E \cap A_i$ has a first element x for the well-ordering R_i of A_i. Take $y \in E$. For some $j \in I$, $y \in A_j$. If in fact $y \in A_i$, then xR_iy and hence xSy. If $y \notin A_i$, then $A_i \subsetneq A_j$, $(A_i,R_i) \leq (A_j,R_j)$, xR_jy by condition (iii), and hence xSy again. Thus x is a first element of E for S.

We have shown that S well-orders B, that is, $(B,S) \in \mathcal{E}$. It is easily checked that $(A_i,R_i) \leq (B,S)$ for each $i \in I$.

Zorn's lemma may now be applied to furnish a maximal element (M,R) of $\mathcal{E}$. It remains only to show $M = X$. Just suppose $M \neq X$. Choose $x \in X \setminus M$. Let $E = M \cup \{x\}$ and let S be the extension of R making x the last element. Then $(E,S) \in \mathcal{E}$ with $(M,R) < (E,S)$, a contradiction to the maximality of (M,R). ☐

The proof above is a typical application of Zorn's lemma: To show a set X has a certain property, form the collection $\mathcal{E}$ of all subsets of X having this property. Suitably partially order $\mathcal{E}$, obtain a maximal member M of $\mathcal{E}$, and conclude that $M = X$ by showing that otherwise $M \cup A \in \mathcal{E}$ for some $A \neq \varnothing$.

There are three things the well-ordering theorem does *not* say. First, it does not assert that any given total ordering is a well-ordering. Second, it does not guarantee the uniqueness of a well-ordering of a given set; if $n \in \omega$ with $n \geq 2$, then the usual order relation in n and its opposite are distinct well-orderings of n. Third, it does not furnish a prescription for constructing any specific well-ordering, for the statement of the theorem is purely existential. In fact, at the present time it is not known how to exhibit explicitly a well-ordering of any uncountable set.

One can simultaneously choose a well-ordering of every set:

17.5. COROLLARY

There exists $F: \mathcal{U} \to \mathcal{U}$ such that $F(X)$ is a well-ordering of X for each set X.

Proof

By 17.4, for each set X the set

$$W(X) = \{R \mid R \text{ is a well-ordering of } X\}$$

is nonempty. Then there exists a choice function F for the family $(W(X) \mid X \in \mathcal{U})$. ☐

TRANSFINITE INDUCTION

17.6. DEFINITION

Let x be a member of a partially ordered class X. An element $y \in X$ is called an *immediate successor* (respectively, *immediate predecessor*) *of x in X* if y is a strict successor (respectively, strict predecessor) of x in X and if no element of X is strictly between x and y.

If a member x of a well-ordered class X has some strict successor, then it has an immediate successor in X, namely the first element of the class of all strict successors of x in X. The dual statement for predecessors is false. For example, the last element ω of ω^+ has denumerably many strict predecessors but no immediate predecessor in ω^+. In contrast, each element of ω other than the first has an immediate predecessor.

Let X be the well-ordered set ω. Then the principle of mathematical induction (6.7) may be stated: Let $E \subset X$. Suppose the first element of X belongs to E. Suppose, further, for each $x \in X$ having an immediate predecessor y in X, if $y \in E$ then $x \in E$. Then $E = X$.

According to Chapter 6, Exercise I, this result holds also when X is the well-ordered set n for some $n \in \omega$.

If we try to extend the principle of mathematical induction to arbitrary well-ordered classes X, we immediately face a counterexample, namely, $X = \omega^+$ and $E = \omega \subsetneq X$. The trouble here is simply the absence of any immediate predecessor of ω in ω^+. Nonetheless, it is still possible to prove an extended version of induction. We generalize not the principle of mathematical induction (6.7), but rather the principle of complete induction (Chapter 6, Exercise G). Some notation will simplify the exposition.

17.7. DEFINITION

Let x be a member of a preordered class X. The *initial segment of X at x*, denoted $X]\!\leftarrow,x[$ or simply $]\!\leftarrow,x[$, is defined to be the subclass

$$\{y \mid y \in X \,\&\, y < x\}$$

of X consisting of all strict predecessors of x in X. (The notation suppresses mention of the particular preordering of X in question.)

As an example of this notation,

$$\omega]\!\leftarrow,n[\;=\; n$$

for each $n \in \omega$.

If X is a partially ordered class, then

$$x \leq y \Rightarrow]\!\leftarrow, x[\subset]\!\leftarrow, y[$$

for all $x, y \in X$. For later use we note next two other elementary properties of initial segments, of which the second is simply a restatement of 15.19.

17.8. LEMMA

Let X be a totally ordered class, and let $x, y \in X$. Then:

(1) Either $]\!\leftarrow, x[\subset]\!\leftarrow, y[$ or $]\!\leftarrow, y[\subset]\!\leftarrow, x[$.
(2) If $]\!\leftarrow, x[\subset]\!\leftarrow, y[$, then $x \leq y$.

17.9. THEOREM (PRINCIPLE OF TRANSFINITE INDUCTION)

Let X be a well-ordered class, and let $E \subset X$. Assume

$$x \in X \,\&\,]\!\leftarrow, x[\subset E \Rightarrow x \in E.$$

Then $E = X$.

Proof

Just suppose $E \neq X$. Then $X \setminus E$ is nonempty and therefore has a first element x. For each $y < x$, $y \notin X \setminus E$, so $]\!\leftarrow, x[\subset E$. By hypothesis $x \in E$, a contradiction. ▯

ORDER-HOMOMORPHISMS

If $f: \omega \to \omega$ is either of the maps

$$n \mapsto n + 1$$

$$n \mapsto n^2$$

then $f(m) \leq f(n)$ whenever $m, n \in \omega$ with $m \leq n$. We next study such maps abstractly.

17.10. DEFINITION

Let X and Y be preordered classes. A map $f: X \to Y$ is said to be *order-preserving* and is called an *order-homomorphism* for the given preorderings of X and Y if for all $x_1, x_2 \in X$,

$$x_1 \leq x_2 \Rightarrow f(x_1) \leq f(x_2).$$

(The occurrence of the symbol '$\leq$' on the left of the implication sign refers to the preordering of X, the occurrence on the right to the preordering of Y.)

Order-preserving maps play the same role in the study of preordered sets as do continuous maps in the study of metric spaces and linear maps in the study of vector spaces. In each study one considers sets with additional "structure," and then one investigates maps between these sets "preserving" the additional structure.

17.11. PROPOSITION

Let $f: X \to Y$ be an order-preserving bijection from a totally ordered class X to a preordered class Y. Then

$$f\langle]\leftarrow,x[\rangle =]\leftarrow, f(x)[$$

for each $x \in X$.

Proof

Let $x \in X$, and let $y = f(x)$. If $a \in]\leftarrow,x[$, then $a < x$, $f(a) < f(x) = y$ since f is an order-preserving injection, and $f(a) \in]\leftarrow, f(x)[$. Now let $z \in]\leftarrow,y[$. Choose $a \in X$ with $z = f(a)$. If $a \notin]\leftarrow,x[$, then $x \leq a$, and $y = f(x) \leq f(a) = z$, which is impossible since $z < y$. □

We consider next certain maps playing the same role in the study of preordered classes as do isomorphisms in the study of vector spaces.

17.12. DEFINITION

Let X and Y be preordered classes. One calls a map $f: X \to Y$ an *order-isomorphism* and writes

$$f: X \sim Y$$

if f is a bijection and if the maps $f: X \to Y$ and $f^{-1}: Y \to X$ are both order-preserving. One says that X is *order-isomorphic to* Y and writes

$$X \sim Y$$

to mean there exists some order-isomorphism $f: X \sim Y$.

A bijection $f: X \simeq Y$ from one preordered class X to another Y is an order-isomorphism if and only if

$$x_1 \leq x_2 \Leftrightarrow f(x_1) \leq f(x_2),$$

or equivalently,

$$x_1 < x_2 \Leftrightarrow f(x_1) < f(x_2),$$

for all $x_1, x_2 \in X$.

An order-isomorphism $f: X \sim Y$ is a very special kind of one-to-one correspondence between X and Y: If $y_1, y_2 \in Y$ correspond under f to $x_1, x_2 \in X$ respectively, then y_1 precedes y_2 in Y if and only if x_1 precedes x_2

in X. Hence the assertion '$X \sim Y$' signifies that as preordered classes X and Y are essentially the same, or as is sometimes said, X and Y *have the same order type.*

17.13. EXAMPLES

(1) The bijection

$$\omega \to \omega \setminus \{0\}$$
$$n \mapsto n+1 \qquad (n \in \omega)$$

is an order-isomorphism.

(2) The bijection $f: \omega^+ \simeq \omega$ given by

$$f(x) = \begin{cases} 0 & \text{if } x = \omega, \\ x+1 & \text{if } x \in \omega \end{cases}$$

is not an order-isomorphism. In fact, ω^+ is not order-isomorphic to ω, since ω^+ has a last element but ω does not (compare 17.16 below).

(3) Let X be a set, and let

$$\varphi: \mathcal{P}(X) \simeq \text{Map } (X,2)$$

be the bijection

$$E \mapsto \chi_E \qquad (E \in \mathcal{P}(X))$$

constructed in the proof of 13.4. Partially order $\mathcal{P}(X)$ by inclusion, and define a partial ordering $\leq$ of Map $(X,2)$ pointwise, so that

$$f \leq g \Leftrightarrow (\forall x \in X)(f(x) \leq g(x))$$

for all $f, g \in$ Map $(X,2)$. Then φ is an order-isomorphism.

For preordered classes X and Y, clearly

$$X \sim Y \Rightarrow X \simeq Y,$$

so the statement "X is order-isomorphic to Y" is stronger than the statement "X is equipollent to Y." However, $X \simeq Y$ need not imply $X \sim Y$: see Example (2) above.

For preordered classes X, Y, Z, clearly

(i) $1_X: X \sim X$;
(ii) if $f: X \sim Y$, then $f^{-1}: Y \sim X$;
(iii) if $f: X \sim Y$ and $g: Y \sim Z$, then $g \circ f: X \sim Z$.

Hence the relation $\sim$ is an equivalence relation on the class of all preordered sets.

ORDER-ISOMORPHIC WELL-ORDERED CLASSES

When the preorderings of preordered classes are actually well-orderings, only certain order-isomorphisms can subsist between them.

17.14. PROPOSITION

Let $f: X \to X$ be an order-preserving injection on a well-ordered class X to itself. Then

$$x \leq f(x) \qquad (x \in X).$$

Proof

Define

$$E = \{x \mid x \in X \,\&\, x \leq f(x)\}.$$

We use transfinite induction (17.9) to prove $E = X$. Let $x \in X$ and suppose $]\leftarrow, x[\subset E$. Then $y < x$ implies $y \leq f(y) < f(x)$ and $y < f(x)$. By 17.8(2), $x \leq f(x)$, that is, $x \in E$. ☐

17.15. COROLLARY

Let X be a well-ordered class which is order-isomorphic to the well-ordered class Y. Then there exists a *unique* order-isomorphism $f: X \sim Y$.

Proof

Suppose

$$f_1: X \sim Y, \qquad f_2: X \sim Y.$$

We show $f_1 = f_2$. Since $f_2^{-1} \circ f_1: X \sim X$,

$$x \leq (f_2^{-1} \circ f_1)(x) \qquad (x \in X)$$

by 17.14, and

$$f_2(x) \leq f_1(x) \qquad (x \in X).$$

Similarly,

$$f_1(x) \leq f_2(x) \qquad (x \in X).$$

Hence $f_1(x) = f_2(x)$ for all $x \in X$. ☐

17.16. COROLLARY

Let X be a well-ordered class. Then X is not order-isomorphic to $]\leftarrow, x[$ for any $x \in X$.

Proof

Just suppose there exists $x \in X$ and an order-isomorphism

$$f: X \sim \,]\!\leftarrow, x[.$$

By 17.14, $x \leq f(x)$. This is impossible since f sends x to some element of $]\!\leftarrow, x[$. □

This corollary and the next easy lemma prepare the way for 17.18, the second of the two main results of this chapter (the first being 17.4).

17.17. LEMMA

Let X be a well-ordered class, and let $E \subset X$. Assume

$$x \in E \Rightarrow \,]\!\leftarrow, x[\, \subset E.$$

Then either $E = X$ or else $E = \,]\!\leftarrow, z[$ where z is the first element of $X \setminus E$.

17.18. THEOREM (TRICHOTOMY LAW FOR WELL-ORDERINGS)

Let X and Y be well-ordered classes. Then exactly one of the following three alternatives holds:

(i) $X \sim Y$;
(ii) $X \sim \,]\!\leftarrow, y[$ for some $y \in Y$;
(iii) $]\!\leftarrow, x[\, \sim Y$ for some $x \in X$.

Proof

By 17.16, (i) and (ii) cannot both hold, nor can both (i) and (iii). If (ii) and (iii) both hold, then from 17.11 it follows that X is order-isomorphic to an initial segment of itself, in contradiction to 17.16. Thus at most one of the three alternatives holds.

We show that at least one of the three alternatives holds. Define

$$A = \{x \mid x \in X \,\&\, (\exists\, y \in Y)(]\!\leftarrow, x[\, \sim \,]\!\leftarrow, y[)\}.$$

Since any two initial segments of Y are comparable under inclusion [17.8(1)], it follows from 17.16 that

$$(\forall x \in A)(\exists!\, y \in Y)(]\!\leftarrow, x[\, \sim \,]\!\leftarrow, y[).$$

Let

$$B = \{y \mid y \in Y \,\&\, (\exists\, x \in A)(]\!\leftarrow, x[\, \sim \,]\!\leftarrow, y[)\}.$$

Then the map

$$f: A \rightarrow B$$

such that $f(x)$ is the unique $y \in Y$ with $]\leftarrow,x[\sim]\leftarrow,y[$ for each $x \in A$ is a bijection. In fact, f is an order-isomorphism, as is easily checked.

To complete the proof, we show that $A = X$ or $B = Y$. Just suppose $A \subsetneq X$ and $B \subsetneq Y$. Clearly $]\leftarrow,x[\subset A$ for each $x \in A$. By 17.17, $A =]\leftarrow,a[$, where a is the first element of $X \setminus A$. Similarly, $B =]\leftarrow,b[$, where b is the first element of $Y \setminus B$. Then the restriction of f to $]\leftarrow,a[$ and $]\leftarrow,b[$ is an order-isomorphism of $]\leftarrow,a[$ onto $]\leftarrow,b[$. By definition of A, $a \in A$, an impossibility. □

With the aid of this trichotomy law and the well-ordering theorem, the deferred proof of the comparability theorem is easy.

17.19. COROLLARY (COMPARABILITY THEOREM 13.14)

Let X and Y be sets. Then $X \preceq Y$ or $Y \preceq X$.

Proof

By 17.4, there exist well-orderings of X and Y. Choose a well-ordering of X and one of Y, and consider the resulting well-ordered sets. One of the alternatives in 17.18 holds. Then $X \simeq Y$ in case (i), $X \preceq Y$ in case (ii), and $Y \preceq X$ in case (iii). □

TRANSFINITE RECURSION

As was the case with induction, ordinary recursion (10.1) does not generalize directly from ω to arbitrary well-ordered sets. Recall that for a map f on ω constructed by ordinary recursion, the value of f at each $x \in \omega$ other than the first element of ω is determined by the value of f at the immediate predecessor of x. For a map f, on a well-ordered class W, constructed by the generalized recursion theorem below, the value of f at each $x \in W$ is determined by the values of f at all strict predecessors of x.

17.20. THEOREM (TRANSFINITE RECURSION)

Let W be a well-ordered class such that $]\leftarrow,x[$ is a set for each $x \in W$. Let X be a class and let

$$G: \mathcal{P}(X) \to X$$

be a given map. Then there exists a *unique* map

$$f: W \to X$$

such that

$$f(x) = G(f\langle]\leftarrow,x[\rangle) \qquad (x \in W).$$

Proof

Uniqueness

Suppose $f_1, f_2 \colon W \to X$ are maps satisfying

$$f_1(x) = G(f_1\langle]\leftarrow,x[\rangle), \qquad f_2(x) = G(f_2\langle]\leftarrow,x[\rangle)$$

for all $x \in W$. To show $f_1 = f_2$, define

$$E = \{x \mid x \in W \,\&\, f_1(x) = f_2(x)\}.$$

Then an easy transfinite induction shows $E = W$.

Existence

Define $\mathcal{H}$ to be the class of all maps h with codomain X such that:

(i) $\operatorname{dmn} h \subset W$;
(ii) if $x \in \operatorname{dmn} h$, then $]\leftarrow,x[\subset \operatorname{dmn} h$ and $h(x) = G(h\langle]\leftarrow,x[\rangle)$.

If $h_1, h_2 \in \mathcal{H}$, then by transfinite induction one shows

$$h_1 \mid (\operatorname{dmn} h_1 \cap \operatorname{dmn} h_2) = h_2 \mid (\operatorname{dmn} h_1 \cap \operatorname{dmn} h_2).$$

By 8.28, there exists a (unique) map

$$f \colon \bigcup\{\operatorname{dmn} h \mid h \in \mathcal{H}\} \to X$$

such that

$$f \mid \operatorname{dmn} h = h \qquad (h \in \mathcal{H}).$$

Then $x \in \operatorname{dmn} f$ implies $x \in \operatorname{dmn} h$ for some $h \in \mathcal{H}$, and by (ii),

$$f(x) = h(x) = G(h\langle]\leftarrow,x[\rangle) = G(f\langle]\leftarrow,x[\rangle).$$

It remains to show $\operatorname{dmn} f = W$. Just suppose $\operatorname{dmn} f \neq W$. In view of (ii), $x \in \operatorname{dmn} f$ implies $]\leftarrow,x[\subset \operatorname{dmn} f$. By 17.17,

$$\operatorname{dmn} f =]\leftarrow,z[$$

for some $z \in W$. Define a map

$$h \colon \{x \mid x \in W \,\&\, x \leq z\} \to X$$

by

$$h \mid]\leftarrow,z[= f,$$
$$h(z) = G(f\langle]\leftarrow,z[\rangle).$$

Then $h \in \mathcal{H}$ and $\operatorname{dmn} f \subsetneq \operatorname{dmn} h$. This is impossible since $\operatorname{dmn} h \subset \operatorname{dmn} f$ by definition of f. □

Notice that if z is the first element of W, then $]\leftarrow,z[= \varnothing$ and $f\langle]\leftarrow,z[\rangle = f\langle\varnothing\rangle = \varnothing$, so that

$$f(z) = G(\varnothing).$$

Notice also that if $W = \omega$, then G need not be defined on all of $\mathcal{P}(X)$, but only on the class of all finite subsets of X.

When a map f is obtained by applying 17.20, one says f is "defined by transfinite recursion." Several examples of transfinite recursion will be given in the next chapter.

EXERCISES

A. Show that each total ordering of a finite set is a well-ordering.

B. (a) If X is a finite set, prove there is a well-ordering R of X such that R^{-1} also well-orders X.
(b) If X is a class in which a relation R and its opposite R^{-1} both well-order X, must X be finite?

C. Construct a nonempty preordered set in which each subset having more than one element has more than one first element.

D. Let $(A_i \mid i \in I)$ be a disjoint family of well-ordered sets indexed by a well-ordered set I. Show that $\bigcup_{i \in I} A_i$ is well-ordered by the relation $\leq$ defined as follows: for $x \in A_i$ and $y \in A_j$, $x \leq y$ in A if and only if either $i < j$ in I or else $i = j$ and $x \leq y$ in A_j. [This generalizes 17.2(4).]

E. Let A and B be well-ordered sets.
(a) Show that $A \times B$ is well-ordered by the relation $\leq$ defined as follows: for $a_1, a_2 \in A$ and $b_1, b_2 \in B$,

$$(a_1,b_1) \leq (a_2,b_2) \Leftrightarrow (b_1 < b_2) \vee (b_1 = b_2 \,\&\, a_1 \leq a_2).$$

With this well-ordering, $A \times B$ is called the *ordinal product of A and B*.
(b) Obtain the ordinal product of A and B as a special case of the construction in Exercise D.
(c) Is the dictionary ordering (Chapter 15, Exercise H) of $A \times B$ a well-ordering?

F. Let X be a totally ordered set having a first element. Assume that each element of X having some strict successor in X has an immediate successor in X. Must X be well-ordered?

G. Let $\leq$ be a well-ordering of a set X, and let $\sim$ be an equivalence relation on X with which $\leq$ is compatible. Is the total ordering of $X/\sim$ induced by $\leq$ also a well-ordering?

H. Formulate and justify a proof rule based on the principle of transfinite induction analogous to the method of induction (6.8).

I. Show that well-ordered classes are the only totally ordered classes in which transfinite induction is possible. More precisely, prove:

Let X be totally ordered. Assume that if $E \subset X$ and if

$$x \in X \,\&\,]\!\leftarrow, x[\,\subset E \Rightarrow x \in E,$$

then $E = X$. Then X is well-ordered.

J. Let A and B be well-ordered sets.
(a) Suppose $A \cap B = \varnothing$. Show that S is an initial segment of the ordinal sum of A and B if and only if either S is an initial segment of A or $S = A \cup D$ for some initial segment D of B.
(b) Determine all initial segments of the ordinal product (Exercise E) of A and B.

K. Without using the axiom of choice, deduce from the well-ordering theorem the following statement: If $\mathcal{A}$ is a *set* of nonempty sets, then there exists a choice function for $\mathcal{A}$.

L. Let A and B be well-ordered classes. Construct *disjoint* well-ordered classes A' and B' with $A \sim A'$ and $B \sim B'$.

M. (a) Let X and Y be preordered classes with $X \sim Y$. Show that X is partially ordered (respectively, totally ordered, well-ordered) if and only if Y is.
(b) Let X be a preordered class, let Y be a class, and let $f: X \simeq Y$. Show there is a unique preordering of Y making f an order-isomorphism.

N. Let X and Y be partially ordered classes, and let $f: X \simeq Y$ be order-preserving.
(a) Show by example that f^{-1} need not be order-preserving, so that f need not be an order-isomorphism.
(b) Find a sufficient condition on X for f to be an order-isomorphism.

P. Prove the following converse of 17.11: Let $f: X \simeq Y$ be a bijection from one totally ordered class X to another Y. If

$$f\langle]\!\leftarrow, x[\rangle =]\!\leftarrow, f(x)[$$

for all $x \in X$, then f is order-preserving.

Q. Let X and Y be two denumerable partially ordered sets each of which has neither a first nor a last element and each of which is order-dense in itself (Chapter 15, Exercise E). Prove that $X \sim Y$. (Note that the partially ordered set $\mathbb{Q}$ of all rational numbers is of this type, so every partially ordered set of this kind is order-isomorphic to $\mathbb{Q}$.) [*Hint*: Write

$$X = \{x_n \mid n \in \omega\}, \qquad Y = \{y_n \mid n \in \omega\}.$$

Let $a_0 = x_0$, $b_0 = y_0$, $b_1 = y_1$. If $b_1 \le b_0$, let a_1 be the x_n with least subscript n such that $x_n \le a_0$; if $b_0 \le b_1$, let a_1 be the x_n with least subscript n such that $a_0 \le x_n$. In other words, a_1 is the element of X with least subscript which has the same order relative to a_0 as does b_1 to b_0. Let a_2 be the element of X with least subscript which is different from a_0 and a_1. Let b_2 be the element of Y with least subscript which has the same order relative to b_0 and b_1 as does a_2 relative to a_0 and a_1. Continue. Assign a_i to b_i for each i.]

R. Let $\langle X, \leq \rangle$ be a nonempty preordered set. Suppose each subset of X which is well-ordered by $\leq$ is bounded above in X. Prove that X has a maximal member. [*Hint*: Let

$$\mathcal{W} = \{E \mid E \subset X \;\&\; \leq \text{ well-orders } E\}.$$

Define a preordering $\leq'$ of $\mathcal{W}$ by

$$A \leq' B \Leftrightarrow (\forall x \in A)(B]\!\leftarrow, x[\; \subset A).$$

Obtain a maximal $M \in \mathcal{W}$ for $\leq'$, and consider an upper bound m of M in X.]

S. Prove the *Knaster-Tarski fixed point theorem*: Let X be a complete partially ordered set (Chapter 15, Exercise G), and let f be an order-homomorphism from X into X. Then there exists $x \in X$ for which $f(x) = x$.

18. Ordinal Numbers

A certain proper class of well-ordered sets, the ordinal numbers, will be defined below which has the property that each well-ordered set is order-isomorphic to exactly one ordinal number.

INTRODUCTION

Let $\mathcal{W}$ denote the class of all well-ordered sets. The relation "is order-isomorphic to" defined in Chapter 17 induces an equivalence relation $\sim$ on the class $\mathcal{W}$. Given $X, Y \in \mathcal{W}$, $X \sim Y$ if and only if X and Y have the same order type.

Our aim is to abstract from any given $X \in \mathcal{W}$ some object $\operatorname{ord}(X)$, to be called the "ordinal of X," which represents the "type" of the given well-ordering of X while disregarding the particular nature of the elements of X. In other words, we require

$$(*) \qquad X \sim Y \Leftrightarrow \operatorname{ord}(X) = \operatorname{ord}(Y) \qquad (X, Y \in \mathcal{W}).$$

The problem is how to define $\operatorname{ord}(X)$.

One obvious candidate for the ordinal of an $X \in \mathcal{W}$ is the equivalence class

$$\operatorname{ord}(X) = \{Y \mid Y \in \mathcal{W} \;\&\; X \sim Y\}$$

of X in $\mathcal{W}$ modulo $\sim$. This is essentially the definition used in Cantor's naive theory. It trivially satisfies requirement $(*)$. The trouble is that under this definition ord(X) is a proper class unless X is empty (see Exercise A). In order that ordinals of well-ordered sets can themselves be elements of classes, we insist that ord(X) shall always be a set.

To obtain a satisfactory definition, imagine the elements of an $X \in \mathcal{W}$ arranged in order by the given well-ordering of X. Let us "count" the elements of X one-by-one by a map $f\colon X \to \mathcal{U}$ which assigns 0 to the first element, 1 to the second, and so on. If x_0 is the first element of X, let

$$f(x_0) = 0.$$

If x_1 is the first element of $X \setminus \{x_0\}$, let

$$f(x_1) = 1 = \{0\} = \{f(x_0)\}.$$

Continuing this way, if x_n is the first element of $X \setminus \{x_0,\ldots,x_{n-1}\}$, let

$$f(x_n) = n = \{0,\ldots,n-1\} = \{f(x_0),\ldots,f(x_{n-1})\}$$
$$= f\langle\{x_0,\ldots,x_{n-1}\}\rangle = f\langle]\!\leftarrow,x_n[\rangle.$$

Of course we need not be able to exhaust all the elements of X in this way (consider $X = \omega^+$).

In order to define $f(x)$ for all $x \in X$ in a manner consistent with the above, we want

$$f(x) = f\langle]\!\leftarrow,x[\rangle \qquad (x \in X).$$

That a unique map

$$f\colon X \to \mathcal{U}$$

with this property exists follows by transfinite recursion [in 17.20 take W to be X, X to be $\mathcal{U}$, and G to be the identity map of $\mathcal{U} = \mathcal{P}(\mathcal{U})$]. We propose to take

$$\text{ord}(X) = f\langle X\rangle$$

for this map f. Clearly ord(X) will then be a set. We must verify requirement (*).

First we observe

$$(**) \qquad t < x \Leftrightarrow f(t) \in f(x) \qquad (t,\, x \in X).$$

In fact, $t < x$ implies $f(t) \in f\langle]\!\leftarrow,x[\rangle = f(x)$. Conversely, assume $t \not< x$. If $t = x$, then $f(t) = f(x)$; if $x < t$, then $f(x) \in f(t)$. In either case, $f(t) \notin f(x)$.

We verify (*). Let Y be another well-ordered set and let

$$g\colon Y \to \mathcal{U}$$

be the map such that

$$g(y) = g\langle]\!\leftarrow,y[\rangle \qquad (y \in Y),$$

so that

$$\text{ord}(Y) = g\langle Y\rangle.$$

If $X \sim Y$, say $h: X \sim Y$, then

$$]\leftarrow,h(x)[\; = h\langle]\leftarrow,x[\rangle \qquad (x \in X),$$

an easy transfinite induction shows

$$g(h(x)) = f(x) \qquad (x \in X),$$

and $\operatorname{ord}(X) = f\langle X\rangle = g\langle h\langle X\rangle\rangle = g\langle Y\rangle = \operatorname{ord}(Y)$. Conversely, assume $\operatorname{ord}(X) = \operatorname{ord}(Y)$. By (**), f is injective and so defines a bijection from X to $f\langle X\rangle$ which we denote again by f; similarly we obtain a bijection g from Y to $f\langle Y\rangle$. Then

$$g^{-1} \circ f: X \to Y$$

is a bijection. Moreover, by (**) and its analog for Y, $g^{-1} \circ f$ is an order-isomorphism. Hence $X \sim Y$.

Now that our definition of the ordinal of a well-ordered set has been shown to satisfy (*), we ask what sort of set can be the ordinal of a well-ordered set. In other words, we seek an intrinsic characterization of those sets α for which $\alpha = \operatorname{ord}(X)$ for some well-ordered set X. Let α be such a set, and choose some well-ordered set X with $\alpha = \operatorname{ord}(X)$. Let $f: X \to \mathfrak{U}$ be the map satisfying (**) as before. First, the elementhood relation connects α, for the relation $<$ in X connects X. Second, α is full, for if $y \in \alpha$ and $z \in y$, then $y = f(x)$ and $z = f(t)$ with $t < x$, so

$$z = f(t) \in f(x) = f\langle]\leftarrow,x[\rangle \subset f\langle X\rangle = \alpha.$$

Conversely, if α is a set which is full and connected by the elementhood relation, then $\alpha = \operatorname{ord}(X)$ for some X (see Exercise B).

We are going to reverse the foregoing considerations. We shall study abstractly the class of all sets which are full and connected by the elementhood relation. Then we shall define anew the ordinal of a well-ordered set to be one of these. The discussion above is purely motivational.

THE CLASS OF ORDINAL NUMBERS

18.1. DEFINITION

A set α is called an *ordinal number* if it is full and if the elementhood relation connects it, in other words, if α has the two properties:

(i) If $x \in \alpha$, then $x \subset \alpha$.
(ii) If $x \in \alpha$ and $y \in \alpha$, then $x = y$ or $x \in y$ or $y \in x$ (by the axiom of foundation, the three alternatives $x = y$, $x \in y$, $y \in x$ are mutually exclusive).

The class

$$\{\alpha \mid \alpha \text{ is an ordinal number}\}$$

of all ordinal numbers is denoted by Ord.

18.2. EXAMPLES

(1) If $n \in \omega$, then $n \in$ Ord by 6.10 and 6.18. Thus ordinal numbers are a generalization of natural numbers; see also 18.9(1).

(2) By 6.21 and 6.18, $\omega \in$ Ord.

(3) The set $\omega^+ = \omega \cup \{\omega\} \in$ Ord.

(4) By ordinary recursion, there is a sequence $(\alpha_n \mid n \in \omega)$ such that $\alpha_0 = \omega$ and $\alpha_{n+1} = \alpha_n{}^+$ for each $n \in \omega$. An easy induction shows $\alpha_n \in$ Ord for each n.

(5) In the notation of (4), let $\alpha = \bigcup_{n=0}^{\infty} \alpha_n$. Then $\alpha \in$ Ord.

The three lemmas which follow allow us to define a well-ordering of Ord which automatically induces a natural well-ordering on each ordinal number.

18.3. LEMMA

Let X be a class. Define R to be the relation in X such that

$$xRy \Leftrightarrow x \in y \vee x = y \qquad (x, y \in X).$$

Then R well-orders X if and only if R connects X.

Proof

Obviously R well-orders X only if it connects X. Conversely, assume R connects X. Let $\varnothing \neq E \subset X$. By the axiom of foundation, there exists $z \in E$ with $z \cap E = \varnothing$. If $x \in E$, then $x \notin z$, so zRx since R connects X. Thus z is a first element of E for R. If y is also a first element of E for R with $y \neq z$, then $y \in z$ since yRz, and this is impossible since $y \in E$. Hence E has a unique first element. ☐

18.4. LEMMA

The class Ord is full.

Proof

Let $\beta \in \alpha \in$ Ord. We must show $\beta \in$ Ord. Since α is full, $\beta \subset \alpha$. Then the elementhood relation connects β since it connects α. It remains to show that β is full.

Let $y \in x \in \beta$. We show $y \in \beta$. Note that $y \in \alpha$ since $y \in x \in \beta \subset \alpha$ and α is full. In 18.3 take $X = \alpha$. Since the elementhood relation connects

X, so does the relation R. By 18.3, the subset $\{x,y,\beta\}$ of X has a first element z for R. Now $z \neq x$ since $y \in x$, and $z \neq \beta$ since $x \in \beta$. Hence $z = y$. Then $yR\beta$, that is, $y = \beta$ or $y \in \beta$. But $y \neq \beta$ since $y \in x \in \beta$. Hence $y \in \beta$. ☐

18.5. LEMMA

Let $\alpha, \beta \in$ Ord. Then

$$\alpha \in \beta \Leftrightarrow \alpha \subsetneq \beta.$$

Proof

If $\alpha \in \beta$, then $\alpha \subset \beta$ since β is full, so $\alpha \subsetneq \beta$.
Conversely, assume $\alpha \subsetneq \beta$. By 18.3, there exists $x \in \beta \setminus \alpha$ such that

$$(*) \qquad\qquad y \in \beta \setminus \alpha \,\&\, y \neq x \Rightarrow x \in y$$

(take $X = \beta$ in 18.3—then $\beta \setminus \alpha$ has a first element x for the well-ordering R). To show $\alpha \in \beta$, it suffices to show $x = \alpha$. Now $x \subset \alpha$, since $y \in x$ implies $x \notin y \neq x$ whence $y \in \alpha$ by $(*)$. To prove the opposite inclusion, let $y \in \alpha$. Then $x \in \beta$ and $y \in \beta$, so $x = y$ or $x \in y$ or $y \in x$. Now $x \neq y$ since $x \notin \alpha$ and $y \in \alpha$; $x \notin y$ since $y \in \alpha$ and α is full. Hence $y \in x$. ☐

18.6. DEFINITION

The inclusion relation in Ord is denoted by $\leq$. Then for $\alpha, \beta \in$ Ord,

$$\alpha \leq \beta \Leftrightarrow \alpha \subset \beta$$

and by 18.5,

$$\alpha \leq \beta \Leftrightarrow \alpha \in \beta \lor \alpha = \beta.$$

Any future reference to a partial ordering of Ord is to this relation $\leq$.

If $\alpha \in$ Ord, then $\alpha \subset$ Ord by 18.4, so the partial ordering of Ord induces a partial ordering on α which is again denoted by $\leq$. Any future reference to an ordinal number as a partially ordered set is understood to mean with respect to the induced partial ordering.

It was noted in 18.2 that $\omega \subset$ Ord. Observe that for natural numbers m and n the meaning of the statement $m \leq n$ given by 18.6 is precisely the same as the meaning given earlier by 6.14. The partial ordering of Ord induces on ω its usual order relation.

Lemma 18.3 says that the partial ordering of each $\alpha \in$ Ord is in fact a well-ordering. Thus *each ordinal number is a well-ordered set.*

If $\alpha \in$ Ord, then in the partially ordered class Ord

$$]\leftarrow,\alpha[\,= \alpha.$$

In fact, each element of α is an ordinal number since both α and Ord are full, so

$$\alpha = \{x \mid x \in \alpha\} = \{x \mid x \in \text{Ord} \,\&\, x \in \alpha\}$$
$$= \{x \mid x \in \text{Ord} \,\&\, x < \alpha\} = \,]\!\leftarrow,\alpha[.$$

We come now to the principal result of this chapter.

18.7. THEOREM

(1) The class Ord is a well-ordered proper class.

(2) If $\mathcal{W}$ denotes the class of all well-ordered sets, then there exists a unique map

$$\text{ord}: \mathcal{W} \to \text{Ord}$$

such that

$$X \sim \text{ord}(X) \qquad (X \in \mathcal{W}).$$

Proof

(1) To show $\leq$ well-orders Ord, by 18.3 it suffices to show $\leq$ connects Ord. Let α, $\beta \in$ Ord. We show $\alpha \leq \beta$ or $\beta \leq \alpha$. Define

$$\gamma = \alpha \cap \beta.$$

Then $\gamma \in$ Ord, $\gamma \leq \alpha$, and $\gamma \leq \beta$. Then $\gamma = \alpha$ or $\gamma = \beta$, for $\gamma \neq \alpha$ and $\gamma \neq \beta$ implies $\gamma \in \alpha \cap \beta = \gamma$, an impossibility.

Just suppose Ord is a set. By 18.4, Ord is full. By the preceding paragraph, the elementhood relation connects Ord. Then Ord is an ordinal number, that is, Ord $\in$ Ord. This is absurd.

(2) Let $X \in \mathcal{W}$. By 17.18, one of the following alternatives holds: (i) $X \sim$ Ord; (ii) $]\!\leftarrow,x[\, \sim$ Ord for some $x \in X$; (iii) $X \sim \,]\!\leftarrow,\alpha[$ for some $\alpha \in$ Ord. Since X is a set and Ord is a proper class, (i) and (ii) are excluded. Hence

$$X \sim \,]\!\leftarrow,\alpha[\, = \alpha$$

for some $\alpha \in$ Ord.

To prove the uniqueness of the $\alpha \in$ Ord for which $X \sim \alpha$, let $\beta \in$ Ord with $\alpha \neq \beta$. We show α is not order-isomorphic to β. Without loss of generality we may suppose $\alpha \leq \beta$. Then α is the initial segment $]\!\leftarrow,\alpha[$ of β, and by 17.16, α is not order-isomorphic to β.

Thus we have a map ord: $\mathcal{W} \to$ Ord such that

$$\text{ord}(X) = \alpha \Leftrightarrow \alpha \in \text{Ord} \,\&\, X \sim \alpha \qquad (X \in \mathcal{W}).$$

The uniqueness of this map is immediate from the preceding paragraph. ☐

In Cantor's unaxiomatized set theory, it was tacitly assumed that Ord is a set. In 1897, C. Burali-Forti derived the very contradiction Ord $\in$ Ord obtained in the proof of 18.7, thereby discovering the first paradox in informal set theory.

18.8. DEFINITION

Let X be a well-ordered set. Then $\text{ord}(X)$, the unique $\alpha \in \text{Ord}$ such that $X \sim \alpha$, is called the *ordinal of* X.

If X is a well-ordered set, not only is $\text{ord}(X)$ the unique ordinal number isomorphic to X, but by 17.15 there is a unique order-isomorphism from X to $\text{ord}(X)$.

For each ordinal number α,

$$\text{ord}(\alpha) = \alpha.$$

Moreover, for ordinal numbers α and β,

$$\alpha \sim \beta \Leftrightarrow \alpha = \beta.$$

Hence for well-ordered sets X and Y,

$$X \sim Y \Leftrightarrow \text{ord}(X) = \text{ord}(Y).$$

This means that for a well-ordered set X, $\text{ord}(X)$ is a member of the equivalence class of X modulo $\sim$ in the class of all well-ordered sets, and $\text{ord}(X)$ is order-isomorphic to every member of this equivalence class.

SOME PROPERTIES AND KINDS OF ORDINALS

The next two propositions elucidate Examples 18.2.

18.9. PROPOSITION

(1) An ordinal number α is finite if and only if $\alpha \in \omega$.
(2) The set ω is the first infinite ordinal number.

Proof

(1) By 18.2, each natural number is a finite ordinal number. Let $\alpha \in \text{Ord}$ with α finite. Define $n = \#(\alpha)$. We show $\alpha = n$ by proving $\alpha \sim n$. By 17.18, either $\alpha \sim n$, or $\alpha \sim m$ for some $m \in n$, or $\beta \sim n$ for some $\beta \in \alpha$. The last two alternatives are impossible since $\alpha \simeq n$.

(2) By 18.2, $\omega \in \text{Ord}$. Let α be an infinite ordinal number. By 17.18, either $\omega \sim \alpha$, or $\omega \sim \beta$ for some $\beta \in \alpha$, or $n \sim \alpha$ for some $n \in \omega$. The third alternative is impossible since α is infinite. If $\omega \sim \alpha$, then $\omega = \alpha$. If $\omega \sim \beta$ for some $\beta \in \alpha$, then $\omega = \beta < \alpha$. □

18.10. PROPOSITION

(1) If $\alpha \in \text{Ord}$, then α^+ is the first ordinal number greater than α.
(2) If E is a set of ordinal numbers, then $\bigcup E$ is the first ordinal number β such that $\beta \geq \alpha$ for each $\alpha \in E$.

Proof

(1) Trivial.

(2) Let E be a subset of Ord, and let $\gamma = \bigcup E$. Then γ is a set. Since the union of any class of full sets is full, γ is full. That the elementhood relation connects γ follows easily from the fact that any two members of E are comparable. Hence $\gamma \in$ Ord. Clearly $\gamma \geq \alpha$ for each $\alpha \in E$. If $\beta \in$ Ord with $\beta \geq \alpha$ for each $\alpha \in E$, then $\beta \supset \alpha$ for each $\alpha \in E$, $\beta \supset \gamma = \bigcup E$, and $\beta \geq \gamma$. □

All the ordinal numbers listed in 18.2, as well as ordinal numbers constructed from these by use of 18.10, are countable. On the other hand, there certainly exist uncountable ordinal numbers. In fact, if X is any uncountable set, there exists a well-ordering of X, and the ordinal of the resulting well-ordered set is an uncountable ordinal number. This justifies the next definition.

18.11. DEFINITION

The first uncountable ordinal number is denoted by Ω.

The following result partially extends 13.13 from natural numbers to arbitrary ordinal numbers.

18.12. PROPOSITION

Let $\alpha, \beta \in$ Ord. Then:

(1) If $\alpha \prec \beta$, then $\alpha < \beta$.
(2) If $\alpha < \beta$, then $\alpha \preceq \beta$.
(3) If $\alpha \leq \beta$, then $\alpha \preceq \beta$.

Proof

(1) If $\alpha \nless \beta$, then $\beta \leq \alpha$, $\beta \subset \alpha$, $\beta \preceq \alpha$, and it is not the case that $\alpha \prec \beta$.

(2) If $\alpha < \beta$, then $\alpha \subset \beta$ and so $\alpha \preceq \beta$.

(3) follows from (2). □

The two ordinal numbers ω and ω^+ satisfy

$$\omega < \omega^+, \qquad \omega \simeq \omega^+.$$

Hence (2) above cannot be strengthened to read, "If $\alpha < \beta$, then $\alpha \prec \beta$," and the converses of (1)–(3) all fail. In Chapter 19 we shall define a subclass of Ord, containing ω, to which 13.13 extends *in toto*.

A rough classification of ordinal numbers is suggested by the observation that ω has no immediate predecessor in Ord, but ω^+ is the immediate successor of ω.

18.13. DEFINITION

An ordinal number α is called a *limit ordinal* if $\alpha \neq 0$ and if $\alpha \neq \beta^+$ for all $\beta \in \text{Ord}$. The class of all limit ordinals is denoted by Lim. Elements of Ord $\setminus$ Lim are called *nonlimit ordinals.*

Of the ordinal numbers in 18.2, only ω and α are limit ordinals.

18.14. PROPOSITION

Let $\alpha \in \text{Ord}$ with $\alpha \neq 0$. Then each of the following conditions is necessary and sufficient for $\alpha \in \text{Lim}$:

(1) $\beta < \alpha$ implies $\beta^+ < \alpha$;
(2) α has no last element;
(3) $\bigcup \alpha = \alpha$.

Proof

Clearly each of (1), (2) is equivalent to $\alpha \in \text{Lim}$.
Assume (2). We show (3). Since α is full, $\bigcup \alpha \subset \alpha$. To prove the opposite inclusion, let $x \in \alpha$. Then $x < \alpha$, by (2) there exists $\beta \in \alpha$ with $x \in \beta$, and so $x \in \bigcup \alpha$.
Assume (3), and just suppose $\alpha = \beta^+$ for some $\beta \in \text{Ord}$. Since $\beta \in \beta^+ = \alpha$, $\beta \in \gamma$ for some $\gamma \in \beta^+$, contradicting 18.10(1). ☐

ORDINAL RECURSION

In order to provide names for particular ordinal numbers, we discuss briefly the arithmetic of ordinal numbers.
First, we shall define an operation $\oplus$ of addition of ordinal numbers which extends ordinary addition of natural numbers. (To distinguish this operation on Ord from addition of cardinal numbers as defined in Chapter 19, we use the notation $\oplus$ here.) Certainly the operation $\oplus$ in Ord should satisfy

$$\alpha \oplus 0 = \alpha$$

$$\alpha \oplus \beta^+ = (\alpha \oplus \beta)^+$$

for all $\alpha, \beta \in \text{Ord}$ if it is to generalize ordinary addition in ω. These two conditions do not specify $\alpha \oplus \gamma$ for all $\gamma \in \text{Ord}$, but only for nonlimit ordinals γ. For $\gamma \in \text{Lim}$ and $\alpha \in \text{Ord}$, it is reasonable to expect $\alpha \oplus \gamma > \alpha \oplus \beta$ for each $\beta < \gamma$, and so guided by example 18.2(5), we also require

$$\alpha \oplus \gamma = \bigcup_{\beta \in \gamma} \alpha \oplus \beta \qquad (\alpha \in \text{Ord}, \gamma \in \text{Lim}).$$

Of course, the existence of a unique binary operation satisfying the three requirements must be established. For this we prove a special version of transfinite recursion for the well-ordered class Ord.

18.15. THEOREM (TRANSFINITE RECURSION—ORDINAL FORM)

Let X be a class such that $\bigcup E \in X$ for each nonempty subset E of X. Let

$$G: X \to X, \qquad z \in X$$

be given. Then there exists a unique map

$$f: \text{Ord} \to X$$

such that

$$\begin{aligned} f(0) &= z, \\ f(\alpha^+) &= G(f(\alpha)) && (\alpha \in \text{Ord}), \\ f(\alpha) &= \textstyle\bigcup_{\beta \in \alpha} f(\beta) && (\alpha \in \text{Lim}). \end{aligned}$$

Proof

Uniqueness

Suppose f_1, f_2: Ord $\to X$ both have the stated properties. Define

$$E = \{\alpha \mid \alpha \in \text{Ord} \;\&\; f_1(\alpha) = f_2(\alpha)\}.$$

By transfinite induction one shows E = Ord (in the inductive step the three cases $\alpha = 0$, $\alpha = \beta^+$ for some $\beta \in$ Ord, $\alpha \in$ Lim must be considered separately).

Existence

For technical reasons it will be convenient to construct not the map f: Ord $\to X$ itself, but rather its graph f. For each $\alpha \in$ Ord, we denote by

$$F(\alpha,X)$$

the class of all functional relations h such that

$$\text{dmn } h = \alpha, \qquad \text{rng } h \subset X.$$

The functional relation f we seek is to have domain Ord, and Ord $\supset \alpha$ for each $\alpha \in$ Ord. The idea of the proof below is to construct, for each $\alpha \in$ Ord, the piece $g_\alpha \in F(\alpha,X)$ of f whose domain is α. If $\alpha \in$ Ord and if the pieces $g_\beta \in F(\beta,X)$ have been constructed for all $\beta < \alpha$, then the ordered pairs $(\gamma,x) \in f$ are known for all $\gamma < \beta < \alpha$; to obtain the piece $g_\alpha \in F(\alpha,X)$, we must specify the ordered pairs $(\beta,x) \in f$ for all $\beta < \alpha$. Hence g_α should depend on whether $\alpha = \beta^+$ for some $\beta < \alpha$, or $\alpha \in$ Lim, or $\alpha = 0$.

We now proceed with the actual construction of f. Let

$$H: \mathcal{P}(\text{Ord} \times X) \to \text{Ord} \times X$$

be the map defined by

$$H(h) = \begin{cases} (\alpha, G(h(\beta)) & \text{if } \beta \in \text{Ord}, \alpha = \beta^+, h \in F(\alpha,X), \\ (\alpha, \bigcup_{\beta\in\alpha} h(\beta)) & \text{if } \alpha \in \text{Lim}, h \in F(\alpha,X), \\ (0,z) & \text{otherwise.} \end{cases}$$

By transfinite recursion (17.20), there exists a map

$$g: \text{Ord} \to \text{Ord} \times X$$

such that

$$g(\alpha) = H(g\langle\alpha\rangle) \qquad (\alpha \in \text{Ord}).$$

For simplicity of notation, define

$$g_\alpha = g\langle\alpha\rangle \qquad (\alpha \in \text{Ord}).$$

We use transfinite induction to prove

$$g_\alpha \in F(\alpha,X) \qquad (\alpha \in \text{Ord}).$$

Let $\alpha \in$ Ord, and assume

$$g_\beta \in F(\beta,X) \qquad (\beta < \alpha).$$

We distinguish three cases.

Case (1)

$\alpha = 0$. Then

$$g_\alpha = g\langle\varnothing\rangle = \varnothing \in F(\alpha,X).$$

Case (2)

$\alpha = \beta^+$ for some $\beta \in$ Ord. Then

$$g_\alpha = g\langle\beta \cup \{\beta\}\rangle = g\langle\beta\rangle \cup \{g(\beta)\} = g_\beta \cup \{H(g_\beta)\}.$$

Now $g_\beta \in F(\beta,X)$ and $H(g_\beta) = (\beta,x)$ for some $x \in X$. Then g_β and $\{H(g_\beta)\}$ are functional relations having disjoint domains β and $\{\beta\}$. Hence g_α is a functional relation having domain $\beta \cup \{\beta\} = \alpha$. Thus $g_\alpha \in F(\alpha,X)$.

Case (3)

$\alpha \in$ Lim. Then

$$g_\alpha = g\langle\bigcup_{\beta\in\alpha} \beta\rangle = \bigcup_{\beta\in\alpha} g\langle\beta\rangle = \bigcup_{\beta\in\alpha} g_\beta.$$

Now $g_\beta \in F(\beta,X)$ for each $\beta \in \alpha$, and $\{g_\beta \mid \beta \in \alpha\}$ is totally ordered under inclusion since α is. By 8.28 and Chapter 8, Exercise DD, g_α is a functional relation such that

$$\operatorname{dmn} g_\alpha = \bigcup_{\beta \in \alpha} \operatorname{dmn} g_\beta = \bigcup_{\beta \in \alpha} \beta = \alpha.$$

Thus $g_\alpha \in F(\alpha, X)$.

We have just proved that $g_\alpha \in F(\alpha, X)$ for every $\alpha \in \mathrm{Ord}$. By 8.28 and Chapter 8, Exercise DD, there is a functional relation f with

$$\operatorname{dmn} f = \mathrm{Ord}, \qquad \operatorname{rng} f \subset X$$

such that

$$f(\beta) = g_\alpha(\beta) \qquad (\alpha \in \mathrm{Ord}, \beta \in \alpha).$$

If $\alpha \in \mathrm{Ord}$, then

$$f(\alpha) = g_{\alpha^+}(\alpha) = g\langle \alpha^+ \rangle(\alpha),$$

and since for some $x \in X$

$$g(\alpha) = H(g\langle \alpha \rangle) = (\alpha, x)$$

is the unique element of $g\langle \alpha^+ \rangle$ having first coordinate α,

$$(\alpha, f(\alpha)) = g(\alpha).$$

An easy computation now shows that f has the desired properties. $\square$

ORDINAL ADDITION

Addition of ordinal numbers can now be defined in the way suggested earlier. Let $\alpha \in \mathrm{Ord}$. Taking for X in 18.15 the class Ord, for z the element α of X, and for G the map $\beta \mapsto \beta^+$, we obtain a unique map

$$s_\alpha : \mathrm{Ord} \longrightarrow \mathrm{Ord}$$

such that

$$s_\alpha(0) = \alpha,$$

$$s_\alpha(\beta^+) = (s_\alpha(\beta))^+ \qquad (\beta \in \mathrm{Ord}),$$

$$s_\alpha(\beta) = \bigcup_{\gamma \in \beta} s_\alpha(\gamma) \qquad (\beta \in \mathrm{Lim}).$$

18.16. DEFINITION

The binary operation

$$(\alpha, \beta) \mapsto s_\alpha(\beta)$$

on Ord is called *ordinal addition* and is denoted by $\oplus$. Thus $\alpha \in \mathrm{Ord}$ implies

$$\alpha \oplus 0 = \alpha,$$

$$\alpha \oplus \beta^+ = (\alpha \oplus \beta)^+ \qquad (\beta \in \mathrm{Ord}),$$

$$\alpha \oplus \beta = \bigcup_{\gamma \in \beta} \alpha \oplus \gamma \qquad (\beta \in \mathrm{Lim}),$$

and in particular,

$$\alpha \oplus 1 = \alpha^+.$$

If $\alpha, \beta \in \text{Ord}$, one calls $\alpha \oplus \beta$ the *ordinal sum of* α *and* β.

It is hardly to be expected that all the familiar laws of addition in ω carry over to Ord. For example,

$$1 \oplus \omega = \bigcup_{n \in \omega} 1 \oplus n = \bigcup_{n \in \omega} 1 + n = \omega,$$

but $\omega \oplus 1 = \omega^+$, so

$$1 \oplus \omega \neq \omega \oplus 1.$$

Also, $0 < 1$, but $0 \oplus \omega = \omega = 1 \oplus \omega$, so

$$0 \oplus \omega \nless 1 \oplus \omega.$$

Some properties of addition in ω do generalize to addition in Ord. To establish these, a concrete representation of $\alpha \oplus \beta$ is especially convenient. Recall the definition of the ordinal sum of disjoint well-ordered sets: 17.2(4). Note that if $\alpha, \beta \in \text{Ord}$, then by Chapter 17, Exercise L there are disjoint well-ordered sets A, B with $\text{ord}(A) = \alpha$, $\text{ord}(B) = \beta$.

18.17. PROPOSITION

Let $\alpha, \beta \in \text{Ord}$. If A and B are disjoint well-ordered sets with

$$\text{ord}(A) = \alpha, \qquad \text{ord}(B) = \beta,$$

then

$$\alpha \oplus \beta = \text{ord}(A \cup B),$$

where $A \cup B$ stands for the ordinal sum of A and B.

Proof

The union of any pair of disjoint well-ordered sets below is understood to be provided with the ordinal sum of their given well-orderings.

Consider first the special case $\beta = 1$. Then B is a singleton $\{b\}$, $A \cup B = A \cup \{b\} \sim \alpha \cup \{\alpha\} = \alpha^+ = \alpha \oplus \beta$, and hence $\text{ord}(A \cup B) = \alpha \oplus \beta$.

The general case is treated by transfinite induction on β. Assume that if $\gamma < \beta$ and if C is a well-ordered set disjoint from A with $\text{ord}(C) = \gamma$, then $\alpha \oplus \gamma = \text{ord}(A \cup C)$.

If $\beta = 0$, then $B = \varnothing$, and

$$\alpha \oplus \beta = \alpha = \text{ord}(A) = \text{ord}(A \cup B).$$

Suppose $\beta = \gamma^+$ for some $\gamma \in \text{Ord}$. There exists $f: \beta \sim B$. Let $C = f\langle\gamma\rangle$ and $S = \{f(\gamma)\}$. Then C is an initial segment of B, and B is the ordinal sum of C and S, so

$$\beta = \gamma \oplus 1 = \text{ord}(C \cup S).$$

By the inductive hypothesis,

$$\alpha \oplus \gamma = \operatorname{ord}(A \cup C).$$

By the special case treated in the first paragraph, this yields

$$\alpha \oplus \beta = (\alpha \oplus \gamma) \oplus 1 = \operatorname{ord}((A \cup C) \cup S).$$

But $(A \cup C) \cup S = A \cup B$ as ordinal sums.

Suppose $\beta \in \text{Lim}$. Let $f\colon \beta \sim B$. For each $\gamma \in \beta$, let $C_\gamma = f\langle\gamma\rangle$. By the inductive hypothesis, for each $\gamma \in \beta$ there exists a (necessarily unique) order-isomorphism

$$g_\gamma\colon \alpha \oplus \gamma \sim A \cup C_\gamma.$$

Now

$$\alpha \oplus \beta = \bigcup_{\gamma\in\beta} \alpha \oplus \gamma,$$

$$A \cup B = \bigcup_{\gamma\in\beta} A \cup C_\gamma,$$

so there exists a bijection

$$g\colon \alpha \oplus \beta \simeq A \cup B$$

with

$$g \mid (\alpha \oplus \gamma) = g_\gamma \qquad (\gamma \in \beta).$$

For each $\gamma \in \beta$, the ordinal sum well-ordering of $A \cup C_\gamma$ is induced by the well-ordering of $A \cup B$. It follows that g is an order-isomorphism. □

Thus $\alpha \oplus \beta$ has the same order type as the well-ordered set obtained by placing a copy of β after a copy of α. For example,

$$1 \oplus \omega \sim \{0\} \cup \{1,2,\ldots\} = \omega$$

since $\omega \sim \omega \setminus \{0\}$. Taking $A = \varnothing$ in 18.17, we obtain

$$0 \oplus \beta = \beta \qquad (\beta \in \text{Ord}).$$

Further use of 18.17 is illustrated by the next proof.

18.18. PROPOSITION

Let $\alpha, \beta, \gamma \in \text{Ord}$. Then:

(1) $(\alpha \oplus \beta) \oplus \gamma = \alpha \oplus (\beta \oplus \gamma)$.
(2) If $\beta < \gamma$, then $\alpha \oplus \beta < \alpha \oplus \gamma$ and $\beta \oplus \alpha \leq \gamma \oplus \alpha$.

Proof

(2) Assume $\beta < \gamma$. Choose disjoint well-ordered sets A, C with $\operatorname{ord}(A) = \alpha$, $\operatorname{ord}(C) = \gamma$. Since β is an initial segment of γ, there is an initial segment B of C such that $\operatorname{ord}(B) = \beta$.

By 18.17,

$$\alpha \oplus \beta = \text{ord}(A \cup B), \qquad \alpha \oplus \gamma = \text{ord}(A \cup C).$$

Then $\alpha \oplus \beta$ is order-isomorphic to the initial segment $A \cup B$ of $A \cup C$, so $\alpha \oplus \beta < \alpha \oplus \gamma$.

By 18.17 again,

$$\beta \oplus \alpha = \text{ord}(B \cup A), \qquad \gamma \oplus \alpha = \text{ord}(C \cup A).$$

Then $\beta \oplus \alpha$ is order-isomorphic to the subset $B \cup A$ of $C \cup A$. Now $x \in B \cup A$ implies

$$(C \cup A)]\leftarrow,x[\subset B \cup A.$$

By 17.17, either $B \cup A = C \cup A$, whence $\beta \oplus \alpha = \gamma \oplus \alpha$, or else $B \cup A$ is an initial segment of $C \cup A$, whence $\beta \oplus \alpha < \gamma \oplus \alpha$. $\square$

ORDINAL MULTIPLICATION AND EXPONENTIATION

We treat ordinal multiplication and exponentiation more summarily than addition.

18.19. DEFINITION

The unique operation $\cdot$ on Ord such that

$$\alpha \cdot 0 = 0 \qquad (\alpha \in \text{Ord}),$$

$$\alpha \cdot (\beta \oplus 1) = (\alpha \cdot \beta) \oplus \alpha \qquad (\alpha, \beta \in \text{Ord}),$$

$$\alpha \cdot \beta = \bigcup_{\gamma \in \beta} \alpha \cdot \gamma \qquad (\alpha \in \text{Ord}, \beta \in \text{Lim})$$

is called *ordinal multiplication*, and for α, $\beta \in$ Ord, $\alpha \cdot \beta$ is called the *ordinal product of* α *and* β.

The unique binary operation

$$(\alpha,\beta) \mapsto \alpha^\beta$$

on Ord such that

$$\alpha^0 = 1 \qquad (\alpha \in \text{Ord}),$$

$$\alpha^{\beta \oplus 1} = \alpha^\beta \cdot \alpha \qquad (\alpha, \beta \in \text{Ord}),$$

$$\alpha^\beta = \begin{cases} \bigcup_{\gamma \in \beta} \alpha^\gamma & \text{if } \alpha \neq 0 \\ 0 & \text{if } \alpha = 0 \end{cases} \qquad (\alpha \in \text{Ord}, \beta \in \text{Lim})$$

is called *ordinal exponentiation*, and for α, $\beta \in$ Ord, α^β is called the *ordinal power of* α *to the* β.

When restricted to $\omega \times \omega$, these two operations in Ord yield multiplication and exponentiation as defined in Chapter 10.

Ordinal multiplication is not commutative, for

$$\omega \cdot 2 = (\omega \cdot 1) \oplus \omega = \omega \oplus \omega \neq \omega = \bigcup_{n \in \omega} 2n = 2 \cdot \omega.$$

However, it is associative, obeys the distributive law

$$\alpha \cdot (\beta \oplus \gamma) = \alpha \cdot \beta \oplus \alpha \cdot \gamma,$$

and has the properties

$$\alpha < \beta \Rightarrow \alpha \cdot \gamma \leq \beta \cdot \gamma,$$

$$0 < \alpha < \beta \,\&\, \gamma \neq 0 \Rightarrow \gamma \cdot \alpha < \gamma \cdot \beta.$$

Ordinal exponentiation satisfies

$$1 < \alpha \,\&\, \beta < \gamma \Rightarrow \alpha^{\beta} < \alpha^{\gamma}.$$

For these and additional facts about ordinal multiplication and exponentiation see the exercises.

The following representation theorem is an analog of the usual representation of natural numbers in base 10. It will not be used to prove any subsequent theorem, and its proof is accordingly omitted (see, for example, Rubin [34, Theorem 9.1.2] for a proof).

18.20. THEOREM

Let $\alpha \in$ Ord with $\alpha \neq 0$. Then there exists a unique natural number n, a unique family $(a_i \mid i = 0,1,\ldots,n)$ of nonzero natural numbers, and a unique family $(\alpha_i \mid i = 0,1,\ldots,n)$ of ordinal numbers such that

$$\alpha_0 < \alpha_1 < \ldots < \alpha_n$$

and

$$(*) \qquad \alpha = \omega^{\alpha_n} \cdot a_n \oplus \omega^{\alpha_{n-1}} \cdot a_{n-1} \oplus \ldots \oplus \omega^{\alpha_0} \cdot a_0.$$

The representation (*) of α is called the *normal form of* α.

Theorem 18.20 can be used, for example, to list systematically all ordinal numbers α strictly less than ω^ω. Let $\alpha \in$ Ord with $0 < \alpha$ and consider the normal form (*) of α. We claim that $\alpha < \omega^\omega$ if and only if $\alpha_n < \omega$. In fact, if $\alpha_n \geq \omega$, then $\alpha \geq \omega^{\alpha_n} \geq \omega^\omega$. Conversely, if $\alpha_n < \omega$, then $\alpha_n \oplus 1 = \alpha_n + 1 < \omega$, so that

$$\begin{aligned}
\alpha &= \omega^{\alpha_n} \cdot a_n \oplus \omega^{\alpha_{n-1}} \cdot a_{n-1} \oplus \ldots \oplus \omega^{\alpha_0} \cdot a_0 \\
&\leq \omega^{\alpha_n} \cdot a_n \oplus \omega^{\alpha_n} \cdot a_{n-1} \oplus \ldots \oplus \omega^{\alpha_n} \cdot a_0 \\
&= \omega^{\alpha_n} \cdot (a_n + a_{n-1} + \ldots + a_0) \\
&< \omega^{\alpha_n} \cdot \omega \\
&= \omega^{\alpha_n \oplus 1} \\
&< \omega^\omega.
\end{aligned}$$

Thus the ordinal numbers less than ω^ω are, in order:

$$\begin{aligned}
&0, 1, 2, \ldots, \omega,\\
&\omega \oplus 1, \omega \oplus 2, \ldots, \omega \cdot 2,\\
&\omega \cdot 2 \oplus 1, \omega \cdot 2 \oplus 2, \ldots, \omega \cdot 3, \ldots,\\
&\omega \cdot 4, \ldots, \omega \cdot 5, \ldots, \omega^2,\\
&\omega^2 \oplus 1, \omega^2 \oplus 2, \ldots, \omega^2 \oplus \omega,\\
&\omega^2 \oplus \omega \oplus 1, \omega^2 \oplus \omega \oplus 2, \ldots, \omega^2 \oplus \omega \cdot 2,\\
&\omega^2 \oplus \omega \cdot 2 \oplus 1, \ldots, \omega^2 \oplus \omega \cdot 3, \ldots,\\
&\omega^2 \cdot 2, \ldots, \omega^2 \cdot 3, \ldots, \omega^3, \ldots,\\
&\omega^4, \ldots, \omega^5, \ldots.
\end{aligned}$$

These are all countable! (See Exercise N.)

EQUIPOLLENCE OF PROPER CLASSES

The remainder of this chapter is devoted to proving that each proper class is equipollent to Ord. The method of proof is to show that $\mathcal{U}$, like Ord, can be well-ordered in such a way that each of its initial segments is a set.

The transfinite recursion theorem 17.20 justifies the following definition.

18.21. DEFINITION

We define

$$\tau\colon \text{Ord} \to \mathcal{U}$$

to be the unique map such that

$$\tau(\alpha) = \mathcal{P}(\bigcup \tau\langle\alpha\rangle) \qquad (\alpha \in \text{Ord}).$$

This map has the property

$$\alpha \leq \beta \Rightarrow \tau(\alpha) \subset \tau(\beta) \qquad (\alpha, \beta \in \text{Ord}).$$

18.22. LEMMA

If $x \in \mathcal{U}$, then $x \in \tau(\alpha)$ for some $\alpha \in \text{Ord}$.

Proof

Define

$$X = \{x \mid (\forall \alpha \in \text{Ord})(x \notin \tau(\alpha))\},$$

and just suppose $X \neq \varnothing$. By the axiom of foundation we can choose $u \in X$ with $u \cap X = \varnothing$. If $y \in u$, then $y \in \tau(\alpha)$ for some $\alpha \in \text{Ord}$ since $y \notin X$, and we may define α_y to be the first such ordinal number α. Define

$$\beta = \bigcup_{y \in u} \alpha_y.$$

Then $y \in u$ implies $\alpha_y \subset \beta$ and $y \in \tau(\alpha_y) \subset \tau(\beta)$. Hence $u \subset \tau(\beta)$. But then

$$u \in \mathcal{P}(\tau(\beta)) \subset \mathcal{P}(\bigcup_{\gamma \in \beta^+} \tau(\gamma)) = \tau(\beta^+),$$

which contradicts the fact that $u \in X$. □

Lemma 18.22 justifies the following definition.

18.23. DEFINITION

The map

$$\rho: \mathfrak{U} \to \mathrm{Ord}$$

is defined by

$$\rho(x) = \text{first element of } \{\alpha \mid \alpha \in \mathrm{Ord} \ \& \ x \in \tau(\alpha)\} \qquad (x \in \mathfrak{U}).$$

For $x \in \mathfrak{U}$, $\rho(x)$ is called the *rank of* x.

18.24. LEMMA

Let $\alpha \in \mathrm{Ord}$. Then

$$\{x \mid \rho(x) \leq \alpha\}$$

is a set.

Proof

From the definition of ρ it follows at once that

$$\{x \mid \rho(x) \leq \alpha\} = \tau(\alpha),$$

and $\tau(\alpha) \in \mathfrak{U}$. □

18.25. LEMMA

There exists a well-ordering R of $\mathfrak{U}$ with respect to which each initial segment is a set.

Proof

By 17.5, there exists a map

$$F: \mathfrak{U} \to \mathfrak{U}$$

such that $F(x)$ is a well-ordering of x for each $x \in \mathfrak{U}$. If $x \in \mathfrak{U}$, then

$$\{y \mid \rho(y) = \rho(x)\} \in \mathfrak{U}$$

by 18.24, and we define the well-ordering W_x by

$$W_x = F(\{y \mid \rho(y) = \rho(x)\}).$$

Define R to be the relation in $\mathfrak{U}$ such that

$$xRy \Leftrightarrow \rho(x) < \rho(y) \vee (\rho(x) = \rho(y) \;\&\; xW_x y).$$

It is easy to verify that R totally orders $\mathfrak{U}$.

Let E be a nonempty subset of $\mathfrak{U}$. We show that E has a first element for R. The class

$$\{\alpha \mid \alpha \in \text{Ord} \;\&\; (\exists x \in E)(\rho(x) = \alpha)\}$$

is nonempty since $E \neq \varnothing$, and so it has a first element α_0. Choose $x \in E$ with $\rho(x) = \alpha_0$. Define

$$E_0 = \{y \mid y \in E \;\&\; \rho(y) = \alpha_0\}.$$

Now $E_0 \neq \varnothing$ since $x \in E_0$, and E_0 is well-ordered by W_x and, a fortiori, by R. Hence E_0 has a first element z. By definition of α_0,

$$\rho(y) \geq \alpha_0 = \rho(z) \qquad (y \in E).$$

Hence z is a first element of E.

Let $y \in \mathfrak{U}$. To see that the initial segment $]\!\leftarrow, y[$ in $\mathfrak{U}$ with respect to R is a set, note that

$$]\!\leftarrow, y[\; \subset \{x \mid \rho(x) \leq \rho(y)\}$$

by definition of R, and apply 18.24. ☐

The well-ordering R of $\mathfrak{U}$ constructed in the preceding proof has the interesting property

$$x \in y \Rightarrow xRy \qquad (x, y \in \mathfrak{U});$$

this follows from Chapter 18, Exercise V.

18.26. THEOREM

Let X be a proper class. Then

$$X \simeq \text{Ord}.$$

Proof

Since $X \subset \mathfrak{U}$, there exists by 18.25 a well-ordering R of X with respect to which each initial segment of X is a set. Now Ord is well-ordered by $\leq$, and each initial segment of Ord is a set. Since X and Ord are both proper classes, it follows from 17.18 that X is order-isomorphic to Ord. Hence $X \simeq \text{Ord}$. ☐

18.27. COROLLARY

(1) If X and Y are proper classes, then $X \simeq Y$.
(2) If X is a set and Y is a proper class, then $X \prec Y$.

Proof

(1) is an immediate consequence of 18.26.

(2) Assume X is a set and Y is a proper class. By 18.26, $Y \simeq \text{Ord}$. By the well-ordering theorem, we can choose a well-ordering of X. Let $\alpha = \text{ord}(X)$. Then $X \simeq \alpha \subset \text{Ord}$, so $X \preceq \text{Ord}$. But $X \not\simeq \text{Ord}$ since Ord is not a set. $\square$

In view of 18.27, the theory of equipollence for proper classes is utterly trivial.

EXERCISES

A. Let X be a nonempty set.
   (a) Show that the class of all sets equipollent to X is a proper class. [*Hint*: $X \times \{Y\} \simeq X$ for each set Y.]
   (b) If a well-ordering of X is given, prove that the class of all well-ordered sets order-isomorphic to X is a proper class. [*Hint*: Use (a) and Chapter 17, Exercise M(b).]

B. (a) Let X be a well-ordered set and let $f: X \to \mathcal{U}$ be the map constructed in the Introduction. Prove that $f\langle X\rangle$, the ordinal of X in the sense of the Introduction, equals $\text{ord}(X)$, the ordinal of X in the sense of Definition 18.8. [*Hint*: We know $f\langle X\rangle \in \text{Ord}$. Show that the map $f^*: X \to f\langle X\rangle$ having the same graph as f is an order-isomorphism.]
   (b) Deduce that each $\alpha \in \text{Ord}$ is the ordinal, in the sense of the Introduction, of some well-ordered set.

C. If x is a subset of Ord, show that $x \in \text{Ord}$ if and only if x is full.

D. Let $\mathcal{P}$ be the class of all full sets α having the property:

$$\varnothing \neq E \subset \alpha \Rightarrow (\exists x \in E)(\forall y \in E)(y \neq x \Rightarrow x \in y).$$

   Prove that $\mathcal{P} = \text{Ord}$.

E. For each ordinal number $\alpha > 0$, construct a well-ordered set X_α such that $\text{ord}(X_\alpha) = \alpha$ but $X_\alpha \notin \text{Ord}$.

F. Let x be a set. If $x^+ \in \text{Ord}$, is $x \in \text{Ord}$?

G. If $\varnothing \neq E \subset \text{Ord}$, show that $\bigcap E$ is the first element of E.

H. In unaxiomatized set theory, the statement "$X \in X$" is not expressly prohibited. Why then, in the absence of the axiom of foundation, does the statement "Ord $\in$ Ord" lead to a contradiction (the *Burali-Forti paradox*)? [*Hint*: Consider ord(Ord).]

I. Let $\alpha \in \text{Ord}$ with $\alpha \neq 0$. Show that α is a nonlimit ordinal if and only if $(\bigcup\alpha)^+ = \alpha$.

J. Is Lim a set?

K. (a) For which subsets E of Ord is $\bigcup E > \alpha$ for each $\alpha \in E$?
(b) If $E \subset$ Ord and E is a proper class, show $\bigcup E =$ Ord.

L. Let Y be a class, and define

$$X = \{h \mid h: \alpha \to Y \text{ for some } \alpha \in \text{Ord}\}.$$

Let $F: X \to X$ be given. By using 18.15 (or otherwise), show there exists a unique map g: Ord $\to Y$ such that

$$g(\alpha) = F(g \mid \alpha) \qquad (\alpha \in \text{Ord}).$$

M. Find a counterexample to each of the following statements about ordinal numbers α, β, γ:
(a) $(\alpha \oplus \beta) \cdot \gamma = \alpha \cdot \gamma \oplus \beta \cdot \gamma$.
(b) $\alpha < \beta \;\&\; \gamma \neq 0 \Rightarrow \alpha \cdot \gamma < \beta \cdot \gamma$.
(c) $(\alpha \cdot \beta)^\gamma = \alpha^\gamma \cdot \beta^\gamma$.
(d) $\alpha < \beta \;\&\; \gamma \neq 0 \Rightarrow \alpha^\gamma < \beta^\gamma$.

N. Show that ω^ω is denumerable.

P. Let $\alpha, \beta \in$ Ord. If A, B are well-ordered sets with $\text{ord}(A) = \alpha$, $\text{ord}(B) = \beta$, show that $\alpha \cdot \beta = \text{ord}(A \times B)$, where $A \times B$ is the ordinal product (Chapter 17, Exercise E) of A and B.

Q. Use Exercise P and 18.17 to prove the left distributive law

$$\alpha \cdot (\beta \oplus \gamma) = \alpha \cdot \beta \oplus \alpha \cdot \gamma \qquad (\alpha, \beta, \gamma \in \text{Ord}).$$

R. Use transfinite induction to prove each of the following:
(a) $0 \cdot \alpha = 0$.
(b) $1 \cdot \alpha = \alpha = \alpha \cdot 1$.
(c) $(\alpha \cdot \beta) \cdot \gamma = \alpha \cdot (\beta \cdot \gamma)$.
(d) $\alpha < \beta \Rightarrow \alpha \cdot \gamma \leq \beta \cdot \gamma$.
(e) $0 < \alpha < \beta \;\&\; \gamma \neq 0 \Rightarrow \gamma \cdot \alpha < \gamma \cdot \beta$.
(f) $\alpha^1 = \alpha$.
(g) $\alpha^2 = \alpha \cdot \alpha$.
(h) $(\alpha^\beta)^\gamma = \alpha^{\beta \cdot \gamma}$.
(i) $1 < \alpha \;\&\; \beta < \gamma \Rightarrow \alpha^\beta < \alpha^\gamma$.

S. (a) If $\alpha \in$ Ord and $0 < \beta \in$ Ord, show

$$\alpha \oplus \beta \in \text{Lim} \Leftrightarrow \beta \in \text{Lim}.$$

(b) When is $\alpha \cdot \beta \in$ Lim?

T. In the notation of 18.21, prove $\alpha < \beta \in \text{Ord} \Rightarrow \tau(\alpha) \in \tau(\beta)$.

U. Exhibit all sets whose rank is α for $\alpha = 0$, $\alpha = 1$, $\alpha = 2$, and $\alpha = 3$.

V. Show that $\rho(x) < \rho(y)$ whenever x and y are sets with $x \in y$.

W. Given $\rho(x) = \alpha$, compute $\rho(\{x\})$.

19. Cardinal Numbers

INTRODUCTION

Suppose X is a set. We want to describe in some way the "size" of X, that is, to answer meaningfully the question, "How many elements does X have?" Of course, we can compare the size of X with the size of other sets by the formula, "X has at most as many elements as Y if and only if $X \preceq Y$." What we are really seeking, however, is an actual measure of the size of X. After all, $\mathcal{P}(\omega)$ is larger than ω, and $\mathcal{P}(\mathcal{P}(\omega))$ is larger than $\mathcal{P}(\omega)$, but just how much larger?

Our approach is to assume we already have such a measure at hand, to decide what features it ought to display, and then to see whether such a measure can actually be defined. Thus we suppose there is assigned to each set X some class $\operatorname{card}(X)$, called the "cardinality of X," which measures the size of X. Since we interpret '$X \simeq Y$' to mean "X has the same number of elements as Y," we demand

$$(*) \qquad X \simeq Y \Leftrightarrow \operatorname{card}(X) = \operatorname{card}(Y).$$

One obvious choice for $\operatorname{card}(X)$ is just the equivalence class

$$\operatorname{card}(X) = \{Y \mid Y \simeq X\}$$

of X in $\mathfrak{U}$ modulo $\simeq$. Roughly speaking, under this definition $\operatorname{card}(X)$ is the property of being equipollent to X: $\operatorname{card}\{1\}$ is "oneness," $\operatorname{card}\{1,2\}$ is "twoness," and so on. Just this definition was introduced by G. Frege and B. Russell to formalize Cantor's informal notion of cardinality. Unfortunately, under the Frege-Russell definition, card(X) is a proper class unless $X = \varnothing$ [see Chapter 18, Exercise A(a)]. We shall insist that card(X) be a set.

The problem of measuring the size of X has already been solved in Chapter 14 for the case of X countable. We can tentatively define

$$\text{card}(X) = \#(X) \text{ if } X \text{ is finite}$$

and

$$\text{card}(X) = \omega \text{ if } X \text{ is denumerable.}$$

Then (*) holds for all countable sets X and Y.

In order to go beyond the countable case, let us observe that for any countable set X these tentative definitions give

$$(**) \qquad \text{card}(X) \in \text{Ord} \;\&\; X \simeq \text{card}(X).$$

It is reasonable to require our definition of card(X) for arbitrary X to satisfy (**). That an ordinal number equipollent to a given set always exists is guaranteed by:

19.1. THEOREM (COUNTING THEOREM)

Let X be a set. Then there exists some $\alpha \in \text{Ord}$ such that $X \simeq \alpha$.

Proof

By the well-ordering theorem (17.4), there exists some well-ordering $\leq$ of X. By 18.7, there exists $\alpha \in \text{Ord}$ such that $\langle X, \leq \rangle$ is order-isomorphic to α. Hence $X \simeq \alpha$. ▯

If $\alpha \in \text{Ord}$ with $X \simeq \alpha$, then α may be used to "count" X: Choose a bijection $f: X \simeq \alpha$ and name each $x \in X$ by the corresponding ordinal number $f(x)$.

An ordinal number whose existence is given by the counting theorem is by no means unique. In Chapter 18 we named infinitely many ordinal numbers which are equipollent to ω. Moreover, if X is any infinite set and if $\alpha \in \text{Ord}$ with $X \simeq \alpha$, then also $X \simeq \alpha^+ \simeq (\alpha^+)^+ \simeq \ldots$.

Among all the ordinal numbers α which count a given X in the sense that $X \simeq \alpha$, there is one which counts X most efficiently, namely, the first one.

These observations suggest a definition for those objects that are to serve as the measures of sets.

THE CLASS OF CARDINAL NUMBERS

19.2. DEFINITION

A *cardinal number* is an ordinal number α having the property

$$\beta \in \text{Ord} \;\&\; \beta < \alpha \Rightarrow \beta \not\simeq \alpha$$

(such an ordinal number is often called an "initial ordinal"). The class

$$\{\alpha \mid \alpha \text{ is a cardinal number}\}$$

is denoted by Card. Following tradition, we denote cardinal numbers by lower case German Fraktur letters (the letters $\mathfrak{c}$ and $\mathfrak{f}$ are reserved for special cardinal numbers).

Since

$$\text{Card} \subset \text{Ord},$$

the class Card is well-ordered by $\leq$.

19.3. EXAMPLES

(1) Each natural number is a cardinal number: $\omega \subset$ Card. In fact, by 18.9, ω is the set of all finite cardinal numbers.

(2) The set ω is the first infinite cardinal number.

(3) The ordinal number ω^+ is not a cardinal number.

The next result generalizes something we already know for finite cardinal numbers. The proof is trivial.

19.4. PROPOSITION

Let $\mathfrak{a}, \mathfrak{b} \in$ Card. Then:

(1) $\mathfrak{a} = \mathfrak{b}$ if and only if $\mathfrak{a} \simeq \mathfrak{b}$.

(2) $\mathfrak{a} < \mathfrak{b}$ if and only if $\mathfrak{a} \prec \mathfrak{b}$.

(3) $\mathfrak{a} \leq \mathfrak{b}$ if and only if $\mathfrak{a} \preceq \mathfrak{b}$.

The next theorem justifies at last a formal definition of cardinality satisfying the conditions (*) and (**) stated in the Introduction.

19.5. THEOREM

There exists a unique map

$$\text{card}\colon \mathfrak{U} \to \text{Card}$$

such that

$$\text{card}(X) \simeq X$$

for each set X.

Proof

Uniqueness follows from 19.4, for if $\mathfrak{a}, \mathfrak{b} \in$ Card with $\mathfrak{a} \simeq X$ and $\mathfrak{b} \simeq X$, then $\mathfrak{a} \simeq \mathfrak{b}$ and hence $\mathfrak{a} = \mathfrak{b}$. We prove existence. If $X \in \mathfrak{U}$, then the counting theorem says that the class

$$\{\alpha \mid \alpha \in \text{Ord} \;\&\; X \simeq \alpha\}$$

of ordinal numbers is nonempty, and we define card(X) to be the first element of this class. $\square$

19.6. DEFINITION

If X is a set, then using the notation of 19.5 we call $\operatorname{card}(X)$ the *cardinal* (or *cardinality*) *of* X; we also denote $\operatorname{card}(X)$ more simply by $\operatorname{card} X$. (Alternate, more classical notations are $\overline{\overline{X}}$ and $|X|$.)

19.7. COROLLARY

If X and Y are sets, then:

(1) $\operatorname{card} X = \operatorname{card} Y$ if and only if $X \simeq Y$.
(2) $\operatorname{card} X < \operatorname{card} Y$ if and only if $X \prec Y$.
(3) $\operatorname{card} X \leq \operatorname{card} Y$ if and only if $X \preceq Y$.

Proof

Use 19.4 and 19.5. □

Two trivial consequences of the definition are:

$$\operatorname{card} \mathfrak{a} = \mathfrak{a} \qquad (\mathfrak{a} \in \mathrm{Card}),$$

$$\operatorname{card}(\operatorname{card} X) = \operatorname{card} X \qquad (X \in \mathfrak{U}).$$

Notice also that $\operatorname{card} X = \#(X)$ if X is finite.

Corollary 19.7 enables us to translate into the language of cardinal numbers our earlier results concerning equipollence and dominance. For example, the Schröder-Bernstein theorem is simply the statement that $\leq$ is antisymmetric in Card. Cantor's theorem (13.7) says

$$\operatorname{card} X < \operatorname{card} \mathcal{P}(X) \qquad (X \in \mathfrak{U}).$$

In particular

$$\mathfrak{a} \in \mathrm{Card} \Rightarrow \mathfrak{a} < \operatorname{card} \mathcal{P}(\mathfrak{a}),$$

so there is no largest cardinal number.

19.8. THEOREM

The class Card is not a set.

Proof

Just suppose Card is a set. Then

$$X = \bigcup \{\mathfrak{a} \mid \mathfrak{a} \in \mathrm{Card}\}$$

is also a set, and hence so is

$$Y = \mathcal{P}(X).$$

Since $\operatorname{card} Y \in \mathrm{Card}$, $\operatorname{card} Y \subset X$. Then

$$\operatorname{card} Y = \operatorname{card}(\operatorname{card} Y) \leq \operatorname{card} X.$$

But

$$\text{card } X < \text{card } Y$$

by Cantor's theorem. ▯

ALEPH

We have already noted that ω is the set of all finite cardinal numbers. Next we name the infinite—or so-called *transfinite*—cardinal numbers.

19.9. THEOREM

There exists a *unique* order-isomorphism

$$\aleph: \text{Ord} \sim \text{Card} \setminus \omega.$$

(The symbol $\aleph$ is the first letter, aleph, of the Hebrew alphabet.)

Proof

Both Ord and Card $\setminus \omega$ are well-ordered proper classes, and each initial segment of Ord, and a fortiori of Card $\setminus \omega$, is a set. Now use 17.18 and 17.15. ▯

19.10. DEFINITION

In the notation of 19.9, for each $\alpha \in$ Ord the value of $\aleph$ at α is denoted by $\aleph_\alpha$.

The smallest transfinite cardinal number is $\aleph_0$ (*aleph null*), so

$$\aleph_0 = \omega.$$

Hence a set X is denumerable if and only if card $X = \aleph_0$. In the language of cardinal numbers, 14.16 and 14.18 translate as:

$$X \text{ is countable} \Leftrightarrow \text{card } X \leq \aleph_0,$$

$$X \text{ is finite} \Leftrightarrow \text{card } X < \aleph_0.$$

Since each countable set is dominated by $\aleph_0$, the first uncountable ordinal number is the transfinite cardinal number $\aleph_1$, so

$$\aleph_1 = \Omega.$$

19.11. DEFINITION

The cardinal number card $\mathcal{P}(\omega)$ is denoted by $\mathfrak{c}$ and is called the *cardinal of the continuum*.

This name is used for $\mathfrak{c}$ since $\mathcal{P}(\omega) \simeq \text{Map}(\omega,2) \simeq \mathbb{R}$ implies

$$\mathfrak{c} = \text{card } \mathbb{R}.$$

Since $\mathfrak{c}$ is uncountable and $\aleph_1$ is the first uncountable cardinal number, we have

$$(*) \qquad \aleph_1 \leq \mathfrak{c}.$$

Cantor conjectured the so-called *continuum hypothesis*:

$$(\text{CH}) \qquad \aleph_1 = \mathfrak{c}.$$

This conjecture is of interest in analysis for it asserts that each uncountable subset of $\mathbb{R}$ is actually equipollent to $\mathbb{R}$.

If $\alpha \in \text{Ord}$, then $\aleph_\alpha < \mathcal{P}(\aleph_\alpha)$ by Cantor's theorem, and since $\aleph_{\alpha \oplus 1}$ is the first cardinal number greater than $\aleph_\alpha$, we have

$$(**) \qquad \aleph_{\alpha \oplus 1} \leq \text{card } \mathcal{P}(\aleph_\alpha) \qquad (\alpha \in \text{Ord}).$$

Since $\mathfrak{c} = \text{card } \mathcal{P}(\aleph_0)$, (**) generalizes (*). Then a generalization of (CH) is the *generalized continuum hypothesis*:

$$(\text{GCH}) \qquad \aleph_{\alpha \oplus 1} = \text{card } \mathcal{P}(\aleph_\alpha) \qquad (\alpha \in \text{Ord}).$$

In other words, (GCH) asserts that for each given ordinal number α, there is no cardinal number $\mathfrak{b}$ satisfying $\aleph_\alpha < \mathfrak{b} < \text{card } \mathcal{P}(\aleph_\alpha)$.

In 1940, Gödel showed that (GCH), and hence (CH), is consistent with the axioms of set theory. In 1963, P. J. Cohen showed that the negations of (CH) and (GCH) are also consistent with the axioms, so that (CH) and (GCH) are independent of the axioms. Thus (CH) and (GCH) have the same metamathematical status as the axiom of choice. Hence the assumption of (CH) or even (GCH) as an additional axiom is a matter of taste and convenience. In contrast to the case with the axiom of choice, essential applications of (CH) and (GCH) do not pervade mathematics, so we do not assume them to be axioms here.

One further metamathematical comment. In 1926, A. Lindenbaum and A. Tarski announced that a weak version of the axiom of choice (each *set* of nonempty sets has a choice function) could be deduced from (GCH). The first proof of this was published in 1947 by W. Sierpinski (see [36, pages 434–436]).

CARDINAL ARITHMETIC

The remainder of this chapter is devoted to the arithmetic of cardinal numbers. Cardinal arithmetic allows one to compute the cardinality of a set constructed from given sets whose cardinalities are known. Many of the proofs below lean heavily on the axiom of choice. Frequent use will be made of results from Chapter 12, usually without explicit reference.

19.12. DEFINITION

Let $\mathfrak{a}$ and $\mathfrak{b}$ be cardinal numbers. The *sum of* $\mathfrak{a}$ *and* $\mathfrak{b}$, denoted $\mathfrak{a} + \mathfrak{b}$, is defined to be

$$\operatorname{card}((\mathfrak{a} \times \{0\}) \cup (\mathfrak{b} \times \{1\})).$$

The *product of* $\mathfrak{a}$ *and* $\mathfrak{b}$, denoted $\mathfrak{a} \cdot \mathfrak{b}$ or simply $\mathfrak{a}\mathfrak{b}$, is defined to be

$$\operatorname{card}(\mathfrak{a} \times \mathfrak{b}).$$

Finally, $\mathfrak{a}$ *to the* $\mathfrak{b}$ *power*, denoted $\mathfrak{a}^{\mathfrak{b}}$, is defined to be

$$\operatorname{card}(\operatorname{Map}(\mathfrak{b},\mathfrak{a})).$$

Because cardinal numbers are ordinal numbers, there is the possibility of confusing the operations of addition, multiplication, and exponentiation just defined with the corresponding ordinal operations. Only in the exceptional case is the sum, product, or power of two cardinal numbers equal to the corresponding ordinal sum, product, or power. For example, if $\mathfrak{a}$ is a transfinite cardinal, then $\mathfrak{a} \times \{0\} \simeq \mathfrak{a}$ and

$$1 \times \{1\} = \{(0,1)\} \simeq \{\mathfrak{a}\},$$

$$(\mathfrak{a} \times \{0\}) \cup \{1 \times \{1\}\} \simeq \mathfrak{a} \cup \{\mathfrak{a}\},$$

$\mathfrak{a} \cup \{\mathfrak{a}\} \simeq \mathfrak{a}$ since $\mathfrak{a}$ is infinite, and hence

$$\mathfrak{a} + 1 = \operatorname{card} \mathfrak{a} = \mathfrak{a} \neq \mathfrak{a}^{+} = \mathfrak{a} \oplus 1.$$

No ambiguity should arise, for in this chapter we never use ordinal multiplication and exponentiation, and ordinal addition is always denoted by $\oplus$.

The binary operations of addition and multiplication can be extended to infinitary operations.

19.13. DEFINITION

Let $(\mathfrak{a}_i \mid i \in I)$ be a family of cardinal numbers whose index class is a set. Then the *sum* of this family, denoted $\sum_{i \in I} \mathfrak{a}_i$, is defined to be

$$\operatorname{card}(\bigcup_{i \in I} (\mathfrak{a}_i \times \{i\}),$$

and the *product* of this family, denoted $\prod_{i \in I} \mathfrak{a}_i$, is defined to be

$$\operatorname{card}(\times_{i \in I} \mathfrak{a}_i).$$

If $\mathfrak{a}$ and $\mathfrak{b}$ are two cardinal numbers, then

$$\mathfrak{a} + \mathfrak{b} = \sum_{i \in I} \mathfrak{a}_i,$$

$$\mathfrak{a} \cdot \mathfrak{b} = \prod_{i \in I} \mathfrak{a}_i,$$

where $I = \{1,2\}$ and $\mathfrak{a}_1 = \mathfrak{a}$, $\mathfrak{a}_2 = \mathfrak{b}$. Hence the definitions of sum and product given in 19.12 are really special cases of 19.13.

Applications of cardinal arithmetic to computing cardinalities of sets rest on:

19.14. PROPOSITION

(1) If $(\mathfrak{a}_i \mid i \in I)$ is a family in Card indexed by a set, and if $(X_i \mid i \in I)$ is a disjoint family of sets such that

$$\operatorname{card} X_i = \mathfrak{a}_i \qquad (i \in I),$$

then

$$\operatorname{card}(\textstyle\bigcup_{i\in I} X_i) = \sum_{i\in I} \mathfrak{a}_i.$$

(2) If $(\mathfrak{a}_i \mid i \in I)$ is as in (1), and if $(X_i \mid i \in I)$ is any family of sets such that

$$\operatorname{card} X_i = \mathfrak{a}_i \qquad (i \in I),$$

then

$$\operatorname{card}(\textstyle\times_{i\in I} X_i) = \prod_{i\in I} \mathfrak{a}_i.$$

(3) If $\mathfrak{a}, \mathfrak{b} \in$ Card and if X, Y are sets such that

$$\operatorname{card} X = \mathfrak{a}, \qquad \operatorname{card} Y = \mathfrak{b},$$

then

$$\operatorname{card} \operatorname{Map}(X,Y) = \mathfrak{b}^{\mathfrak{a}}.$$

Proof

(1) Under the hypothesis,

$$X_i \simeq \mathfrak{a}_i \simeq \mathfrak{a}_i \times \{i\} \qquad (i \in I),$$

and since $(\mathfrak{a}_i \times \{i\} \mid i \in I)$ as well as $(X_i \mid i \in I)$ is disjoint

$$\textstyle\bigcup_{i\in I} X_i \simeq \bigcup_{i\in I} \mathfrak{a}_i \times \{i\}. \quad \square$$

This proposition says, in particular, that if $\mathfrak{a}, \mathfrak{b} \in$ Card and if X, Y are sets with

$$\operatorname{card} X = \mathfrak{a}, \qquad \operatorname{card} Y = \mathfrak{b},$$

then

$$\operatorname{card}(X \times Y) = \mathfrak{a}\mathfrak{b},$$

and if in addition $X \cap Y = \varnothing$, then

$$\operatorname{card}(X \cup Y) = \mathfrak{a} + \mathfrak{b}.$$

Since $\operatorname{card} X = \#(X)$ for any finite set X, it follows from 19.14 together with results in Chapter 14 that for finite cardinal numbers m and n their cardinal sum, product, and power are just the usual natural numbers $m + n$, mn, and m^n as defined in Chapter 10.

In 19.14(1) it is of course essential to assume that $(X_i \mid i \in I)$ be disjoint. Without this assumption, one has only an inequality:

19.15. PROPOSITION

If $(X_i \mid i \in I)$ is a family of sets indexed by a set, then

$$\operatorname{card}(\textstyle\bigcup_{i\in I} X_i) \leq \sum_{i\in I} \operatorname{card} X_i.$$

Proof

The map

$$f\colon \textstyle\bigcup_{i\in I} X_i \times \{i\} \to \bigcup_{i\in I} X_i$$

such that $f(x,i) = x$ if $i \in I$ and $x \in X_i$ is surjective, so its range is dominated by its domain. Now the family $(X_i \times \{i\} \mid i \in I)$ is disjoint, so

$$\textstyle\sum_{i\in I} \operatorname{card} X_i = \operatorname{card}(\operatorname{dmn} f)$$

by 19.14 (1). ▯

For any set X, $X^1 \simeq X$ and $X^2 \simeq X \times X$. Hence

$$\mathfrak{a}^1 = \mathfrak{a}, \qquad \mathfrak{a}^2 = \mathfrak{a} \cdot \mathfrak{a} \qquad (\mathfrak{a} \in \text{Card}).$$

Similarly,

$$1 \cdot \mathfrak{a} = \mathfrak{a}, \qquad 2 \cdot \mathfrak{a} = \mathfrak{a} + \mathfrak{a} \qquad (\mathfrak{a} \in \text{Card}).$$

More generally, multiplication is repeated addition, and exponentiation is repeated multiplication:

19.16. PROPOSITION

Let $\mathfrak{a}, \mathfrak{b} \in \text{Card}$. Then

$$\mathfrak{a} \cdot \mathfrak{b} = \textstyle\sum_{i\in I} \mathfrak{b}_i$$

and

$$\mathfrak{b}^{\mathfrak{a}} = \textstyle\prod_{i\in I} \mathfrak{b}_i$$

where $I = \mathfrak{a}$ and $\mathfrak{b}_i = \mathfrak{b}$ for each $i \in I$.

Proof

We have

$$\mathfrak{a} \times \mathfrak{b} = \textstyle\bigcup_{i\in \mathfrak{a}} \{i\} \times \mathfrak{b},$$

and so on. ▯

A number of results in Chapter 14 concerning countable sets may now be concisely summarized.

(1) We have

$$\aleph_0 + \aleph_0 = \aleph_0^2 = \aleph_0.$$

(2) If $\mathfrak{a} \in$ Card with $\mathfrak{a} < \aleph_0$, then

$$\mathfrak{a} + \aleph_0 = \aleph_0,$$

and if in addition $0 < \mathfrak{a}$, then

$$\mathfrak{a} \cdot \aleph_0 = \aleph_0^{\mathfrak{a}} = \aleph_0.$$

(3) If $(\mathfrak{a}_n \mid n \in \omega)$ is a sequence in Card with $0 < \mathfrak{a}_n \leq \aleph_0$ for each n, then

$$\textstyle\sum_{n\epsilon\omega} \mathfrak{a}_n = \aleph_0.$$

A concrete example of exponentiation is:

$$\text{card}\, \mathcal{P}(X) = 2^{\text{card } X} \qquad (X \in \mathfrak{U}).$$

Recall that ordinal addition, multiplication, and exponentiation violate many of the familiar laws which hold for the corresponding operations on natural numbers. In contrast, cardinal operations are rather well-behaved.

19.17. PROPOSITION

Let $\mathfrak{a} \in$ Card, and let $(\mathfrak{a}_i \mid i \in I)$, $(\mathfrak{b}_i \mid i \in I)$ be families in Card indexed by a nonempty set I. Then:

(1) (associative laws) If $(I_j \mid j \in J)$ is a partition of I, then

$$\textstyle\sum_{i\epsilon I} \mathfrak{a}_i = \sum_{j\epsilon J} (\sum_{i\epsilon I_j} \mathfrak{a}_i),$$

$$\textstyle\prod_{i\epsilon I} \mathfrak{a}_i = \prod_{j\epsilon J} (\prod_{i\epsilon I_j} \mathfrak{a}_i).$$

(2) (commutative laws) If σ is a permutation of I, then

$$\textstyle\sum_{i\epsilon I} \mathfrak{a}_i = \sum_{i\epsilon I} \mathfrak{a}_{\sigma(i)},$$

$$\textstyle\prod_{i\epsilon I} \mathfrak{a}_i = \prod_{i\epsilon I} \mathfrak{a}_{\sigma(i)}.$$

(3) (distributive law)

$$\textstyle\mathfrak{a} \cdot \sum_{i\epsilon I} \mathfrak{b}_i = \sum_{i\epsilon I} \mathfrak{a} \cdot \mathfrak{b}_i.$$

(4) (monotonicity laws) If

$$\mathfrak{a}_i \leq \mathfrak{b}_i \qquad (i \in I),$$

then

$$\textstyle\sum_{i\epsilon I} \mathfrak{a}_i \leq \sum_{i\epsilon I} \mathfrak{b}_i,$$

$$\textstyle\prod_{i\epsilon I} \mathfrak{a}_i \leq \prod_{i\epsilon I} \mathfrak{b}_i.$$

(5) (laws of exponentiation)

$$\mathfrak{a}^{\Sigma_{i\in I}\mathfrak{b}_i} = \prod_{i\in I} \mathfrak{a}^{\mathfrak{b}_i},$$

$$(\prod_{i\in I} \mathfrak{a}_i)^{\mathfrak{b}} = \prod_{i\in I} \mathfrak{a}_i^{\mathfrak{b}}.$$

Proof

(1) and (2) follow from the associativity and commutativity of (arbitrary) unions and cartesian products.

(3) If $(X_i \mid i \in I)$ is a disjoint family of sets with card $X_i = \mathfrak{a}_i$ for each $i \in I$, then $(\mathfrak{a} \times X_i \mid i \in I)$ is also disjoint, and

$$\mathfrak{a} \times \bigcup_{i\in I} X_i = \bigcup_{i\in I} \mathfrak{a} \times X_i.$$

(4) Assume $\mathfrak{a}_i \leq \mathfrak{b}_i$ for all $i \in I$. Then $\mathfrak{a}_i \subset \mathfrak{b}_i$ for all $i \in I$, so

$$\bigcup_{i\in I} (\mathfrak{a}_i \times \{i\}) \subset \bigcup_{i\in I} (\mathfrak{b}_i \times \{i\}).$$

Also,

$$\times_{i\in I} \mathfrak{a}_i \preceq \times_{i\in I} \mathfrak{b}_i,$$

because each element of $\times_{i\in I} \mathfrak{a}_i$ is a map from I to $\bigcup_{i\in I} \mathfrak{a}_i$ whose graph is the graph of a unique map from I to $\bigcup_{i\in I} \mathfrak{b}_i$ which is an element of $\times_{i\in I} \mathfrak{b}_i$.

(5) Let $(Y_i \mid i \in I)$ be a disjoint family of sets with card $Y_i = \mathfrak{b}_i$ for each $i \in I$. By Chapter 12, Exercise N,

$$\times_{i\in I} \operatorname{Map}(Y_i,\mathfrak{a}) \simeq \operatorname{Map}(\bigcup_{i\in I} Y_i, \mathfrak{a}).$$

This proves the first assertion of (5). The second follows from 12.12. □

19.18. COROLLARY

Let $\mathfrak{a}, \mathfrak{b}, \mathfrak{p}, \mathfrak{q} \in$ Card. Then:

(1) $(\mathfrak{a} + \mathfrak{b}) + \mathfrak{p} = \mathfrak{a} + (\mathfrak{b} + \mathfrak{p})$
$(\mathfrak{a} \cdot \mathfrak{b}) \cdot \mathfrak{p} = \mathfrak{a} \cdot (\mathfrak{b} \cdot \mathfrak{p})$.

(2) $\mathfrak{a} + \mathfrak{b} = \mathfrak{b} + \mathfrak{a}$
$\mathfrak{a} \cdot \mathfrak{b} = \mathfrak{b} \cdot \mathfrak{a}$.

(3) $\mathfrak{a} \cdot (\mathfrak{b} + \mathfrak{p}) = (\mathfrak{a} \cdot \mathfrak{b}) + (\mathfrak{a} \cdot \mathfrak{p})$.

(4) If $\mathfrak{a} \leq \mathfrak{p}$ and $\mathfrak{b} \leq \mathfrak{q}$, then
$\mathfrak{a} + \mathfrak{b} \leq \mathfrak{p} + \mathfrak{q}$
$\mathfrak{a} \cdot \mathfrak{b} \leq \mathfrak{p} \cdot \mathfrak{q}$.

(5) $\mathfrak{a}^{\mathfrak{p}+\mathfrak{q}} = \mathfrak{a}^{\mathfrak{p}} \cdot \mathfrak{a}^{\mathfrak{q}}$
$(\mathfrak{a} \cdot \mathfrak{b})^{\mathfrak{q}} = \mathfrak{a}^{\mathfrak{q}} \cdot \mathfrak{b}^{\mathfrak{q}}$.

(6) If $\mathfrak{a} \leq \mathfrak{p}$ and $\mathfrak{b} \leq \mathfrak{q}$, then
$\mathfrak{a}^{\mathfrak{b}} \leq \mathfrak{p}^{\mathfrak{q}}$.

(7) $(\mathfrak{a}^{\mathfrak{p}})^{\mathfrak{q}} = \mathfrak{a}^{\mathfrak{p}\cdot\mathfrak{q}}$.

Proof

(1)–(5) follow from 19.17. Statement (7) follows from the fact that

$$\text{Map}\ (Z,\ \text{Map}\ (Y,X)) \simeq \text{Map}\ (Y \times Z,\ X)$$

for all sets X, Y, Z (see Chapter 12, Exercise H). We leave (6) as an exercise. ☐

Recall that

$$n + \aleph_0 = \aleph_0 = n \cdot \aleph_0 \qquad (0 < n \in \omega).$$

Hence there can be no universally valid cancellation laws for cardinal addition and multiplication: neither $\mathfrak{a} + \mathfrak{p} = \mathfrak{b} + \mathfrak{p}$ nor $\mathfrak{a} \cdot \mathfrak{q} = \mathfrak{b} \cdot \mathfrak{q}$ $(\mathfrak{q} > 0)$ need imply $\mathfrak{a} = \mathfrak{b}$. See, however, Exercise M.

THE FUNDAMENTAL THEOREM OF CARDINAL ARITHMETIC

We now come to the main result of this chapter which drastically simplifies the arithmetic of transfinite cardinals. Given two nonzero cardinal numbers at least one of which is transfinite, their sum as well as their product is just the larger of the two given numbers!

19.19. THEOREM (FUNDAMENTAL THEOREM OF CARDINAL ARITHMETIC)

Let $\mathfrak{a}$ and $\mathfrak{b}$ be cardinal numbers such that $\mathfrak{b}$ is *transfinite* and

$$0 < \mathfrak{a} \leq \mathfrak{b}.$$

Then

$$\mathfrak{a} + \mathfrak{b} = \mathfrak{a} \cdot \mathfrak{b} = \mathfrak{b}$$

and if $1 < \mathfrak{a}$

$$\mathfrak{a}^{\mathfrak{b}} = 2^{\mathfrak{b}}.$$

The computation of $\mathfrak{a}^{\mathfrak{b}}$ for $\mathfrak{a} > \mathfrak{b}$ is of course not made trivial by this theorem.

For the proof of 19.19 we need two preparatory lemmas. Because the second of these uses Zorn's lemma, the proof of 19.19 depends ultimately on the axiom of choice. (Actually, Theorem 19.19 and Lemma 19.21 are each equivalent to the weak form of the axiom of choice stating that each set of nonempty sets has a choice function.)

19.20. LEMMA

Let $\mathfrak{a} \in \text{Card}$. Assume

$$\mathfrak{a}^2 = \mathfrak{a}.$$

Then

$$\mathfrak{a} = 2\mathfrak{a} = 3\mathfrak{a} = \mathfrak{a}^2.$$

Proof

We may suppose $\mathfrak{a} > 0$. Now $\mathfrak{a} \geq \aleph_0$, for $\mathfrak{a}^2 > \mathfrak{a}$ for each nonzero finite cardinal $\mathfrak{a}$. Then $3 < \mathfrak{a}$, so

$$\mathfrak{a} = \mathfrak{a} + 0 \leq \mathfrak{a} + \mathfrak{a} = 2 \cdot \mathfrak{a} \leq 3 \cdot \mathfrak{a} \leq \mathfrak{a} \cdot \mathfrak{a} = \mathfrak{a}^2 = \mathfrak{a}. \quad \square$$

We now come to the crucial lemma, which generalizes the equation $\aleph_0{}^2 = \aleph_0$.

19.21. LEMMA

Let $\mathfrak{a}$ be a transfinite cardinal. Then

$$\mathfrak{a}^2 = \mathfrak{a}.$$

Proof

Define

$$\mathcal{A} = \{(A, f) \mid A \subset \mathfrak{a} \;\&\; f\colon A \simeq A \times A\}.$$

By means of Zorn's lemma it will be shown that $(M, f) \in \mathcal{A}$ for some f and some $M \subset \mathfrak{a}$ satisfying card $M = \mathfrak{a}$. This will prove the lemma, for then

$$\mathfrak{a}^2 = (\text{card } M) \cdot (\text{card } M) = \text{card } M = \mathfrak{a}.$$

The set $\mathcal{A}$ is nonempty. In fact, since $\mathfrak{a}$ is infinite, there exists some denumerable subset A of $\mathfrak{a}$, and then $A \simeq A \times A$.

Partially order $\mathcal{A}$ by defining

$$(A, f) \leq (B, g) \Leftrightarrow A \subset B \;\&\; g \mid A = f$$

(by $g \mid A$ we mean actually the restriction of $g\colon B \to B \times B$ to A and $A \times A$, so that $g \mid A\colon A \to A \times A$).

Let $((A_i, f_i) \mid i \in I)$ be a family in $\mathcal{A}$ such that $\{(A_i, f_i) \mid i \in I\}$ is a chain in $\mathcal{A}$. We construct an upper bound of this chain in $\mathcal{A}$. Let

$$A = \bigcup_{i \in I} A_i.$$

Consider the map

$$f\colon A \to A \times A$$

such that

$$f(x) = f_i(x) \qquad (x \in A_i,\ i \in I).$$

We claim that f is bijective, so that $(A, f) \in \mathcal{A}$ and hence (A, f) is an upper bound of the given chain in $\mathcal{A}$. Surely f is injective. To see that it is also surjective, let $(x,y) \in A \times A$. There exist $i, j \in I$ with $x \in A_i$, $y \in A_j$. We may suppose $(A_i, f_i) \leq (A_j, f_j)$. Then $x \in A_j$ also. Since $f_j\colon A_j \to A_j \times A_j$ is surjective, $(x,y) = f_j(a)$ for some $a \in A_j$, and then $(x,y) = f(a)$.

By Zorn's lemma, $\mathfrak{A}$ has a maximal member (M, f). We shall prove card $M = \mathfrak{a}$. Suppose not. Then

$$\text{card } M < \mathfrak{a}.$$

Since $M \simeq M \times M$, 19.20 implies

$$2 \cdot \text{card } M = \text{card } M.$$

If $\text{card}(\mathfrak{a} \setminus M) \leq \text{card } M$, then

$$\mathfrak{a} = \text{card } M + \text{card}(\mathfrak{a} \setminus M) \leq \text{card } M + \text{card } M$$
$$= 2 \cdot \text{card } M = \text{card } M.$$

Hence $\text{card } M < \text{card}(\mathfrak{a} \setminus M)$, and there exists

$$E \subset \mathfrak{a} \setminus M$$

with

$$\text{card } M = \text{card } E.$$

By 19.20 again,

$$3 \cdot \text{card } E = \text{card } E = (\text{card } E)^2.$$

We shall contradict the maximality of (M, f) by showing that $(M \cup E, g) \in \mathfrak{A}$ for some extension g of f. Since $E \subset \mathfrak{a} \setminus M$, we have

$$(M \cup E) \times (M \cup E) = (M \times M) \cup (M \times E) \cup (E \times M) \cup (E \times E)$$

with the four products on the right pairwise disjoint. Now

$$\begin{aligned}\text{card}((M \times E) \cup (E \times M) \cup (E \times E)) \\ &= \text{card}(M \times E) + \text{card}(E \times M) + \text{card}(E \times E) \\ &= (\text{card } M) \cdot (\text{card } E) + (\text{card } E) \cdot (\text{card } M) + (\text{card } E)^2 \\ &= 3 \cdot (\text{card } E)^2 \\ &= \text{card } E\end{aligned}$$

so there exists

$$h \colon E \simeq (M \times E) \cup (E \times M) \cup (E \times E).$$

Then the map

$$g \colon M \cup E \to (M \cup E) \times (M \cup E)$$

such that

$$g(x) = f(x) \qquad (x \in M), \qquad g(x) = h(x) \qquad (x \in E)$$

is bijective, and $(M \cup E, g) \in \mathfrak{A}$. $\square$

19.22. COROLLARY

If $\mathfrak{a}$ is a transfinite cardinal and $0 < n \in \omega$, then

$$\mathfrak{a}^n = \mathfrak{a}.$$

Proof of Theorem 19.19

Lemma 19.21 yields

$$\mathfrak{b} = 1 \cdot \mathfrak{b} \leq \mathfrak{a} \cdot \mathfrak{b} \leq \mathfrak{b} \cdot \mathfrak{b} = \mathfrak{b}^2 = \mathfrak{b},$$

and $\mathfrak{a} \cdot \mathfrak{b} = \mathfrak{b}$. Next,

$$2 \cdot \mathfrak{b} = \mathfrak{b}^2 = \mathfrak{b}$$

by 19.20 and 19.21,

$$\mathfrak{b} \leq \mathfrak{a} + \mathfrak{b} \leq \mathfrak{b} + \mathfrak{b} = 2 \cdot \mathfrak{b} = \mathfrak{b},$$

and $\mathfrak{a} + \mathfrak{b} = \mathfrak{b}$. Finally, $\mathfrak{a} < 2^{\mathfrak{a}}$,

$$2^{\mathfrak{b}} \leq \mathfrak{a}^{\mathfrak{b}} \leq (2^{\mathfrak{a}})^{\mathfrak{b}} = 2^{\mathfrak{a} \cdot \mathfrak{b}} = 2^{\mathfrak{b}},$$

and $\mathfrak{a}^{\mathfrak{b}} = 2^{\mathfrak{b}}$. □

Theorem 19.19 has many important consequences. The first says that subtraction and division of transfinite cardinals are possible—and trivial.

19.23. PROPOSITION

Let $\mathfrak{a}$ and $\mathfrak{b}$ be transfinite cardinals with

$$\mathfrak{a} < \mathfrak{b}.$$

Then:

(1) There is a unique $\mathfrak{p} \in$ Card such that

$$\mathfrak{a} + \mathfrak{p} = \mathfrak{b},$$

namely, $\mathfrak{p} = \mathfrak{b}$.

(2) There is a unique $\mathfrak{q} \in$ Card such that

$$\mathfrak{a} \cdot \mathfrak{q} = \mathfrak{b},$$

namely, $\mathfrak{q} = \mathfrak{b}$.

Proof

(1) By 19.19, $\mathfrak{a} + \mathfrak{b} = \mathfrak{b}$. Now let $\mathfrak{p} \in$ Card with

$$\mathfrak{a} + \mathfrak{p} = \mathfrak{b}.$$

We show $\mathfrak{p} = \mathfrak{b}$. If $\mathfrak{p} < \mathfrak{a}$, then since $\mathfrak{a}$ is transfinite,

$$\mathfrak{a} + \mathfrak{p} = \mathfrak{p} + \mathfrak{a} = \mathfrak{a} < \mathfrak{b}.$$

Hence $\mathfrak{a} \leq \mathfrak{p}$. But then $\mathfrak{p}$ is also transfinite and

$$\mathfrak{b} = \mathfrak{a} + \mathfrak{p} = \mathfrak{p}. \quad \square$$

19.24. DEFINITION

If $\mathfrak{a}$ and $\mathfrak{b}$ are transfinite cardinals with $\mathfrak{a} < \mathfrak{b}$, then the unique cardinal number $\mathfrak{p}$ for which $\mathfrak{a} + \mathfrak{p} = \mathfrak{b}$ is denoted by $\mathfrak{b} - \mathfrak{a}$.

If Y is an infinite subset of a set X, then

$$\operatorname{card}(X \setminus Y) = \operatorname{card} X - \operatorname{card} Y,$$

hence

$$\operatorname{card} Y < \operatorname{card} X \Rightarrow \operatorname{card}(X \setminus Y) = \operatorname{card} X.$$

19.25. COROLLARY

If $\mathfrak{a}$ is a transfinite cardinal, then

$$2^{\mathfrak{a}} - \mathfrak{a} = 2^{\mathfrak{a}}.$$

Proof

We have $\mathfrak{a} < 2^{\mathfrak{a}}$. ▯

A proof of 19.25 not involving the use of 19.19 has been given by Sierpinski, but Sierpinski's proof is quite difficult.

Corollary 19.25 has the interesting consequence that the set of all irrational numbers, and even the set of all transcendental numbers, is equipollent to $\mathbb{R}$. More generally, if D is any denumerable subset of $\mathbb{R}$, then

$$\begin{aligned}\operatorname{card}(\mathbb{R} \setminus D) &= \operatorname{card} \mathbb{R} - \operatorname{card} D \\ &= 2^{\aleph_0} - \aleph_0 \\ &= 2^{\aleph_0}.\end{aligned}$$

As a final application of 19.19, we estimate certain infinite sums.

19.26. PROPOSITION

Let $\mathfrak{a}$ be a transfinite cardinal, and let $(\mathfrak{a}_i \mid i \in I)$ be a family of cardinal numbers indexed by a set I satisfying

$$\operatorname{card} I \leq \mathfrak{a}.$$

Suppose

$$\mathfrak{a}_i \leq \mathfrak{a} \qquad (i \in I).$$

Then

$$\textstyle\sum_{i \in I} \mathfrak{a}_i \leq \mathfrak{a}.$$

If in addition $\mathfrak{a}_j = \mathfrak{a}$ for some $j \in I$, then

$$\textstyle\sum_{i \in I} \mathfrak{a}_i = \mathfrak{a}.$$

Proof

Let $\mathfrak{b} = \text{card } I$. By 19.17(4) and 19.16,

$$\sum_{i \in I} \mathfrak{a}_i \leq \mathfrak{a} \cdot \mathfrak{b} \leq \mathfrak{a}^2 = \mathfrak{a}.$$

If $\mathfrak{a}_j = \mathfrak{a}$ for some $j \in I$, then also

$$\mathfrak{a} = \mathfrak{a}_j \leq \sum_{i \in I} \mathfrak{a}_i. \quad \square$$

We conclude with an inequality between infinite sums and products.

19.27. THEOREM (KÖNIG)

Let $(\mathfrak{a}_i \mid i \in I)$ and $(\mathfrak{b}_i \mid i \in I)$ be families of cardinal numbers indexed by a set I. Suppose

$$0 < \mathfrak{a}_i < \mathfrak{b}_i \qquad (i \in I).$$

Then

$$\sum_{i \in I} \mathfrak{a}_i < \prod_{i \in I} \mathfrak{b}_i.$$

Proof

Let $(Y_i \mid i \in I)$ be a disjoint family of sets such that

$$\text{card } Y_i = \mathfrak{b}_i \qquad (i \in I).$$

Using the axiom of choice we can choose a family $(X_i \mid i \in I)$ with

$$X_i \subset Y_i, \qquad \text{card } X_i = \mathfrak{a}_i \qquad (i \in I).$$

By hypothesis, $Y_i \setminus X_i \neq \varnothing$ for each $i \in I$, so we may choose some

$$c \in \mathsf{X}_{i \in I} (Y_i \setminus X_i).$$

Let

$$g \colon \bigcup_{i \in I} X_i \to I$$

be the map such that $i \in I$ and $x \in X_i$ implies $g(x) = i$ (note that the X_i's are disjoint from each other). Let

$$Y = \mathsf{X}_{i \in I} Y_i.$$

Define a map

$$f \colon \bigcup_{i \in I} X_i \to \mathsf{X}_{i \in I} Y_i$$

as follows: if $x \in \bigcup_{i \in I} X_i$, then

$$f(x) = (f(x)_i \mid i \in I),$$

where

$$f(x)_i = \begin{cases} x & \text{if } i = g(x), \\ c_i & \text{if } i \neq g(x). \end{cases}$$

It is easy to see that f is injective. Hence

$$\sum_{i \in I} \mathfrak{a}_i \leq \prod_{i \in I} \mathfrak{b}_i.$$

Just suppose equality holds in this inequality. Then there exists a partition $(E_i \mid i \in I)$ of Y such that

$$\text{card } E_i = \mathfrak{a}_i \qquad (i \in I).$$

Consider the projection maps

$$\text{pr}_i\colon Y \to Y_i.$$

For each $i \in I$, pr_i is surjective, so

$$\text{card}(\text{pr}_i\langle E_i \rangle) \leq \text{card } E_i = \mathfrak{a}_i < \mathfrak{b}_i = \text{card } Y_i$$

whence $Y_i \setminus \text{pr}_i\langle E_i \rangle$ is nonempty. Choose some

$$y \in \mathsf{X}_{i \in I} (Y_i \setminus \text{pr}_i\langle E_i \rangle).$$

Then $y \in Y$, and so $y \in E_j$ for some $j \in I$. But

$$\text{pr}_j(y) = y_j \notin \text{pr}_j\langle E_j \rangle,$$

an absurdity. □

EXERCISES

A. Prove or disprove:
   (a) Each transfinite cardinal is a limit ordinal.
   (b) Each limit ordinal is a transfinite cardinal.

B. In Cantor's unaxiomatized set theory, $\mathfrak{U}$ was tacitly assumed to be a set, and in 1899 Cantor himself derived a paradox concerning $\text{card}(\mathfrak{U})$. Deduce a contradiction concerning $\text{card}(\mathfrak{U})$ from the assumption that $\mathfrak{U}$ is a set.

C. Prove: For each $\alpha \in \text{Ord}$,

$$\text{card } \alpha \leq \alpha \leq \aleph_\alpha.$$

D. (a) If $\mathfrak{a} \in \text{Card}$, show there exists no $\mathfrak{b} \in \text{Card}$ for which $\mathfrak{a} < \mathfrak{b} < \mathfrak{a} + 1$. Do not use 19.19.
   (b) If $\mathfrak{a}, \mathfrak{b} \in \text{Card}$, show that $\mathfrak{a} + 1 = \mathfrak{b} + 1$ implies $\mathfrak{a} = \mathfrak{b}$ without using 19.19.

E. If $\alpha, \beta \in \text{Ord}$, show that

$$\text{card}(\alpha \oplus \beta) = \text{card } \alpha + \text{card } \beta.$$

Does a similar result hold for the cardinality of the ordinal product of two ordinal numbers?

F. Let $n \in \omega$. Prove *without* using 19.19:

(a) $n + \mathfrak{c} = \aleph_0 + \mathfrak{c} = \mathfrak{c} + \mathfrak{c} = \mathfrak{c}$.

(b) If $n > 0$, then

$$n \cdot \mathfrak{c} = \aleph_0 \cdot \mathfrak{c} = \mathfrak{c}^2 = \mathfrak{c}^n = \mathfrak{c}.$$

(c) If $n > 1$, then

$$n^{\aleph_0} = \aleph_0^{\aleph_0} = \mathfrak{c}^{\aleph_0} = \mathfrak{c}.$$

G. Let $C(\mathbb{R})$ denote the set of all continuous $f \in \text{Map}(\mathbb{R},\mathbb{R})$. What is card $C(\mathbb{R})$? [*Hint*: If $f, g \in C(\mathbb{R})$, then $f \mid \mathbb{Q} = g \mid \mathbb{Q}$ implies $f = g$.]

H. Let H be the Hilbert sequence space, that is, the set of all sequences $(x_n \mid n \in \omega)$ of real numbers such that $\sum_{n=0}^{\infty} | x_n |^2$ converges. Show that H is equipollent to $\mathbb{R}$.

I. If $(\mathfrak{a}_n \mid n \in \omega)$ is a sequence of cardinal numbers with $1 < \mathfrak{a}_n < \aleph_0$ for each n, compute $\prod_{n \in \omega} \mathfrak{a}_n$.

J. Let

$$\mathfrak{f} = \text{card Map}(\mathbb{R},\mathbb{R}).$$

Prove *without* using 19.19:

(a) If $n \in \omega$, then

$$n + \mathfrak{f} = \aleph_0 + \mathfrak{f} = \mathfrak{c} + \mathfrak{f} = \mathfrak{f} + \mathfrak{f} = \mathfrak{f}.$$

(b) If $0 < n \in \omega$, then

$$n \cdot \mathfrak{f} = \aleph_0 \cdot \mathfrak{f} = \mathfrak{c} \cdot \mathfrak{f} = \mathfrak{f} \cdot \mathfrak{f} = \mathfrak{f}.$$

(c) If $0 < n \in \omega$, then

$$\mathfrak{f}^n = \mathfrak{f}^{\aleph_0} = \mathfrak{f}^{\mathfrak{c}} = \mathfrak{f}.$$

(d) If $1 < n \in \omega$, then

$$n^{\mathfrak{c}} = \aleph_0^{\mathfrak{c}} = \mathfrak{c}^{\mathfrak{c}} = \mathfrak{f}.$$

K. Show that summands equal to 0 and factors equal to 1 can be deleted from infinite sums and infinite products respectively. More precisely: Let $(\mathfrak{a}_i \mid i \in I)$ be a family in Card indexed by a set I, and let $J \subset I$. Then:

(a) $\mathfrak{a}_i = 0$ for all $i \in I \setminus J$ implies

$$\textstyle\sum_{i \in I} \mathfrak{a}_i = \sum_{j \in J} \mathfrak{a}_j.$$

(b) $\mathfrak{a}_i = 1$ for all $i \in I \setminus J$ implies

$$\textstyle\prod_{i \in I} \mathfrak{a}_i = \prod_{j \in J} \mathfrak{a}_j.$$

L. Generalize the distributive law 19.17(3) to the case of the product of a family of sums.

M. Let $\mathfrak{a}, \mathfrak{b}, \mathfrak{p}, \mathfrak{q}$ be transfinite cardinals with $\mathfrak{a} < \mathfrak{p}$ and $\mathfrak{b} < \mathfrak{q}$. Show that

$$\mathfrak{a} + \mathfrak{b} < \mathfrak{p} + \mathfrak{q}, \qquad \mathfrak{a} \cdot \mathfrak{b} < \mathfrak{p} \cdot \mathfrak{q}.$$

N. Let X be an infinite set. Show there exists a partition $(X_i \mid i \in I)$ of X such that card I = card X and each X_i is denumerable.

P. Let X and Y be sets, and let $f: X \to Y$ be a surjection. If card $Y = \mathfrak{a}$ and

$$\operatorname{card} f^{-1}\langle\{y\}\rangle = \mathfrak{b} \qquad (y \in Y)$$

for some $\mathfrak{b} \in$ Card, compute card X.

Q. Let X be an infinite set. Show that the set of all finite subsets of X is equipollent to X.

R. Let X be an infinite set. Show that the set of all finite families in X is equipollent to X.

S. Let X be an infinite set. Show that the set of all permutations of X is equipollent to $\mathcal{P}(X)$.

T. Let X be an infinite set with card $X = \mathfrak{b}$, and let $\mathfrak{a}$ be a transfinite cardinal with $\mathfrak{a} < \mathfrak{b}$. Show that the set

$$\{E \mid E \subset X \;\&\; \operatorname{card} E = \mathfrak{a}\}$$

has cardinality $\mathfrak{b}^{\mathfrak{a}}$.

U. If $\mathfrak{a}$ is a transfinite cardinal, show that

$$\{\alpha \mid \alpha \in \operatorname{Ord} \;\&\; \alpha \simeq \mathfrak{a}\}$$

is a set and has cardinality $2^{\mathfrak{a}}$.

V. Let $\alpha \in$ Ord.

(a) Prove that

$$\textstyle\sum_{\beta \leq \alpha} \aleph_\beta = \aleph_\alpha.$$

(b) If α is a limit ordinal, prove that

$$\textstyle\sum_{\beta < \alpha} \aleph_\beta = \aleph_\alpha.$$

W. Assume (GCH). Let α be a nonlimit ordinal and β an ordinal number with $\alpha > \beta$. Show that

$$\aleph_\alpha^{\aleph_\beta} = \aleph_\alpha.$$

X. (a) Prove: $\aleph_\omega < \aleph_\omega^{\aleph_0}$.

(b) Is $2^{\aleph_0} = \aleph_\omega$?

Y. (a) Let V be a vector space over the field K of scalars. Recall that dim V, the dimension of V, is the cardinality of some (hence any) basis B of V. Express card V in terms of dim V

and card K. [*Hint*: If B is finite, each $v \in V$ has a unique representation

$$v = \sum_{x \in B} \alpha_x x$$

for a family $(\alpha_x \mid x \in B)$ in K. If B is infinite, each nonzero $v \in V$ has a unique representation

$$v = \sum_{x \in E} \alpha_x x$$

for a nonempty finite subset E of B and a family $(\alpha_x \mid x \in E)$ in $K \setminus \{0\}$; consider separately the cases K finite, K infinite and use Exercise Q.]

(b) Compute dim $\mathbb{R}$ when $\mathbb{R}$ is considered in the standard way as a vector space over the field $\mathbb{Q}$. [*Hint*: Show that dim $\mathbb{R}$ is infinite and apply (a). The continuum hypothesis is not needed.]

Appendix: Construction of the Integers and Rational Numbers

As an application of several set-theoretic ideas, we construct below the algebraic systems $\mathbb{Z}$ of integers and $\mathbb{Q}$ of rational numbers, starting from the system $\mathbb{N}$ of natural numbers. Recall that the notation $\mathbb{N}$ for ω is used in contexts where we wish to emphasize the algebraic structure of ω supplied by the operations of addition and multiplication.

THE ADDITIVE GROUP OF INTEGERS

Below *the symbol* $\mathbb{N}^*$ *denotes the set* $\mathbb{N} \setminus \{0\}$ of all nonzero natural numbers.

For given $m, n \in \mathbb{N}$, the equation

$$(*) \qquad x + n = m$$

has the solution $m - n \in \mathbb{N}$ in case $m \geq n$, but it has no solution in $\mathbb{N}$ in case $m < n$. In particular, for $n \in \mathbb{N}^*$ no $x \in \mathbb{N}$ satisfies the equation

$$(**) \qquad x + n = 0.$$

We propose to associate with each $n \in \mathbb{N}^*$ a new object, $-n$, which we decree to be a solution of (**). By adjoining all these new objects to $\mathbb{N}$ we shall obtain a set $\mathbb{Z}$ containing $\mathbb{N}$. Moreover, we propose to extend addition in $\mathbb{N}$ to a binary operation in the larger set $\mathbb{Z}$ in such a way that $-n$ is *really* a solution of (**), but where the sign $+$ now refers to the extended addition in $\mathbb{Z}$. As a bonus, we shall have for any $m, n \in \mathbb{Z}$ a unique solution of (*).

The minimal requirements upon the sets $-n$ to be adjoined to $\mathbb{N}$ are that $-n \notin \mathbb{N}$ for any $n \in \mathbb{N}^*$ and that $-n = -m$ precisely when $n = m$. Hence we shall define $-n$ to be the ordered pair $(-,n)$, where $-$ is some fixed set. The particular choice for $-$ is immaterial, and we take it to be ω.

A.1. DEFINITION

Let $-$ denote the set ω. The set $\{-\} \times \mathbb{N}^*$ is denoted by $\mathbb{Z}^-$. For each $n \in \mathbb{N}^*$ the ordered pair $(-,n)$ is denoted by $-n$. The set $\mathbb{N} \cup \mathbb{Z}^-$ is denoted by $\mathbb{Z}$. An element of $\mathbb{Z}$ is called an *integer*. (The letter $\mathbb{Z}$ is suggested by the German *Ziffer*, meaning "number.")

The precise goal of our construction is summarized by the following theorem.

A.2. THEOREM

There exists a unique binary operation $\oplus$ on $\mathbb{Z}$ having the following properties:

(0) The operation $\oplus$ on $\mathbb{Z}$ is an extension of the operation $+$ on $\mathbb{N}$.
(1) For all $x, y, z \in \mathbb{Z}$, $(x \oplus y) \oplus z = x \oplus (y \oplus z)$.
(2) For all $x, y \in \mathbb{Z}$, $x \oplus y = y \oplus x$.
(3) For every $x \in \mathbb{Z}$, $x \oplus 0 = x$.
(4) For each $n \in \mathbb{N}^*$, $-n \oplus n = 0$.

Property (0) means, of course, that

$$m \oplus n = m + n \qquad (m, n \in \mathbb{N}).$$

Properties (1)–(3) say that the associative, commutative, and zero laws for addition in $\mathbb{N}$ hold for addition in $\mathbb{Z}$ as well.

From (0)–(4) we deduce:

(5) For each $x \in \mathbb{Z}$, there exists a unique $x' \in \mathbb{Z}$ for which

$$x' \oplus x = 0.$$

In fact, let $x \in \mathbb{Z}$. To show existence of x', take $x' = -x$ if $x \in \mathbb{N}^*$, $x' = 0$ if $x = 0$, and $x' = n$ if $x = -n \in \mathbb{Z}^-$ for $n \in \mathbb{N}^*$. To show uniqueness of x', suppose $y, z \in \mathbb{Z}$ with

$$y \oplus x = 0, \qquad z \oplus x = 0.$$

Then

$$\begin{aligned} y &= y \oplus 0 = y \oplus (y \oplus x) = y \oplus (x \oplus y) \\ &= (y \oplus x) \oplus y = (z \oplus x) \oplus y = z \oplus (x \oplus y) \\ &= z \oplus (y \oplus x) = z \oplus 0 = z, \end{aligned}$$

so $y = z$.

For a given $x \in \mathbf{Z}$, the unique $x' \in \mathbf{Z}$ given by (5) will be denoted by $-x$. In view of (4), this notation does not conflict with the notation $-n$ for a typical element of $\mathbf{Z}^-$.

Properties (1)–(3) and (5) together express the fact that the operation $\oplus$ makes $\mathbf{Z}$ into a *commutative group.*

Proof of Theorem A.2

Deduction of (5) from (0)–(4) began our proof of the uniqueness of $\oplus$ assuming its existence. We continue the uniqueness proof in earnest.

The three sets $\mathbf{N}^*$, $\{0\}$, $\mathbf{Z}^-$ form a partition of $\mathbf{Z}$. Hence for $x, y \in \mathbf{Z}$ exactly one of the following seven possibilities holds:

(a)	$x = 0, y = 0$		
(b)	$x \in \mathbf{N}^*, y = 0$	(f)	$x = 0, y \in \mathbf{N}^*$
(c)	$x \in \mathbf{N}^*, y \in \mathbf{N}^*$		
(d)	$x \in \mathbf{N}, y \in \mathbf{Z}^-$	(g)	$x \in \mathbf{Z}^-, y \in \mathbf{N}$
(e)	$x \in \mathbf{Z}^-, y \in \mathbf{Z}^-$.		

In each case we shall express $x \oplus y$ solely in terms of x, y, and addition in $\mathbf{N}$. Since $x \oplus y = y \oplus x$, we need not explicitly treat cases (f) and (g) once cases (b) and (d) have been treated. Since $x \oplus 0 = x$, cases (a) and (b) are disposed of. In case (c) we have $x \oplus y = x + y$ by property (0).

Case (d)

We have $y = -n$ for a unique $n \in \mathbf{N}^*$. Denote x by m. Suppose first $n \leq m$. Then

$$m \oplus (-n) = m - n.$$

In fact,

$$(m \oplus (-n)) \oplus n = m \oplus (-n \oplus n) = m \oplus 0 = m$$
$$= (m - n) + n = (m - n) \oplus n.$$

If u, v denote $m \oplus (-n)$, $m - n$, then $u \oplus n = v \oplus n$, so

$$u = u \oplus 0 = u \oplus (n \oplus (-n)) = (u \oplus n) \oplus (-n)$$
$$= (v \oplus n) \oplus (-n) = v \oplus (n \oplus (-n)) = v \oplus 0 = v.$$

Now suppose $n > m$. Then

$$m \oplus (-n) = -(n - m).$$

In fact, the preceding paragraph shows

$$n - m = n \oplus (-m),$$

so

$$(m \oplus (-n)) \oplus (n - m) = 0.$$

From this, property (4) together with the uniqueness part of (5) allows us to conclude $m \oplus (-n) = -(n - m)$.

Case (e)

For unique $m, n \in \mathbf{N}^*$, $x = -m$, $y = -n$. Then

$$(-m) \oplus (-n) = -(m + n),$$

for

$$((-m) \oplus (-n)) \oplus (m + n) = ((-m) \oplus (-n)) \oplus (m \oplus n) = 0.$$

Proof of the uniqueness of $\oplus$ is completed. To establish existence of $\oplus$, we propose to turn things around and use the expression for $x \oplus y$ found above in each case as the definition of $x \oplus y$. Unfortunately, this makes verification of (0)–(4) very tedious due to the necessity of considering many separate cases. We seek a single formula for $x \oplus y$ encompassing all cases at once.

The expressions for $m \oplus (-n)$ found above suggest defining a map

$$d: \mathbf{N} \times \mathbf{N} \to \mathbf{Z}$$

by

$$d(m,n) = \begin{cases} m - n & \text{if } m \geq n, \\ -(n - m) & \text{if } m < n. \end{cases}$$

This map d is surjective. In fact,

$$d(m,0) = m \qquad (m \in \mathbf{N}),$$
$$d(0,n) = -n \qquad (n \in \mathbf{N}^*).$$

Hence d has as a right inverse the map

$$e: \mathbf{Z} \to \mathbf{N} \times \mathbf{N}$$

defined by

$$e(x) = \begin{cases} (x,0) & \text{if } x \in \mathbf{N}, \\ (0,n) & \text{if } x = -n,\ n \in \mathbf{N}^*. \end{cases}$$

We can now express $\oplus$ in terms of d, e, and addition in $\mathbf{N}$. Define a binary operation $+$ on $\mathbf{N} \times \mathbf{N}$ by

$$(m,n) + (m',n') = (m + m', n + n') \qquad ((m,n),(m',n') \in \mathbf{N} \times \mathbf{N}).$$

Then

$$\text{(A)} \qquad x \oplus y = d(e(x) + e(y)) \qquad (x, y \in \mathbf{Z}),$$

as is easily verified by checking separately each of the cases (a)–(g) enumerated earlier.

We now *define* the binary operation $\oplus$ on $\mathbb{Z}$ by means of formula (A). To show this $\oplus$ satisfies (0)–(4) of A.2, we need a preliminary result.

A.3. LEMMA

The operation $+$ on $\mathbb{N} \times \mathbb{N}$ satisfies

(i) $(u + v) + w = u + (v + w)$,
(ii) $u + v = v + u$,
(iii) $u + (0,0) = u$,

for all $u, v, w \in \mathbb{N} \times \mathbb{N}$.

Verification of A.2 (0)–(4) is now a triviality. For (0), for example, let $x, y \in \mathbb{N}$; then

$$x \oplus y = d(e(x) + e(y)) = d((x,0) + (y,0))$$
$$= d((x + y, 0)) = x + y.$$

For (2), let $x, y \in \mathbb{Z}$; by A.3(ii),

$$e(x) + e(y) = e(y) + e(x),$$

hence

$$x \oplus y = d(e(x) + e(y)) = d(e(y) + e(x)) = y + x.$$

With the proof of A.2 completed, we abandon the notation $\oplus$ in favor of the more familiar $+$.

Given $y, z \in \mathbb{Z}$, the unique solution of

$$x + y = z$$

in $\mathbb{Z}$ is $z + (-y)$, which we denote as usual by $z - y$.

MULTIPLICATION AND ORDER FOR INTEGERS

Besides addition, $\mathbb{N}$ has the binary operation of multiplication. This operation too extends to one on $\mathbb{Z}$.

A.4. THEOREM

There exists a unique binary operation $\cdot$ on $\mathbb{Z}$ which extends the operation of multiplication in $\mathbb{N}$ such that:

(1) For all $x, y, z \in \mathbb{Z}$, $(x \cdot y) \cdot z = x \cdot (y \cdot z)$.
(2) For all $x, y \in \mathbb{Z}$, $x \cdot y = y \cdot x$.
(3) For all $x \in \mathbb{Z}$, $1 \cdot x = x$.
(4) For all $x, y, z \in \mathbb{Z}$, $x \cdot (y + z) = (x \cdot y) + (x \cdot z)$.

Proof

Uniqueness

Assume the existence of $\cdot$. We develop an expression for it in terms of multiplication and addition in $\mathbf{N}$.

If $x \in \mathbf{Z}$, then

$$x \cdot 0 = x \cdot (0 + 0) = (x \cdot 0) + (x \cdot 0)$$

so that

$$(5) \qquad x \cdot 0 = 0.$$

Then $x, y \in \mathbf{Z}$ imply

$$0 = x \cdot 0 = x \cdot (y + (-y)) = (x \cdot y) + (x \cdot (-y)),$$

so that

$$(6) \qquad x \cdot (-y) = -(x \cdot y).$$

Since $-(-x) = x$, then $x, y \in \mathbf{Z}$ implies

$$(7) \qquad (-x) \cdot (-y) = x \cdot y.$$

Properties (2), (5)–(7) completely determine the extension $\cdot$ of multiplication in $\mathbf{N}$ through consideration of the cases (a)–(g) enumerated earlier. For example, if $x = m \in \mathbf{N}^*$ and $y = -n$ for $n \in \mathbf{N}^*$, then $x \cdot y = m \cdot (-n) = -(m \cdot n) = -(mn)$. Again, however, we seek a single formula encompassing all cases.

For $m, m', n, n' \in \mathbf{N}$, we find

$$(m - n) \cdot (m' - n') = (mm' + nn') - (mn' + nm').$$

Define a binary operation $\cdot$ on $\mathbf{N} \times \mathbf{N}$ by

$$(m,n) \cdot (m',n') = (mm' + nn', mn' + nm').$$

This operation satisfies

(i) $(u \cdot v) \cdot w = u \cdot (v \cdot w)$
(ii) $u \cdot v = v \cdot u$
(iii) $(1,0) \cdot u = u$
(iv) $u \cdot (v + w) = (u \cdot v) + (u \cdot w)$

for all $u, v, w \in \mathbf{N} \times \mathbf{N}$. For all cases (a)–(g), we find

$$(\mathrm{M}) \qquad x \cdot y = d(e(x) \cdot e(y)).$$

Existence

Define a binary operation $\cdot$ on $\mathbf{Z}$ by formula (M). The verification that $\cdot$ extends multiplication in $\mathbf{N}$ and satisfies (1)–(4) is left to the reader. $\square$

Properties (1)–(4) of A.4 say that multiplication together with addition make $\mathbb{Z}$ into a *commutative ring*. The corollary below says that $\mathbb{Z}$ is actually something more, an *integral domain*.

A.5. COROLLARY

If $x, y \in \mathbb{Z}$ and $x \cdot y = 0$, then $x = 0$ or $y = 0$.

Proof

Use the fact that $m, n \in \mathbb{N}$ and $mn = 0$ implies $m = 0$ or $n = 0$. □

The set of natural numbers carries not only the operations of addition and multiplication, but an order relation as well. So does the set of integers.

A.6. THEOREM

There exists a unique total ordering $\leq$ of $\mathbb{Z}$ such that:

(1) For $x, y, z \in \mathbb{Z}$, $x \leq y$ implies $x + z \leq y + z$.
(2) For $x, y, z \in \mathbb{Z}$, $x \leq y$ and $0 \leq z$ implies $xz \leq yz$.
(3) $0 \leq 1$.

Moreover, the total ordering induced by $\leq$ on $\mathbb{N}$ is the usual order relation in $\mathbb{N}$.

Proof

Uniqueness

By (1) and (3), $n \leq n + 1$ for all $n \in \mathbb{N}$. It follows that $\leq$ induces the usual order relation in $\mathbb{N}$. In particular, $0 \leq n$ for all $n \in \mathbb{N}$. More precisely,

$$0 \leq x \Leftrightarrow x \in \mathbb{N} \qquad (x \in \mathbb{Z}).$$

To see that $0 \leq x$ implies $x \in \mathbb{N}$, suppose $0 \leq x$ but $x \notin \mathbb{N}$; then $x = -n$ with $n \in \mathbb{N}^*$, $0 \leq n$,

$$0 \leq x = -n = -n + 0 \leq -n + n = 0,$$

and $x = 0 \in \mathbb{N}$, a contradiction.

We now establish the uniqueness of $\leq$ by showing

$$(O) \qquad x \leq y \Leftrightarrow y - x \in \mathbb{N} \qquad (x, y \in \mathbb{Z}).$$

If $x \leq y$, then

$$0 = x + (-x) \leq y + (-x) = y - x,$$

$0 \leq y - x$, and so $y - x \in \mathbb{N}$. Conversely, if $y - x \in \mathbb{N}$, then $0 \leq y - x$, so

$$x = 0 + x \leq (y - x) + x = y.$$

Existence

Define a relation $\leq$ in $\mathbb{Z}$ by formula (O).

We show $\leq$ totally orders $\mathbb{Z}$. The reflexivity, antisymmetry, and transitivity of $\leq$ follow from

$$0 \in \mathbb{N},$$

$$u \in \mathbb{N} \,\&\, -u \in \mathbb{N} \Rightarrow u = 0,$$

$$u \in \mathbb{N} \,\&\, v \in \mathbb{N} \Rightarrow u + v \in \mathbb{N},$$

respectively. For example, if $x \leq y$ and $y \leq z$, then $y - x \in \mathbb{N}$ and $z - y \in \mathbb{N}$, so

$$z - x = (z - y) + (y - x) \in \mathbb{N},$$

that is, $x \leq z$. That the partial ordering $\leq$ totally orders $\mathbb{Z}$ follows from

$$(\forall u \in \mathbb{Z})\,(u \in \mathbb{N} \vee -u \in \mathbb{N}).$$

The verification of (1)–(3) is left to the reader. □

THE FIELD OF RATIONAL NUMBERS

Below *the symbol* $\mathbb{Z}^*$ *denotes the set* $\mathbb{Z} \setminus \{0\}$ of all nonzero integers.

The system $\mathbb{Z}$ was so constructed that for given $n \in \mathbb{Z}$, the linear equation $x + n = 0$ has a (unique) solution in $\mathbb{Z}$. Not every linear equation with coefficients in $\mathbb{Z}$ has a solution in $\mathbb{Z}$, however. In fact, for given $n \in \mathbb{Z}^*$, the equation

$$(*) \qquad nx - 1 = 0$$

has a solution in $\mathbb{Z}$ only for $n = 1$ or $n = -1$. More generally, for given $m \in \mathbb{Z}$, $n \in \mathbb{Z}^*$, the linear equation

$$(**) \qquad nx - m = 0$$

has a solution in $\mathbb{Z}$ if and only if $m = kn$ for some $k \in \mathbb{Z}$ (prove this).

We are going to enlarge $\mathbb{Z}$ to a new set $\mathbb{Q}$ having operations of addition and multiplication (extending the corresponding operations on $\mathbb{Z}$) such that for given $n \in \mathbb{Z}^*$ equation (*) has a unique solution n^{-1} in $\mathbb{Q}$. Although we are not yet prepared to identify the set $\mathbb{Q}$, we can state now the properties which addition and multiplication on $\mathbb{Q}$ are to have.

A.7. THEOREM

There exist unique binary operations $+$ and $\cdot$ on $\mathbb{Q}$ extending the operations of addition and multiplication on $\mathbb{Z}$ respectively such that

(1) For all $x, y, z \in \mathbb{Q}$, $(x + y) + z = x + (y + z)$.
(2) For all $x, y \in \mathbb{Q}$, $x + y = y + x$.

(3) For all $x \in \mathbb{Q}$, $x + 0 = x$.
(4) For each $x \in \mathbb{Q}$, there exists a unique element $-x \in \mathbb{Q}$ for which $-x + x = 0$.
(5) For all $x, y, z \in \mathbb{Q}$, $(x \cdot y) \cdot z = x \cdot (y \cdot z)$.
(6) For all $x, y \in \mathbb{Q}$, $x \cdot y = y \cdot x$.
(7) For all $x \in \mathbb{Q}$, $1 \cdot x = x$.
(8) For each $x \in \mathbb{Q}$ with $x \neq 0$, there exists a unique element $x^{-1} \in \mathbb{Q}$ for which $x \cdot x^{-1} = 1$.
(9) For all $x, y, z \in \mathbb{Q}$, $x \cdot (y + z) = (x \cdot y) + (x \cdot z)$.
(10) For each $x \in \mathbb{Q}$, there exist $m, n \in \mathbb{Z}$ with $n \neq 0$ for which $x = m \cdot n^{-1}$.

There is no conflict between the notation $-x$ used in (4) and the notation $-x$ introduced for an integer x earlier. In fact, if $x \in \mathbb{Z}$, the unique $x' \in \mathbb{Q}$ with $x' + x = 0$ given by (4) is just the integer $-x$ defined earlier.

Properties (1)–(7) and (9) say that the ring properties for addition and multiplication on $\mathbb{Z}$ persist for the extensions of these operations to $\mathbb{Q}$. If $x, y, z \in \mathbb{Q}$ with $xz = yz$ and $z \neq 0$, then

$$x = x \cdot 1 = x \cdot (z \cdot z^{-1}) = (x \cdot z) \cdot z^{-1}$$
$$= (y \cdot z) \cdot z^{-1} = y \cdot (z \cdot z^{-1}) = y \cdot 1 = y.$$

Hence $+$ and $\cdot$ actually make $\mathbb{Q}$ into an integral domain. Property (8) asserts in particular the existence in $\mathbb{Q}$ of a solution to equation (*) above. Together (1)–(9) express the fact that $+$ and $\cdot$ make $\mathbb{Q}$ into a (commutative) *field*.

For $m \in \mathbb{Z}$ and $n \in \mathbb{Z}^*$, equation (**) has the unique solution $m \cdot n^{-1}$ in $\mathbb{Q}$; we denote $m \cdot n^{-1}$ by $\dfrac{m}{n}$ or m/n. Property (10) says that *every* element of $\mathbb{Q}$ has the form m/n for some $(m,n) \in \mathbb{Z} \times \mathbb{Z}^*$, in short, that the field $\mathbb{Q}$ is a *field of quotients* of the integral domain $\mathbb{Z}$.

It is to be emphasized that the set $\mathbb{Q}$ itself has not yet been defined. However, the proof of uniqueness of the operations $+$ and $\cdot$ on $\mathbb{Q}$ will suggest a definition of $\mathbb{Q}$!

To proceed with the uniqueness proof, we assume given a set $\mathbb{Q}$, containing $\mathbb{Z}$, on which there are given binary operations $+$ and $\cdot$ having the properties stated in A.7.

Owing to (10), the map

$$f: \mathbb{Z} \times \mathbb{Z}^* \to \mathbb{Q}$$
$$(m,n) \mapsto m \cdot n^{-1}$$

is surjective. For (m,n), $(a,b) \in \mathbb{Z} \times \mathbb{Z}^*$,

$$(m \cdot n^{-1}) \cdot (n \cdot b) = m \cdot b = mb,$$
$$(a \cdot b^{-1}) \cdot (n \cdot b) = a \cdot n = an,$$

so

$$f(m,n) = f(a,b) \Leftrightarrow mb = an.$$

Hence the equivalence kernel of f is the equivalence relation $\sim$ on $\mathbb{Z} \times \mathbb{Z}^*$ given by

$$\text{(E)} \qquad (m,n) \sim (a,b) \Leftrightarrow mb = an.$$

Let

$$Q = (\mathbb{Z} \times \mathbb{Z}^*)/\sim.$$

We obtain the commutative triangle

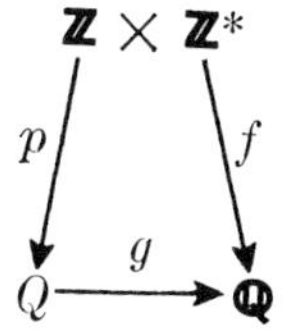

where p is the projection and g is the bijection obtained from f by passing to quotients mod $\sim$.

For each $(m,n) \in \mathbb{Z} \times \mathbb{Z}^*$, let $[m,n]$ denote the equivalence class $p(m,n)$ of (m,n) under $\sim$, so that

$$[m,n] = \{(m',n') \mid (m',n') \in \mathbb{Z} \times \mathbb{Z}^* \ \&\ mn' = m'n\}.$$

Then $(m,n) \in \mathbb{Z} \times \mathbb{Z}^*$ implies

$$g([m,n]) = m \cdot n^{-1}.$$

In view of (E), the equivalence relation $\sim$, and hence the set Q, is determined solely by $\mathbb{Z}$. We are going to show that the operations $+$ and $\cdot$ on $\mathbb{Q}$ are themselves determined by $\mathbb{Z}$.

There are unique operations $+$ and $\cdot$ on Q such that

$$g(x + y) = g(x) + g(y)$$
$$g(x \cdot y) = g(x) \cdot g(y) \qquad (x, y \in Q),$$

namely, the ones defined by the formulas

$$x + y = g^{-1}(g(x) + g(y))$$
$$x \cdot y = g^{-1}(g(x) \cdot g(y)) \qquad (x, y \in Q).$$

Under these operations, elements of Q add and multiply exactly as do the elements of $\mathbb{Q}$ corresponding to them under g (technically speaking, g is an "isomorphism"). Moreover, the operations on $\mathbb{Q}$ are determined by those on Q via

$$u + v = g(g^{-1}(u) + g^{-1}(v))$$
$$u \cdot v = g(g^{-1}(u) \cdot g^{-1}(v)) \qquad (u, v \in \mathbb{Q}).$$

Hence all that remains of our uniqueness proof of A.7 is to show that the operations on Q are determined by $\mathbb{Z}$.

Let $x, y \in Q$. Choose any (m,n), $(a,b) \in \mathbb{Z} \times \mathbb{Z}^*$ with

$$x = [m,n], \qquad y = [a,b].$$

Then in Q

$$[m,n] \cdot [a,b] = [ma,nb],$$

for in $\mathbb{Q}$

$$\begin{aligned} g([m,n] \cdot [a,b]) &= g([m,n]) \cdot g([a,b]) \\ &= (m \cdot n^{-1}) \cdot (a \cdot b^{-1}) \\ &= (m \cdot a) \cdot (n \cdot b)^{-1} \\ &= (ma) \cdot (nb)^{-1} \\ &= g([ma,nb]). \end{aligned}$$

Also,

$$[m,n] + [a,b] = [mb + an, nb],$$

for

$$\begin{aligned} g([m,n] + [a,b]) &= g([m,n]) + g([a,b]) \\ &= m \cdot n^{-1} + a \cdot b^{-1} \\ &= (m \cdot n^{-1} + a \cdot b^{-1}) \cdot ((n \cdot b) \cdot (n \cdot b)^{-1}) \\ &= (m \cdot b + a \cdot n) \cdot (n \cdot b)^{-1} \\ &= (mb + an) \cdot (nb)^{-1} \\ &= g([mb + an, nb]). \end{aligned}$$

Thus $x \cdot y = [ma, nb]$ and $x + y = [mb + an, nb]$ are completely determined by x and y and addition and multiplication in $\mathbb{Z}$. This completes our uniqueness proof.

CONSTRUCTION OF THE FIELD OF RATIONAL NUMBERS

We now drop our earlier assumption of existence of the field $\mathbb{Q}$. It is time to construct the set $\mathbb{Q}$ and the operations on it. The preceding analysis motivates the construction, only now we construct Q and binary operations on Q first.

Define an equivalence relation $\sim$ on $\mathbb{Z} \times \mathbb{Z}^*$ by formula (E) above, and as above take $Q = (\mathbb{Z} \times \mathbb{Z}^*)/\sim$. Again we observe that $\sim$ and Q depend solely on $\mathbb{Z}$.

Define binary operations $+$ and $\cdot$ on $\mathbb{Z} \times \mathbb{Z}^*$ by

$$\left.\begin{aligned} (m,n) + (a,b) &= (mb + an, nb) \\ (m,n) \cdot (a,b) &= (ma,nb) \end{aligned}\right\} \qquad ((m,n),\ (a,b) \in \mathbb{Z} \times \mathbb{Z}^*).$$

These operations are compatible with $\sim$ in the sense that

$$(m',n') \sim (m,n) \ \& \ (a',b') \sim (a,b) \Rightarrow$$
$$(m',n') + (a',b') = (m,n) + (a,b)$$
$$\& \ (m',n') \cdot (a',b') = (m,n) \cdot (a,b).$$

According to Chapter 11, Exercise I(b), there are unique binary operations $+$ and $\cdot$ on Q making the following diagrams commute, where q is given by $q(x,y) = (p(x),p(y))$:

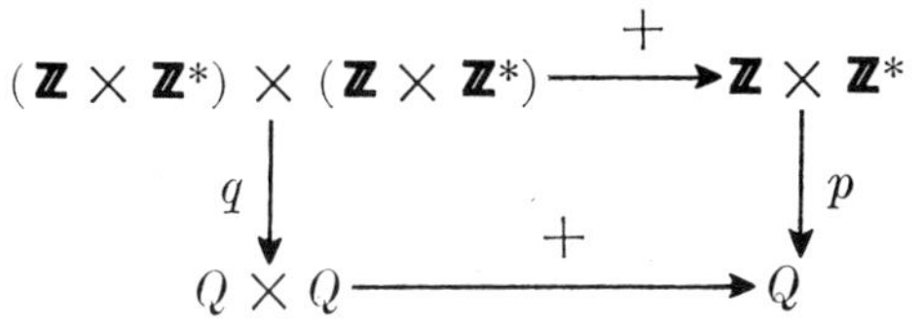

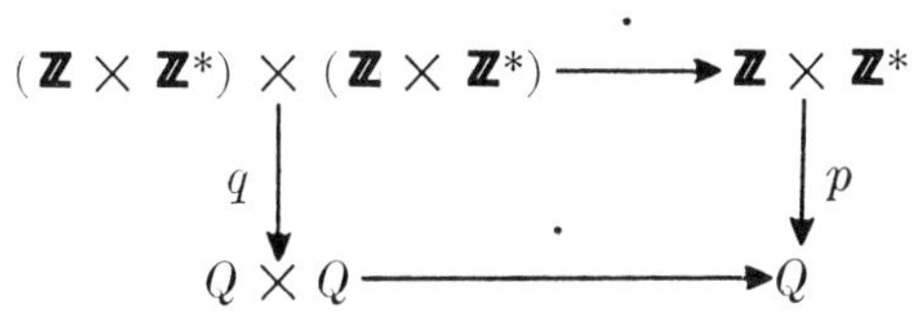

These operations are given explicitly by

$$[m,n] + [a,b] = [mb + an,\ nb],$$
$$[m,n] \cdot [a,b] = [ma,nb].$$

The operations on Q have the following properties making Q a field:

(1′) For all $x, y, z \in Q$, $(x + y) + z = x + (y + z)$.
(2′) For all $x, y \in Q$, $x + y = y + x$.
(3′) For all $x \in Q$, $x + [0,1] = x$.
(4′) For each $x \in Q$, there exists a unique $-x \in Q$ for which $-x + x = [0,1]$, namely, $x = [m,n]$ implies $-x = [-m,n]$.
(5′) For all $x, y, z \in Q$, $(x \cdot y) \cdot z = x \cdot (y \cdot z)$.
(6′) For all $x, y \in Q$, $x \cdot y = y \cdot x$.
(7′) For all $x \in Q$, $x \cdot [1,1] = x$.
(8′) For each $x \in Q$ with $x \neq [0,1]$, there exists a unique $x^{-1} \in Q$ for which $x^{-1} \cdot x = [1,1]$, namely, $x = [m,n]$ implies $x^{-1} = [n,m]$.
(9′) For all $x, y, z \in Q$, $x \cdot (y + z) = (x \cdot y) + (x \cdot z)$.
(10′) For each $x \in Q$, there exist $m \in \mathbb{Z}$ and $n \in \mathbb{Z}^*$ for which $x = [m,1] \cdot [n,1]^{-1}$.

The field Q is not exactly the field $\mathbb{Q}$ we seek, for $\mathbb{Z}$ is not a subset of Q. However, comparison of (1′)–(10′) with A.7(1)–(10) suggests that elements of the form $[n,1]$ of Q play the same role in Q as elements n of $\mathbb{Z}$ are to play in $\mathbb{Q}$. More precisely, consider the injection

$$j: \mathbb{Z} \to Q$$
$$n \mapsto [n,1] \qquad (n \in \mathbb{Z}).$$

Then j defines a one-to-one correspondence between $\mathbb{Z}$ and the subset $j\langle\mathbb{Z}\rangle$ of Q. Moreover,

$$j(m+n) = j(m) + j(n)$$
$$j(mn) = j(m) \cdot j(n) \qquad (m, n \in \mathbb{Z}),$$

so that elements of $j\langle\mathbb{Z}\rangle$ add and multiply exactly as do the elements of $\mathbb{Z}$ corresponding to them under j.

In order to define $\mathbb{Q}$, we simply replace each $[n,1] \in j\langle\mathbb{Z}\rangle$ by the corresponding $n \in \mathbb{Z}$.

A.8. DEFINITION

Define $\mathbb{Q}$ to be the set

$$(Q \setminus j\langle\mathbb{Z}\rangle) \cup \mathbb{Z}.$$

An element of $\mathbb{Q}$ is called a *rational number*.

We have a natural bijection

$$i: \mathbb{Q} \to Q$$

defined by

$$i(x) = \begin{cases} x & \text{if } x \in \mathbb{Q} \setminus j\langle\mathbb{Z}\rangle, \\ j(x) & \text{if } x \in \mathbb{Z}. \end{cases}$$

Operations $+$ and $\cdot$ on Q were defined above. We want to define operations $+$ and $\cdot$ on $\mathbb{Q}$ in such a way that

$$i(x+y) = i(x) + i(y)$$
$$i(x \cdot y) = i(x) \cdot i(y) \qquad (x, y \in \mathbb{Q}).$$

A.9. DEFINITION

Define binary operations $+$ and $\cdot$ on $\mathbb{Q}$ by

$$x + y = i^{-1}(i(x) + i(y))$$
$$x \cdot y = i^{-1}(i(x) \cdot i(y)) \qquad (x, y \in \mathbb{Q}).$$

These operations on $\mathbb{Q}$ extend ordinary addition and multiplication on the subset $\mathbb{Z}$ of $\mathbb{Q}$. In fact, if $m, n \in \mathbb{Z}$, then the sum of m and n in $\mathbb{Q}$ is

$$\begin{aligned} m + n &= i^{-1}(i(m) + i(n)) = i^{-1}(j(m) + j(n)) \\ &= i^{-1}(j(m+n)) \\ &= (i^{-1} \circ i)(m+n) \end{aligned}$$

which is the sum of m and n in $\mathbb{Z}$; similarly for $m \cdot n$.

To show that the operations on $\mathbb{Q}$ satisfy A.7(1)–(10), it remains only to use the bijection i to translate each of the properties (1′)–(10′) into the analogous property for $\mathbb{Q}$. For example, if $x, y \in \mathbb{Q}$, then

$$i(x) + i(y) = i(y) + i(x)$$

by (2′), so

$$x + y = i^{-1}(i(x) + i(y)) = i^{-1}(i(y) + i(x)) = y + x,$$

which gives (2). If $x \in \mathbb{Q}$, then $i(x) \in Q$, by (10′) there exist $m \in \mathbb{Z}$ and $n \in \mathbb{Z}^*$ for which $i(x) = [m,1] \cdot [n,1]^{-1}$, so

$$\begin{aligned} x &= i^{-1}(i(x)) = i^{-1}([m,1] \cdot [n,1]^{-1}) \\ &= i^{-1}(j(m) \cdot (j(n))^{-1}) \\ &= i^{-1}(j(m)) \cdot i^{-1}(j(n)^{-1}) \\ &= i^{-1}(j(m)) \cdot (i^{-1}(j(n)))^{-1} \\ &= m \cdot n^{-1}, \end{aligned}$$

which gives (10). Verification of the remaining parts of A.7 is left to the reader.

With the proof of A.7 completed, we can now forget the actual definition of $\mathbb{Q}$ and base any further discussion of $\mathbb{Q}$ solely on the properties stated in A.7.

ORDERING OF THE RATIONAL NUMBERS

Like $\mathbb{Z}$, the field $\mathbb{Q}$ is an integral domain which can be ordered.

A.10. THEOREM

There exists a unique total ordering $\leq$ of $\mathbb{Q}$ such that:

(1) For $x, y, z \in \mathbb{Q}$, $x \leq y$ implies $x + z \leq y + z$.
(2) For $x, y, z \in \mathbb{Q}$, $x \leq y$ and $0 \leq z$ implies $xz \leq yz$.
(3) $0 \leq 1$.

Moreover, the total ordering of $\mathbb{Z}$ induced by $\leq$ is the usual one (given by A.6).

Proof

Assume the existence of such a total ordering; we deduce its uniqueness. The total ordering induced on $\mathbb{Z}$ satisfies A.6(1)–(3) and hence is the one given by A.6. For this reason,

$$0 \leq n \Leftrightarrow n \in \mathbb{N} \qquad (n \in \mathbb{Z}).$$

We show next that

$$0 < x \Rightarrow 0 < x^{-1} \qquad (x \in \mathbb{Q}).$$

Let $x \in \mathbb{Q}$ with $0 < x$, and suppose $x^{-1} \leq 0$. Then by (2),

$$1 = x^{-1} \cdot x \leq 0 \cdot x = 0,$$

which is impossible in view of (3).

Define

$$P = \{x \mid x \in \mathbb{Q} \;\&\; (\exists m \in \mathbb{N})(\exists n \in \mathbb{N}^*)(x = m \cdot n^{-1})\}.$$

We claim

$$0 \leq x \Leftrightarrow x \in P \qquad (x \in \mathbb{Q}).$$

In fact, assume $x \in P$, say $x = m \cdot n^{-1}$ with $m \in \mathbb{N}$ and $n \in \mathbb{N}^*$. Now $0 \leq m = xn$, and since $0 < n^{-1}$,

$$0 = 0 \cdot n^{-1} \leq m \cdot n^{-1} = x,$$

so $0 \leq x$. Conversely, assume $0 \leq x$. For some $m \in \mathbb{Z}$, $n \in \mathbb{Z}^*$, $x = m \cdot n^{-1}$. Without loss of generality we may assume $n \in \mathbb{N}^*$, for $n \notin \mathbb{N}^*$ implies $-n \in \mathbb{N}^*$ and $x = (-m) \cdot (-n)^{-1}$. Then $0 \leq x = m \cdot n^{-1}$, and since $0 \leq n$,

$$0 = 0 \cdot n \leq x \cdot n = m,$$

so $m \in \mathbb{N}$ and hence $x \in P$.

As in the proof of A.6, one shows

$$x \leq y \Leftrightarrow 0 \leq y - x \qquad (x, y \in \mathbb{Q}).$$

Hence

$$(*) \qquad x \leq y \Leftrightarrow y - x \in P \qquad (x, y \in \mathbb{Q}).$$

Since the definition of P does not depend upon $\leq$, the uniqueness of $\leq$ is established.

To establish the existence of $\leq$, we define a relation $\leq$ in $\mathbb{Q}$ by (*). The verification that $\leq$ is a total ordering of $\mathbb{Q}$ satisfying (1)–(3) proceeds as in the proof of A.6. □

Additional properties of the ordering of $\mathbb{Q}$ appear in the exercises.

The field $\mathbb{Q}$ was constructed in order that each linear equation with coefficients in $\mathbb{Z}$ would have a solution in $\mathbb{Q}$. However, a polynomial equation of degree greater than 1 and with coefficients in $\mathbb{Z}$ need not have any solution in $\mathbb{Q}$. One such is the quadratic equation $x^2 - 2 = 0$.

A.11. PROPOSITION

For all $x \in \mathbb{Q}$, $x^2 \neq 2$.

Proof

Suppose to the contrary that $x^2 = 2$ for some $x \in \mathbb{Q}$. We may assume $x > 0$, for $x < 0$ implies $-x > 0$, and $(-x)^2 = 2$. There exist $m, n \in \mathbb{Z}$

with $m > 0$, $n > 0$, and

$$x = m/n.$$

We may assume m and n are not both even (see Exercise F).

We have $2 = x^2 = m^2/n^2$, so

$$m^2 = 2n^2$$

and m^2 is even. If m were odd, say $m = 2k + 1$ with $k \in \mathbf{N}$, then $m^2 = (2k + 1)^2 = 2(2k^2 + 2k) + 1$ would be odd. It follows that m itself is even.

Since m is even, we may write

$$m = 2k$$

for some $k \in \mathbf{N}$. Then

$$4k^2 = (2k)^2 = m^2 = 2n^2,$$

$$n^2 = 2k^2,$$

and n^2 is even. It follows that n itself is even. Since m is also even, we have reached a contradiction. □

From A.11 we can deduce that *the order relation in* $\mathbf{Q}$ *is not conditionally complete.*

A.12. PROPOSITION

The nonempty subset

$$E = \{x \mid x \in \mathbf{Q} \;\&\; x > 0 \;\&\; x^2 < 2\}$$

of $\mathbf{Q}$ is bounded above in $\mathbf{Q}$ but has no supremum in $\mathbf{Q}$.

Proof

The integer 2 is an upper bound of E in $\mathbf{Q}$, for $x \in E$ implies $0 < x$ and $x^2 < 4$, whence $x < 2$.

Suppose E has a supremum b in $\mathbf{Q}$. We shall show that $b^2 = 2$, thereby contradicting A.11.

We show $2 \leq b^2$. Suppose $b^2 < 2$. Then $2 - b^2 > 0$ and $b > 1$, so $(2 - b^2)/(2b + 1) > 0$. By Exercise I, there exists $\epsilon \in \mathbf{Q}$ with

$$0 < \epsilon < 1, \qquad \epsilon < (2 - b^2)/(2b + 1).$$

Since $\epsilon < 1$,

$$\epsilon < (2 - b^2)/(2b + \epsilon),$$

$$2b\epsilon + \epsilon^2 < 2 - b^2,$$

and

$$(b + \epsilon)^2 = b^2 + 2b\epsilon + \epsilon^2 < 2.$$

Hence $b + \epsilon \in E$ with $b + \epsilon > b$, which contradicts the fact that b is an upper bound of E in $\mathbb{Q}$.

We show $b^2 \leq 2$. Suppose $2 < b^2$. Consider

$$y = b - (b^2 - 2)/2b = (b/2) + (1/b).$$

We have

$$0 < y < b,$$

$$y^2 = b^2 - (b^2 - 2) + ((b^2 - 2)/2b)^2 > b^2 - (b^2 - 2) = 2,$$

and $2 < y^2$. Then $x \in E$ implies $x^2 < 2 < y^2$, and $x < y$. Hence y is an upper bound of E in $\mathbb{Q}$ with $y < b$, which contradicts the fact that b is the supremum of E in $\mathbb{Q}$. $\square$

REAL AND COMPLEX NUMBERS

The proof of A.12 shows that if $\mathbb{Q}$ were conditionally complete, then $\sup E = b$ would be a solution of $x^2 - 2 = 0$. Hence one should try to obtain a solution of this equation by enlarging the ordered field $\mathbb{Q}$ to an ordered field which is conditionally complete. Specifically, one wants to construct a set $\mathbb{R}$, whose elements are called *real numbers,* with the following properties:

(R1) $\mathbb{Q} \subset \mathbb{R}$.

(R2) There exist unique binary operations $+$ and $\cdot$ on $\mathbb{R}$ which make $\mathbb{R}$ a field and which extend the operations $+$ and $\cdot$ on $\mathbb{Q}$.

(R3) There exists a unique total ordering $\leq$ of $\mathbb{R}$ which makes $\mathbb{R}$ an ordered field [that is, the analogs of A.10(1)–(3) hold for $\mathbb{R}$]. Moreover, the total ordering of $\mathbb{Q}$ induced by $\leq$ is the usual one (given by A.10).

(R4) This total ordering of $\mathbb{R}$ is conditionally complete.

In many calculus and analysis texts, the existence of the conditionally complete ordered field $\mathbb{R}$ is postulated. It should come as no surprise to the reader of this book that $\mathbb{R}$ can actually be shown to exist by explicit construction.

One method for constructing $\mathbb{R}$ is due to R. Dedekind (as modified by G. Peano and B. Russell). This method uses "Dedekind cuts": A Dedekind cut is a set A with $\varnothing \neq A \subsetneq \mathbb{Q}$ such that A has no largest element. For example,

$$\{x \mid x \in \mathbb{Q} \;\&\; x^2 < 2\}$$

is such a set. For this method, the reader is referred to Rudin [35].

A second method for constructing $\mathbb{R}$ is due to Cantor. It uses "Cauchy sequences" of rational numbers and identifies as real numbers equivalence classes of Cauchy sequences under a suitable equivalence relation. For this more complicated method, the reader is referred to Suppes [38].

Once the set $\mathbb{R}$ satisfying (R1)–(R4) is constructed, one can show that if $a \in \mathbb{R}$ with $a \geq 0$, then the quadratic equation $x^2 - a = 0$ always has a solution in $\mathbb{R}$.

Construction of number systems does not stop with $\mathbb{R}$, for the equation $x^2 + 1 = 0$ has no solution in $\mathbb{R}$. To obtain a solution, one constructs a set $\mathbb{C}$, whose elements are called *complex numbers,* with the following properties:

(C1) $\mathbb{R} \subset \mathbb{C}$.

(C2) There exist unique binary operations $+$ and $\cdot$ on $\mathbb{C}$ which make $\mathbb{C}$ into a field and which extend the operations $+$ and $\cdot$ on $\mathbb{R}$.

(C3) There exists $i \in \mathbb{C}$ with $i^2 = -1$.

(C4) With i as in (C3), each $z \in \mathbb{C}$ can be written in the form $z = x + i \cdot y$ for some $x, y \in \mathbb{R}$.

In contrast to the constructions of $\mathbb{Z}$, $\mathbb{Q}$, and $\mathbb{R}$, the construction of $\mathbb{C}$ from $\mathbb{R}$ is exceedingly simple. See, for example, Rudin [35] or Landau [21] (Landau constructs all the number systems $\mathbb{N}$, $\mathbb{Z}$, $\mathbb{Q}$, $\mathbb{R}$, and $\mathbb{C}$, starting from the Peano postulates; despite its somewhat old-fashioned flavor, Landau's account is very readable, even in the original German [20]).

With $\mathbb{C}$, the successive enlargements of number systems in which more and more equations can be solved stops. The "fundamental theorem of algebra" asserts that every polynomial equation with coefficients in $\mathbb{C}$ has a solution in $\mathbb{C}$.

EXERCISES

A. Show that the equivalence kernel of the map $d: \mathbb{N} \times \mathbb{N} \to \mathbb{Z}$ used to prove A.2 is the equivalence relation defined in Chapter 11, Exercise J.

B. For $x, y, z \in \mathbb{Z}$, show:

$$x < y \Leftrightarrow -y < -x,$$

$$x < y \;\&\; z < 0 \Rightarrow xz > yz.$$

C. Establish the *division algorithm* for $\mathbb{Z}$: Given $m, n \in \mathbb{Z}$ with $n > 0$, there exist unique $q, r \in \mathbb{Z}$ such that

$$m = qn + r, \qquad 0 \leq r < n.$$

[*Hint*: r is the first element of

$$\{x \mid x \in \mathbb{N} \,\&\, (\exists q \in \mathbb{Z})(qn + x = m)\}.]$$

D. A nonempty subset J of the ring $\mathbb{Z}$ is called an *ideal* in $\mathbb{Z}$ if (i) $m, n \in J$ implies $m + n \in J$, and (ii) $n \in J$ and $m \in \mathbb{Z}$ implies $mn \in J$.
(1) If $k \in \mathbb{Z}$, show that the set

$$(k) = \{nk \mid n \in \mathbb{Z}\}$$

is an ideal in $\mathbb{Z}$; such an ideal is said to be *principal.*
(2) Prove that every ideal J in $\mathbb{Z}$ is principal. [*Hint*: Show that $J = (k)$, where k is the least positive element of J.]

E. For $m, c \in \mathbb{Z}$, one says that c *divides* m and writes $c \mid m$ to mean that $c \neq 0$ and $m = ck$ for some $k \in \mathbb{Z}$. Let $m, n \in \mathbb{Z}^*$. Show there exists $d \in \mathbb{Z}$ such that

$$d \mid m, \qquad d \mid n,$$
$$c \mid m \,\&\, c \mid n \Rightarrow c \mid d \qquad (c \in \mathbb{Z}),$$

and that such a d has the form

$$d = am + bn$$

for some $a, b \in \mathbb{Z}$. [*Hint*: Show that the set

$$\{am + bn \mid a, b \in \mathbb{Z}\}$$

is an ideal in $\mathbb{Z}$ and then apply Exercise D(2).]

F. Let $x \in \mathbb{Q}$. Show there exist $m, n \in \mathbb{Z}$ with $x = m/n$ such that 2 does not divide both m and n.

G. For any $x \in \mathbb{Q}$, $y \in \mathbb{Q}^*$, let $\frac{x}{y}$ denote $x \cdot y^{-1}$. Prove the following hold for all $x, u \in \mathbb{Q}$, $y, v \in \mathbb{Q}^*$:

$$\frac{x}{y} = \frac{u}{v} \Leftrightarrow x \cdot v = u \cdot y,$$

$$\frac{x}{y} + \frac{u}{v} = \frac{xv + uy}{yv},$$

$$\frac{x}{y} \cdot \frac{u}{v} = \frac{xu}{yv},$$

$$\frac{-x}{y} = \frac{x}{-y} = -\frac{x}{y}.$$

H. Using only A.7(1)–(3), prove that if $x \in \mathbb{Q}$, then there is at most one $y \in \mathbb{Q}$ for which $y + x = 0$. Prove an analogous result for multiplication.

I. Prove that the following three assertions about $\mathbb{Q}$ are equivalent:

(i) The set $\mathbb{Z}$ is not bounded above in $\mathbb{Q}$.

(ii) If $x \in \mathbb{Q}$ and $\epsilon \in \mathbb{Q}$ with $\epsilon > 0$, then there exists $n \in \mathbb{N}^*$ such that $n\epsilon > x$.

(iii) If $\epsilon \in \mathbb{Q}$ with $\epsilon > 0$, then there exists $n \in \mathbb{N}^*$ with $1/n < \epsilon$.

Then prove that (i)–(iii) are all true; they say that the ordered field $\mathbb{Q}$ is *archimedean*. (The ordered field $\mathbb{R}$ is also archimedean.)

J. Prove that $\mathbb{Q}$ is order-dense in itself (see Chapter 15, Exercise E). Prove a similar result for $\mathbb{R}$.

K. Prove that the additive group $\mathbb{Z}$ is "universal for groups" in the following sense: Let G be a group, that is, a set on which there is given a binary operation $+$ satisfying:

(i) For all $x, y, z \in G$, $(x + y) + z = x + (y + z)$.

(ii) There exists a unique $u \in G$ such that $x + u = u + x = x$ for all $x \in G$.

(iii) For each $x \in G$, there exists a unique $-x \in G$ for which $(-x) + x = u$.

Let $f: \mathbb{N} \to G$ be a map such that

$$f(m + n) = f(m) + f(n) \qquad (m, n \in \mathbb{N}).$$

Then there exists a unique extension $g: \mathbb{Z} \to G$ of f for which

$$g(m + n) = g(m) + g(n) \qquad (m, n \in \mathbb{Z}).$$

[*Hint*: $g(-n) = -g(n)$ for $n \in \mathbb{N}^*$.]

L. Prove that the additive group $\mathbb{Z}$ is "unique up to isomorphism" in the following sense: Let G be a group as in the preceding exercise. Let $j: \mathbb{N} \to G$ be an injection such that

$$(\forall x \in G)(\exists m, n \in \mathbb{N})(x = j(m) - j(n)).$$

Then there exists a unique bijection $i: \mathbb{Z} \to G$ extending j for which

$$i(m + n) = i(m) + i(n) \qquad (m, n \in \mathbb{Z}).$$

M. With G and j as in Exercise L, what can be said about introducing a multiplication and order relation in G?

N. Formulate and prove analogs of Exercises K and L for $\mathbb{Q}$.

P. (1) Show that the binary operation of exponentiation on $\mathbb{N}$ has a unique extension

$$\mathbb{Q} \times \mathbb{N} \to \mathbb{Q}$$

$$(x,n) \mapsto x^n$$

such that

$$(x \cdot y)^n = x^n \cdot y^n \qquad (x, y \in \mathbb{Q}; n \in \mathbb{N}).$$

Show that this extension satisfies

$$x^{m+n} = x^m \cdot x^n$$

$$(x^m)^n = x^{mn} \qquad (x \in \mathbb{Q}; m, n \in \mathbb{N}).$$

(2) Consider the restriction of the above exponentiation to $\mathbb{Q}^* \times \mathbb{N}$. Show that this restriction has a unique extension

$$\mathbb{Q}^* \times \mathbb{Z} \to \mathbb{Q}$$

$$(x,n) \mapsto x^n$$

such that

$$x^{m+n} = x^m \cdot x^n \qquad (x \in \mathbb{Q}^*; m, n \in \mathbb{Z}).$$

Here $\mathbb{Q}^*$ denotes $\mathbb{Q} / \{0\}$.

Bibliography

[1] P. Bernays, *A system of axiomatic set theory. I–VII*, J. Symbolic Logic **2** (1937), 65–77; **6** (1941), 1–17; **7** (1942), 65–89, 133–145; **8** (1943), 89–106; **13** (1948), 65–79; **19** (1954), 81–96.

[2] P. Bernays and A. A. Fraenkel, *Axiomatic set theory*, North-Holland, Amsterdam, 1958.

[3] N. Bourbaki, *Foundations of mathematics for the working mathematician*, J. Symbolic Logic **14** (1949), 1–8.

[4] N. Bourbaki, *Théorie des ensembles*, 4 vol., chap. 1–2, 3rd ed., 1966; chap. 3, 2nd ed., 1963; chap. 4, 1957; Fascicule de résultats, 4th ed., 1964, Hermann, Paris.

[5] N. Bourbaki, *Theory of sets*, Elements of Math., vol. 3, Addison-Wesley, Reading, Mass., 1968.

[6] G. Cantor, *Beiträge zur Begründung der transfiniten Mengenlehre. I–II*, Math. Ann. **46** (1895), 481–512; **49** (1897), 207–246.

[7] G. Cantor, *Contributions to the founding of the theory of transfinite numbers*, Dover, New York.

[8] P. J. Cohen, *Set theory and the continuum hypothesis*, Benjamin, New York, 1966.

[9] A. A. Fraenkel, *Abstract set theory*, 2nd ed., North-Holland, Amsterdam, 1961.

[10] A. A. Fraenkel and Y. Bar-Hillel, *Foundations of set theory*, North-Holland, Amsterdam, 1958.

[11] K. Gödel, *The consistency of the axiom of choice and of the generalized continuum hypothesis with the axioms of set theory*, Annals of Math. Studies 3, Princeton Univ. Press, Princeton, 1940.

[12] N. Goodman, *Fact, fiction, and forecast*, Harvard Univ. Press, Cambridge, 1955.

[13] W. H. Gottschalk, *The theory of quaternality*, J. Symbolic Logic **18** (1953), 193–196.

[14] P. R. Halmos, *Lectures on Boolean algebras*, Van Nostrand, Princeton, N. J., 1963.

[15] W. S. Hatcher, *Foundations of mathematics*, Saunders, Philadelphia, 1968.

[16] D. Hilbert and W. Ackermann, *Principles of mathematical logic*, Chelsea, New York, 1950.

[17] E. Kamke, *Theory of sets*, Dover, New York, 1950.

[18] J. L. Kelley, *General topology*, Van Nostrand, Princeton, N. J., 1955.

[19] S. C. Kleene, *Mathematical logic*, Wiley, New York, 1967.

[20] E. Landau, *Grundlagen der Analysis*, 1930. Reprint, 4th ed., Chelsea, New York, 1965.

[21] E. Landau, *Foundations of analysis*, 2nd ed., Chelsea, New York, 1960.

[22] F. W. Lawvere, *An elementary theory of the category of sets*, Proc. Nat. Acad. Sci. U.S.A. **52** (1964), 1506–1511.

[23] R. C. Lyndon, *Notes on logic*, Van Nostrand, Princeton, N. J., 1966.

[24] S. MacLane and G. Birkhoff, *Algebra*, Macmillan, New York, 1967.

[25] E. Mendelson, *Introduction to mathematical logic*, Van Nostrand, Princeton, N. J., 1964.

[26] A. P. Morse, *A theory of sets*, Academic Press, New York, 1965.

[27] E. Nagel and J. R. Newman, *Gödel's proof*, New York Univ. Press, New York, 1958.

[28] J. von Neumann, *Eine Axiomatisierung der Mengenlehre*, J. Reine Angew. Math. **154** (1925), 219–240.

[29] J. von Neumann, *Die Axiomatisierung der Mengenlehre*, Math. Z. **27** (1928), 669–752.

[30] J. von Neumann, *Ueber eine Widerspruchsfreiheitsfrage in der axiomatischen Mengenlehre*, J. Reine Angew. Math. **160** (1929), 227–241.

[31] W. V. O. Quine, *Mathematical logic*, rev. ed., Harvard Univ. Press, Cambridge, 1951. Paperback ed., Harper & Row, New York, 1962.

[32] W. V. O. Quine, *Set theory and its logic*, rev. ed., Harvard Univ. Press, Cambridge, 1969.

[33] H. Rubin and J. E. Rubin, *Equivalents of the axiom of choice*, North-Holland, Amsterdam, 1963.

[34] J. E. Rubin, *Set theory for the mathematician*, Holden-Day, San Francisco, 1967.

[35] W. Rudin, *Principles of mathematical analysis*, 2nd ed., McGraw-Hill, New York, 1964.

[36] W. Sierpinski, *Cardinal and ordinal numbers*, Panstwowe Wydawnictwo Naukowe, Warsaw; Hafner, New York, 1958.

[37] P. Suppes, *Introduction to logic*, Van Nostrand, Princeton, N. J., 1957.

[38] P. Suppes, *Axiomatic set theory*, Van Nostrand, Princeton, N. J., 1960.

[39] A. N. Whitehead and B. Russell, *Principia Mathematica*, 2nd ed., 3 vols., Cambridge Univ. Press, Cambridge, 1925–27. Paperback ed., 1962.

Summary of Axiom Schemes

Let **P**, **Q**, **R**, and **C** be formulas, **X** and **Y** be terms, and **x** and **y** be letters. Suppose **x** is free in **P** and in **Q**, **y** does not appear in **P**, and **x** does not appear in **C**. Then the following are axioms:

1. $\mathbf{P} \vee \mathbf{P} \Rightarrow \mathbf{P}$
2. $\mathbf{P} \Rightarrow \mathbf{P} \vee \mathbf{Q}$
3. $\mathbf{P} \vee \mathbf{Q} \Rightarrow \mathbf{Q} \vee \mathbf{P}$
4. $(\mathbf{P} \Rightarrow \mathbf{Q}) \Rightarrow (\mathbf{R} \vee \mathbf{P} \Rightarrow \mathbf{R} \vee \mathbf{Q})$
5. $(\forall \mathbf{x})(\mathbf{P}) \Leftrightarrow (\forall \mathbf{y})([\mathbf{y} \mid \mathbf{x}]\mathbf{P})$
6. $(\forall \mathbf{x})(\mathbf{P}) \Rightarrow [\mathbf{X} \mid \mathbf{x}]\mathbf{P}$
7. $(\mathbf{C} \Rightarrow \mathbf{P}) \Rightarrow (\mathbf{C} \Rightarrow (\forall \mathbf{x})(\mathbf{P}))$
8. $\mathbf{X} = \mathbf{Y} \Rightarrow ([\mathbf{X} \mid \mathbf{x}]\mathbf{P} \Leftrightarrow [\mathbf{Y} \mid \mathbf{x}]\mathbf{P})$
9. $(\iota \mathbf{x})(\mathbf{P}) = (\iota \mathbf{y})([\mathbf{y} \mid \mathbf{x}]\mathbf{P})$
10. $(\forall \mathbf{x})(\mathbf{P} \Leftrightarrow \mathbf{Q}) \Rightarrow (\iota \mathbf{x})(\mathbf{P}) = (\iota \mathbf{x})(\mathbf{Q})$
11. $(\exists ! \mathbf{x})(\mathbf{P}) \Rightarrow (\forall \mathbf{y})([\mathbf{y} \mid \mathbf{x}]\mathbf{P} \Leftrightarrow \mathbf{y} = (\iota \mathbf{x})(\mathbf{P}))$
12. $(\exists \mathbf{y})(\forall \mathbf{x})(\mathbf{x} \in \mathbf{y} \Leftrightarrow \mathbf{x} \text{ is a set } \& \ \mathbf{P})$

The symbolism $[\mathbf{y} \mid \mathbf{x}]\mathbf{P}$ stands for the formula obtained from **P** by replacing each appearance of **x** in **P** by **y**; similarly for $[\mathbf{X} \mid \mathbf{x}]\mathbf{P}$, $[\mathbf{Y} \mid \mathbf{x}]\mathbf{P}$.

Summary of Axioms

1. Extension:

   $(\forall x)(x \in X \Leftrightarrow x \in Y) \Rightarrow X = Y.$

2. Subsets:

   $(\forall X \in \mathfrak{U})(\exists Z \in \mathfrak{U})(\forall Y)(Y \subset X \Rightarrow Y \in Z).$

3. Pairing:

   If X and Y are sets, then $\{X,Y\}$ is a set.

4. Union:

   If $\mathfrak{A}$ is a set, then $\bigcup\mathfrak{A}$ is a set.

5. Infinity:

   There exists an inductive set.

6. Foundation:

   If $X \neq \varnothing$, then there exists $a \in X$ such that $a \cap X = \varnothing$.

7. Replacement:

   If f is a functional relation and if dmn f is a set, then rng f is a set.

8. Choice:

   If $\mathfrak{A}$ is a class of nonempty sets, then there exists a choice function for $\mathfrak{A}$.

List of Symbols

SYMBOL	MEANING	PAGE
$\neg$	not	4, 7
$\vee$	or	4, 7
$(\forall x)$	for every x	4, 7
(ιx)	the x such that	4, 7
$=$	equals	4, 7
$\in$	element of	4, 7
$\Rightarrow$	implies	8
&	and	9, 22
$\Leftrightarrow$	equivalent to	9, 26
$(\exists x)$	there exists an x such that	9, 41
$\mathcal{B} \vdash \mathbf{C}$	$\mathbf{C}$ is deducible from $\mathcal{B}$	10
$\vdash \mathbf{C}$	$\mathbf{C}$ is true	11
$\square$	end of proof	11
$[\mathbf{S} \mid x]\mathbf{P}$	$\mathbf{S}$ replaces x in $\mathbf{P}$	35
$(\exists! x)$	there exists a unique x such that	46
$(\forall x, \mathbf{C})$	for every x satisfying $\mathbf{C}$	47
$(\exists x, \mathbf{C})$	there exists an x satisfying $\mathbf{C}$ such that	47
$\ldots (\mathbf{C})$	with the restriction $\mathbf{C}$	48
$\neq$	unequal to	55
$\notin$	not an element of	55
$\subset, \supset$	contained in, contains	56

SYMBOL	MEANING	PAGE
$\times_{i=0}^{n} X_i$, $\times_{i=0}^{\infty} X_i$, and so on	product	202
$\simeq$	equipollent	216
$\preceq$	dominated by	220
$\prec$	strictly dominated by	220
$\#(X)$	number of elements	229
$<$, $>$, and so on	less than, greater than, and so on (preordering)	248, 249
$\sup Y$	supremum	253
$\inf Y$	infimum	253
$]\leftarrow,x[$, $X]\leftarrow,x[$	initial segment	275
$f: X \sim Y$	order-isomorphism	277
$\sim$	order-isomorphic	277
Ord	class of ordinal numbers	289
$\leq$	less than or equal to (ordinal numbers)	290
$\mathrm{ord}(X)$	ordinal number of X	292
Ω	first uncountable ordinal	293
Lim	class of limit ordinals	294
$\alpha \oplus \beta$	sum of ordinal numbers	297
$\alpha \cdot \beta$	product of ordinal numbers	300
α^{β}	power of ordinal number	300
ρ	rank	303
Card	class of cardinal numbers	309
$\mathrm{card}(X)$	cardinality of X	309
$\aleph$	aleph	311
$\mathfrak{c}$	cardinal of the continuum	311
$\mathfrak{a} + \mathfrak{b}$	sum of cardinal numbers	313
$\mathfrak{a} \cdot \mathfrak{b}$	product of cardinal numbers	313
$\mathfrak{a}^{\mathfrak{b}}$	power of cardinal number	313
$\sum_{i \in I} \mathfrak{a}_i$	sum of family of cardinal numbers	313
$\prod_{i \in I} \mathfrak{a}_i$	product of family of cardinal numbers	313
$\mathfrak{b} - \mathfrak{a}$	difference of cardinal numbers	322
$\mathfrak{f}$	cardinality of Map $(\mathbb{R}, \mathbb{R})$	325
$\mathbb{N}^*$	set of nonzero natural numbers	328
$-n$	negative of natural number	329
$\mathbb{Z}$	set of integers	329
$\mathbb{Z}^*$	set of nonzero integers	335
$\mathbb{Q}$	set of rational numbers	335, 340
$\mathbb{R}$	set of real numbers	344
$\mathbb{C}$	set of complex numbers	345

Index